GIS Applications for Water, Wastewater, and Stormwater Systems

U.M. Shamsi

Taylor & Francis
Taylor & Francis Group

Boca Raton   London   New York   Singapore

A CRC title, part of the Taylor & Francis imprint, a member of the
Taylor & Francis Group, the academic division of T&F Informa plc.

Library of Congress Cataloging-in-Publication Data

Shamsi, U. M. (Uzair M.)
   GIS applications for water, wastewater, and stormwater systems / U.M. Shamsi.
     p. cm.
   Includes bibliographical references and index.
   ISBN 0-8493-2097-6 (alk. paper)
    1. Water—Distribution. 2. Sewage disposal. 3. Runoff—Management. 4. Geographic
information systems. I. Title.

   TD482.S53 2005
   628.1--dc22   2004057108

Visit the CRC Press Web site at www.crcpress.com

© 2005 by CRC Press

No claim to original U.S. Government works
International Standard Book Number 0-8493-2097-6
Library of Congress Card Number 2004057108
Printed in the United States of America   2 3 4 5 6 7 8 9 0
Printed on acid-free paper

Dedication

Dedicated to my beloved wife, Roshi, and my children, Maria, Adam, and Harris

Preface

To fully appreciate the benefits of GIS applications consider the following hypothetical scenario. On March 10, 2004, following a heavy storm event, a sewer customer calls the Sewer Authority of the City of Cleanwater to report minor basement flooding without any property damage. An Authority operator immediately starts the GIS and enters the customer address. GIS zooms to the resident property and shows all the sewers and manholes in the area. The operator queries the inspection data for a sewer segment adjacent to the customer property and finds that a mini movie of the closed-circuit television (CCTV) inspection dated July 10, 1998, is available. The operator plays the movie and sees light root growth in the segment. A query of the maintenance history for that segment indicates that it has not been cleaned since April 5, 1997. This information indicates that the roots were never cleaned and have probably grown to "heavy" status. The operator highlights the sewer segment, launches the work order module, and completes a work order form for CCTV inspection and root removal, if necessary. The export button saves the work order form and a map of the property and adjacent sewers in a PDF file. The operator immediately sends the PDF file by e-mail to the Authority's sewer cleaning contractor. The entire session from the time the customer called the Authority office took about 30 min. The operator does not forget to call the customer to tell him that a work order has been issued to study the problem. This book presents the methods and examples required to develop applications such as this.

The days of the slide rule are long gone. Word processors are no longer considered cutting-edge technology. We are living in an information age that requires us to be more than visionaries who can sketch an efficient infrastructure plan. This tech-heavy society expects us to be excellent communicators who can keep all the stakeholders — the public, the regulators, or the clients — "informed." New information and decision support systems have been developed to help us to be good communicators. GIS is one such tool that helps us to communicate geographic or spatial information. The real strength of GIS is its ability to integrate information. GIS helps decision makers by pulling together crucial bits and pieces of information as a whole and showing them the "big picture." In the past 10 years, the number of GIS users has increased substantially. Many of us are using GIS applications on the Internet and on wireless devices without even knowing that we are using a GIS. Experts believe that in the near future, most water, wastewater, and stormwater system professionals will be using the GIS in the same way they are now using a word processor or spreadsheet. Except for the computer itself, no technology has so revolutionized the water industry. The time has come for all the professionals involved in the planning, design, construction, and operation of water, wastewater, and stormwater systems to enter one of the most promising and exciting technologies of the millennium in their profession — GIS applications.

According to some estimates, more than 80% of all the information used by water and sewer utilities is geographically referenced.

This book was inspired from a continuing education course that the author has been teaching since 1998 for the American Society of Civil Engineers (ASCE). Entitled "GIS Applications in Water, Wastewater and Stormwater Systems," the seminar course has been attended by hundreds of water, wastewater, and stormwater professionals in major cities of the United States. Many models, software, examples, and case studies described in the book (especially those from Pennsylvania) are based on the GIS projects worked on or managed by the author himself.

This is my second GIS book for water, wastewater, and stormwater systems. The first book, *GIS Tools for Water, Wastewater, and Stormwater Systems*, published by American Society of Civil Engineers (ASCE) Press in 2002, was a huge success. The first printing was sold out, and the book achieved ASCE Press's best-seller status within months of publication. Whereas the first book focused on GIS basics and software and data tools to develop GIS applications, this second book focuses on the practical applications of those tools. Despite the similarity of the titles, both books cover different topics and can be read independent of each other.

STYLE OF THE BOOK

This book has been written using the recommendations of the Accreditation Board for Engineering and Technology (ABET) of the U.S. and the American Society of Civil Engineers' (ASCE) Excellence in Civil Engineering Education (ExCEEd) program. Both of these organizations recommend performance- (or outcome-) based learning in which the learning objectives of each lecture (or chapter) are clearly stated up front, and the learning is measured in terms of achieving these learning objectives. Each chapter of this book accordingly starts with learning objectives for that chapter and ends with a chapter summary and questions. Most technical books are written using the natural human teaching style called *deductive*, in which principles are presented before the applications. In this book, an attempt has been made to organize the material in the natural human learning style called *inductive*, in which examples are presented before the principles. For example, in most chapters, case studies are presented before the procedures are explained. The book has numerous maps and illustrations that should cater well to the learning styles of "visual learners" — GIS, after all is regarded as a visual language.

The primary learning objective of this book is to document GIS applications for water, wastewater, and stormwater systems. This book will show you how to use GIS to make tasks easier to do and increase productivity, and hence, save time and money in your business.

ORGANIZATION OF THE BOOK

There are 17 chapters in this book, organized as follows:

- Chapter 1, GIS Applications: Describes why GIS applications are important and how they are created

- Chapter 2, Needs Analysis: Explains how to avoid potential pitfalls of GIS implementation by starting with a needs analysis study

The next five chapters describe four GIS-related technologies that are very beneficial in developing GIS applications:

- Chapter 3, Remote Sensing Applications: Shows how to use satellite imagery in GIS applications
- Chapter 4, DEM Applications: Describes the methods of incorporating digital elevation model (DEM) data
- Chapter 5, GPS Applications: Discusses how to benefit from global positioning system (GPS) technology
- Chapter 6, Internet Applications: Explains the applications of Internet technology in serving GIS maps on the Internet
- Chapter 7, Mobile GIS: Provides information on using GIS in the field for inspection and maintenance work

The GIS applications that are of particular importance to water industry professionals are: Mapping, Monitoring, Modeling, and Maintenance. These four *M*s define some of the most important activities for efficient management of water, wastewater, and stormwater systems, and are referred to as the "4M applications" in this book. The next ten chapters focus on these four *M*s.

- Chapter 8, Mapping: Describes how to create the first *M* of the 4M applications
- Chapter 9, Mapping Applications: Describes examples of the first *M* of the 4M applications
- Chapter 10, Monitoring Applications: Describes the applications of the second *M* of the 4M applications
- Chapter 11, Modeling Applications: Describes the applications of the third *M* of the 4M applications
- Chapter 12, Water Models: Describes examples of the third *M* of the 4M applications for modeling water distribution systems
- Chapter 13, Sewer Models: Describes examples of the third *M* of the 4M applications for modeling sewage collection systems
- Chapter 14, AM/FM/GIS Applications: Describes automated mapping/facilities management/geographic information system (AM/FM/GIS) software tools for implementing the fourth *M* of the 4M applications
- Chapter 15, Maintenance Applications: Describes the applications of the fourth *M* of the 4M applications
- Chapter 16, Security Planning and Vulnerability Assessment: Discusses GIS applications for protecting water and wastewater systems against potential terrorist attacks
- Chapter 17, Applications Sampler: Presents a collection of recent case studies from around the world

Acknowledgments

Case studies presented in Chapter 17, Applications Sampler, were written specially for publication in this book by 18 GIS and water industry experts from 6 countries (Belgium, Bulgaria, Czech Republic, Denmark, Spain and the United States) in response to my call for case studies distributed to various Internet discussion forums. I thank these case study authors for their contributions to this book:

- Bart Reynaert, Rene Horemans, and Patrick Vercruyssen of Pidpa, Belgium
- Carl W. Chen and Curtis Loeb of Systech Engineering, Inc., San Ramon, California
- Dean Trammel, Tucson Water, Tucson, Arizona
- Ed Bradford, Roger Watson, Eric Mann, Jenny Konwinski of Metropolitan Sewerage District of Buncombe County, North Carolina
- Eric Fontenot of DHI, Inc., Hørsholm, Denmark
- Milan Suchanek and Tomas Metelka of Sofiyska Voda A.D., Sofia, Bulgaria
- Peter Ingeduld, Zdenek Svitak, and Josef Drbohlav of Praûská vodohospodáská spolenost a.s. (Prague stockholding company), Prague, Czech Republic
- Hugo Bartolin and Fernando Martinez of Polytechnic University of Valencia, Spain

I also thank the following organizations and companies for providing information for this book: American Society of Civil Engineers, American Water Works Association, Azteca Systems, *CE Magazine*, CH2M Hill, Chester Engineers, Computational Hydraulics International, Danish Hydraulic Institute (DHI), Environmental Systems Research Institute, *Geospatial Solutions Magazine*, *GEOWorld Magazine*, Haestad Methods, Hansen Information Technology, *Journal of the American Water Resources Association*, *Journal of the American Water Works Association*, MWH Soft, *Professional Surveyor Magazine*, USFilter, *Water Environment Federation, and Water Environment & Technology Magazine*. Some information presented in this book is based on my collection of papers and articles published in peer-reviewed journals, trade magazines, conference proceedings, and the Internet. The authors and organizations of these publications are too numerous to be thanked individually, so I thank them all collectively without mentioning their names. Their names are, of course, included in the Reference section.

Finally, I would like to thank you for buying the book. I hope you will find the book useful in maximizing the use of GIS in your organization to make things easier to do, increase productivity, and save time and money.

About the Author

Uzair (Sam) M. Shamsi, Ph.D., P.E., DEE is director of the GIS and Information Management Technology division of Chester Engineers, Pittsburgh, Pennsylvania, and an adjunct assistant professor at the University of Pittsburgh, where he teaches GIS and hydrology courses. His areas of specialization include GIS applications and hydrologic and hydraulic (H&H) modeling. He has been continuing education instructor for the American Society of Civil Engineers (ASCE) and an Environmental Systems Research Institute (ESRI)-authorized ArcView® GIS instructor since 1998. He has taught GIS courses to more than 500 professionals throughout the United States, including a course on "GIS Applications in Water, Wastewater, and Stormwater Systems" for ASCE. Sam earned his Ph.D. in civil engineering from the University of Pittsburgh in 1988. He has 20 years of GIS and water and wastewater engineering experience in teaching, research, and consulting. His accomplishments include more than 120 projects and over 100 lectures and publications, mostly in GIS applications. His previous book, *GIS Tools for Water, Wastewater, and Stormwater Systems,* was an ASCE Press best seller. He is the recipient of the ASCE's Excellence in Civil Engineering Education (EXCEED) training and is a licensed professional engineer in Pennsylvania, Ohio, and West Virginia. In addition to ASCE, he is a member of the American Water Resources Association, the Water Environment Foundation, and the American Water Works Association.

E-mail: ushamsi@GISApplications.com
Web site: www.GISApplications.com

GIS is an instrument for implementing geographic thinking.

Jack Dangermond (1998)

Iron rusts from disuse; water loses its purity from stagnation and in cold weather becomes frozen; even so does inaction sap the vigors of the mind.

Leonardo da Vinci (1452–1519)

Life is like a sewer…what you get out of it depends on what you put into it.

Tom Lehrer (1928–)

Times of general calamity and confusion create great minds. The purest ore is produced from the hottest furnace, and the brightest thunderbolt is elicited from the darkest storms.

Charles Caleb Colton (1780–1832)

Contents

Chapter 15   Maintenance Applications

Chapter 16   Security Planning and Vulnerability Assessment

Chapter 17   Applications Sampler

CHAPTER 1

GIS Applications

Geographic Information System (GIS) is one of the most promising and exciting technology of the decade in our profession. This book will show you that with GIS the possibilities to manage your water, wastewater, and stormwater systems are almost endless.

GIS applications can take you from work frustration to job satisfaction.

LEARNING OBJECTIVE

The learning objective of this chapter is to understand the importance and scope of geographic information system (GIS) applications for water, wastewater, and storm-water systems.

MAJOR TOPICS

- Definition of GIS applications
- History of GIS applications
- Advantages and disadvantages of GIS applications
- Evolving and future GIS applications and trends
- Methods of developing GIS applications

LIST OF CHAPTER ACRONYMS*

CAD Computer-Aided Drafting/Computer-Aided Design
ESRI Environmental Systems Research Institute
GIS Geographic Information Systems
GPS Global Positioning System
GUI Graphical User Interface
H&H Hydrologic and Hydraulic
LBS Location-Based Services
PC Personal Computer
PDA Personal Digital Assistant

GIS Project Nominated for OCEA Award

American Society of Civil Engineers (ASCE) awards Outstanding Civil Engineering Achievement (OCEA) awards to projects based on their contribution to the well-being of people and communities; resourcefulness in planning and solving design challenges; pioneering in use of materials and methods; innovations in construction; impact on physical environment; and beneficial effects including aesthetic value. The Adam County (Illinois) 2002 GIS Pilot Project was a nominee for the 1997 awards. This project was a 10-year, multiparticipant (Adams County, City of Quincy, Two Rivers Regional Planning Council, and a number of state and local agencies) project to develop an accurate, updated GIS designed to create a more efficient local government.

INTRODUCTION

The water industry** business is growing throughout the world. For example, the U.S. market for water quality systems and services had a total value of $103 billion in 2000. The two largest components of this business are the $31-billion

---

* Each chapter of this book begins with a list of frequently used acronyms in the chapter. Appendix A provides a complete list of acronyms used in the book.
** In this book, the term *water industry* refers to water, wastewater, and stormwater systems.

public wastewater treatment market and the $29-billion water supply market (Farkas and Berkowitz, 2001).

One of the biggest challenges in the big cities with aging water, wastewater, and stormwater infrastructures is managing information about maintenance of existing infrastructure and construction of new infrastructure. Many utilities tackle infrastructure problems on a react-to-crisis basis that, despite its conventional wisdom, may not be the best strategy. Making informed infrastructure improvement decisions requires a large amount of diverse information on a continuing basis. If information is the key to fixing infrastructure problems, the first step of any infrastructure improvement project should be the development of an information system.

An information system is a framework that provides answers to questions, from a data resource. A GIS is a special type of information system in which the data source is a database of spatially distributed features and procedures to collect, store, retrieve, analyze, and display geographic data (Shamsi, 2002).

More than 80% of all the information used by water and wastewater utilities is geographically referenced.

In other words, a key element of the information used by utilities is its location relative to other geographic features and objects. GIS technology that offers the combined power of both geography and information systems is an ideal solution for effective management of water industry infrastructure. Geotechnology and geospatial technology are alias names of GIS technology.

The days of the slide rule are long gone. Word processors are no longer considered cutting-edge technology. We are living in the information age, which requires us to be more than visionaries who can sketch an efficient infrastructure plan. Today's tech-savvy society expects us to be excellent communicators who can keep all the stakeholders — the public, the regulators, or the clients — "informed." New information and decision support systems have been developed to help us become good communicators. GIS is one such tool that helps us to communicate geographic or spatial information. In fact, a carefully designed GIS map can be worth more than a thousand words. Sometimes the visual language of GIS allows us to communicate without saying a single word, which is the essence of effective communication.

WHAT ARE GIS APPLICATIONS?

An application is an applied use of a technology. For example, online shopping is an application of Internet technology, automobile navigation is an application of GPS technology, and printing driving-direction maps is an application of GIS technology. No matter how noble a technology is, without applied use it is just a theoretical development. Applications bridge the gap between pure science and applied use. Highly effective water and wastewater utilities strive for continuous operational improvements and service excellence. GIS applications have the potential to enhance the management of our water, wastewater, and stormwater systems and prepare them for the operational challenges of the 21st century.

HISTORY OF GIS APPLICATIONS

GIS technology was conceived in the 1960s as a digital layering system for coregistered overlays. Started in the mid-1960s and still operating today, Canadian GIS is an example of one of the earliest GIS developments. Civilian GIS in the U.S. got a jump start from the military and intelligence imagery programs of the 1960s. The Internet was started in the 1970s by the U.S. Department of Defense to enable computers and researchers at universities to work together. GIS technology was conceived even before the birth of the Internet.

Just as technology has changed our lifestyles and work habits, it has also changed GIS. Though the art of GIS has been in existence since the 1960s, the science was restricted to skilled GIS professionals. The mid-1990s witnessed the inception of a new generation of user-friendly desktop GIS software packages that transferred the power of GIS technology to the average personal computer (PC) user with entry-level computing skills. In the past decade, powerful workstations and sophisticated software brought GIS capability to off-the-shelf PCs. Today, PC-based GIS implementations are much more affordable and have greatly reduced the cost of GIS applications. Today's GIS users are enjoying faster, cheaper, and easier products than ever before, mainly because of the advent of powerful and affordable hardware and software.

There were only a few dozen GIS software vendors before 1988 (Kindleberger, 1992); in 2001, the number had grown to more than 500. This revolution rightfully steered the GIS industry from a focus on the technology itself toward the applications of the technology (Jenkins, 2002). The strength of GIS software is increasing while its learning curve is decreasing. At this time, GIS is one of the fastest growing market sectors of the software industry and for a good reason: GIS applications are valuable for a wide range of users, from city planners to property tax assessors, law enforcement agencies, and utilities. Once the exclusive territory of cartographers and computer-aided drafting (CAD) technicians, today's GIS is infiltrating almost all areas of the water industry.

A GIS article published in *American City and County* in 1992 predicted faster computers and networks and that efficient database management and software will move GIS applications from property recording, assessing, and taxing functions to much more diverse applications during the 1990s (Kindleberger, 1992). This article anticipated future GIS applications to be rich in their use of multimedia, images, and sound. It expected GIS applications to become more closely linked to the 3D world of CAD as used by architects and engineers. Almost all of the GIS applications predicted in 1992 are now available except interacting with GIS data in a "virtual reality" medium wearing helmets and data gloves.

GIS literature is broad due to the wide variety of areas that utilize geographic data. Likewise, the literature describing GIS applications in the water industry is itself very broad. However, much of this work has been in the area of natural hydrology and large-scale, river-basin hydrology. A recent literature review conducted by Heaney et al. (1999) concluded that GIS applications literature exists in several distinct fields. In the field of water resources, recent conferences focusing on urban stormwater have several papers on GIS. Proceedings from two European conferences on urban stormwater by Butler and Maksimovic (1998), and Seiker and Verworn (1996), have a wealth of current information on GIS. The American Water Resources Association (AWRA) has sponsored specialty conferences on GIS applications in water resources, such as Harlin and Lanfear (1993) and Hallam et al. (1996). These reports have sections devoted to urban stormwater, of which modeling is a recurring theme. The International Association of Hydrological Sciences (IAHS) publishes the proceedings from its many conferences, some of which have dealt specifically with the integration and application of GIS and water resources management (e.g., Kovar and Nachtnebel, 1996).

In the early 1990s, not too many people were very optimistic about the future of GIS applications. This perception was based, in part, on geographic information technologies being relatively new at that time and still near the lower end of the growth curve in terms of (1) applications and (2) their influence as tools on the ways in which scientific inquiries and assessments were conducted (Goodchild, 1996). It was felt that several challenges related to our knowledge of specific processes and scale effects must be overcome to brighten the future of GIS applications (Wilson et al., 2000).

GIS applications for the water industry started evolving in the late 1980s. In the early 1990s, the water industry had started to use GIS in mapping, modeling, facilities management, and work-order management for developing capital improvement programs and operations and maintenance plans (Morgan and Polcari, 1991). In the mid-1990s, GIS started to see wide applicability to drinking water studies. Potential applications identified at that time included (Schock and Clement, 1995):

- GIS can provide the basis for investigating the occurrence of regulated contaminants for estimating the compliance cost or evaluating human health impacts.
- Mapping can be used to investigate process changes for a water utility or to determine the effectiveness of some existing treatment such as corrosion control or chlorination.
- GIS can assist in assessing the feasibility and impact of system expansion.
- GIS can assist in developing wellhead protection plans.

According to the American Water Works Association (AWWA), approximately 90% of the water utilities in the U.S. were using GIS technology by the end of the year 2000.

The use of GIS as a management tool has grown since the late 20th century. In the past 10 years, the number of GIS users has increased substantially. GIS technology has eased previously laborious procedures. Exchange of data between GIS, CAD, supervisory control and data acquisition (SCADA), and hydrologic and hydraulic (H&H) models is becoming much simpler. For example, delineating watersheds and stream networks has been simplified and the difficulty of conducting spatial data management and model parameterization reduced (Miller et al., 2004).

Today GIS is being used in concert with applications such as maintenance management, capital planning, and customer service. Many of us are using GIS applications on the Internet and on wireless devices without even knowing that we are using a GIS. These developments make GIS an excellent tool for managing water, wastewater, and stormwater utility information and for improving the operation of these utilities. Experts believe that in the near future, most water industry professionals will be using GIS in the same way they are now using a word processor or spreadsheet. Except for the computer itself, no technology has so revolutionized the field of water resources (Lanfear, 2000). In the early 1990s, GIS was being debated as the most controversial automation technology for the water industry (Lang, 1992). However, the time has come for all the professionals involved in the planning, design, construction, and operation of water, wastewater, and stormwater systems to enter one of the most promising and exciting technologies of the decade in their profession — GIS applications.

The Environmental Systems Research Institute (ESRI), the leading GIS software company in the world, has been a significant contributor to GIS applications in the water industry. ESRI hosts a large annual international user conference. The proceedings archives from these conferences are available at the ESRI Web site. This Web site also has a homepage for water and wastewater applications.

More information about GIS application books, periodicals, and Internet resources is provided in the author's first GIS book (Shamsi, 2002).

4M APPLICATIONS

Representation and analysis of water-related phenomena by GIS facilitates their management. GIS applications that are of particular importance to water industry professionals are: mapping, monitoring, modeling, and maintenance. These four *M*s define some of the most important activities for efficient management of water, wastewater, and stormwater systems, and are referred to as the "4M applications" in this book. With the help of new methods and case studies, the following chapters will show you how a GIS can be used to implement the 4M applications in the water industry. This book will demonstrate that with GIS the possibilities to map, monitor, model, and maintain your water, wastewater, and stormwater systems are almost endless. It will teach you how to apply the power of GIS and how to realize the full potential of GIS technology in solving water-related problems. This book does not train you in the use of a particular GIS software. It is not intended to help you run a GIS map production shop. Simply stated, this book will enable you to identify and apply GIS applications in your day-to-day operations.

ADVANTAGES AND DISADVANTAGES OF GIS APPLICATIONS

As described in the following subsections, GIS applications offer numerous advantages and a few drawbacks.

Advantages

Thanks to recent advances in GIS applications, we are finally within reach of organizing and applying our knowledge of the Earth in our daily lives. Typical advantages of GIS applications are described in the following subsections.

GIS Applications Save Time and Money

The foremost benefit of GIS technology is increased productivity and quicker turnaround. Increased efficiency saves time, which translates into saving money. GIS applications improve the quality of life because they make things easier to do. GIS allows us to perform routine work, such as keeping records of maintenance work or customer complaints, more efficiently. GIS tools have become user-friendly and easier to use. Local governments, utilities, and their consultants are using GIS to analyze problems and recommend solutions in a fraction of the time previously required.

GIS provides a spatial approach to organizing information about customers and the assets of a water or sewer utility, such as pipes, hydrants, pumps, and treatment equipment. GIS applications help a utility to analyze the spatial information about its customers and assets to improve planning, management, operation, and maintenance of its facilities. Municipalities and utilities that have successfully implemented GIS have seen dramatic improvements in the way in which data are retrieved, analyzed, and maintained. These improvements are allowing municipal and utility personnel to collect information more efficiently, better perform routine activities, and make more informed decisions.

GIS applications can significantly reduce time and costs associated with conventional analysis and evaluation methodologies (EPA, 2000).

GIS Applications Are Critical to Sustaining GIS Departments

Continued development of new applications is critical to sustaining the growth of a new technology. GIS, being a new technology itself, might not survive unless people use it in their routine business operations to make things easier to do, to enhance productivity, and to save both time and money. To justify their existence in an organization, GIS departments should use GIS to develop cost-effective solutions that make people's life easier. Simply stated, GIS applications are the key to garnering the management's financial support for GIS departments. This book shows how to put GIS technology to productive use in the water industry.

Although GIS applications in the water industry are not new, getting beyond basic inventory and mapping functions is often challenging. Unfortunately, mapping efforts alone do not always justify the financial support for a GIS group. Unless a GIS is taken to the operational level, it is nothing but a pretty map. That is why GIS emphasis is now shifting from producing high-quality maps to enterprise-wide mission-critical applications. The benefit–cost ratio of GIS increases with its functionality and applications. GIS applications of automated mapping return a 1:1 benefit–cost ratio. Benefit–cost ratios of 4:1 can be attained when GIS use expands to all the departments of an organization (Alston and Donelan, 1993).

GIS Applications Provide the Power of Integration

The typical local government office contains hundreds of maps displaying such information as municipal boundaries, property lines, streets, sewer pipes, water mains, voting district boundaries, zoning areas, flood plains, school bus routes, land use, streams, watersheds, wetlands, topography, geology, and soil types, to name a few. Paper maps, after all, have been the traditional method of storing and retrieving geographically referenced information. The sheer number, range of types, and diversity of maps used by municipalities are evidence of the importance geographically referenced information plays in our day-to-day operations. Unfortunately, the wide variety of maps and diversity of their scales and designs at our disposal make it extremely difficult to access, use, and maximize the value of the information they contain. GIS *integrates* all kinds of information and applications with a geographic component into one manageable system.

The real strength of GIS is its ability to integrate information. This integration power makes the scope of GIS applications almost infinite. A GIS can be whatever we want it to be. GIS can organize the geographic information of a municipality or utility into one seamless environment. The unique integration capability of GIS allows disparate data sets to be brought together (*integrated*) to create a complete picture of a situation. GIS technology illustrates relationships, patterns, and connections that are not necessarily obvious in any one data set but are amazingly apparent once the data sets are integrated. The integration capability of GIS technology empowers organizations to make better and informed decisions based on all relevant factors (ESRI, 2003). GIS offers integrated solutions in the areas of planning and engineering, operation and maintenance, and even finance and administration.

GIS Applications Offer a Decision Support Framework

GIS helps decision makers by pulling together crucial bits and pieces of information as a "whole" and showing them the "big" picture. In this regard, GIS can be used as a consensus-building and decision-making tool. By using geography as the common denominator, GIS permits data from a wide range of disparate sources to be combined and analyzed. Therefore, an important benefit of GIS applications is their inherent ability to integrate and analyze all spatial data to support a decision-making process. GIS provides uniformity of data usage and the flexibility to test and evaluate multiple scenarios. The use of a common database eliminates the differences in presentation, evaluation, and decision making based on using different forms and types of data. For example, civil engineers who need to know the lay of the land to design, build, and maintain projects can learn from the ways in which utilities and municipalities are linking GIS data to every aspect of their computing enterprise (Goldstein, 1997). A GIS provides the opportunity to conduct sensitivity analyses appropriate for the level of accuracy of the input data. This allows engineers, planners, elected officials, and the public to focus on the impacts and analysis of alternatives rather than the accuracy of data. After the planning and decision-making phase has been completed, GIS can continue to support the implementation phase of a project by tracking the success and failures of alternative approaches. Plan performance tracking and testing of new approaches is based on new parameters, new information, and new conditions within or outside the study area (EPA, 2000).

GIS Applications Provide Effective Communication Tools

GIS fosters better communication and cooperation among various stakeholders (e.g., community leaders and the public) of a water industry construction or improvement project. Many people learn better with maps than they do with words or numbers. GIS can be used to communicate with different audiences using visually different views of the same data. For instance, 3D plan views of a water or sewer system improvement project can be used for presentations at town meetings to graphically illustrate necessary improvements. Because GIS is a visual language, it is an excellent communication tool for visual learners. A picture is only worth a thousand words. A map may be worth a thousand numbers. But a GIS is worth a thousand tables.

In the late 1990s, GIS advocates noted that robust citizen participation in ongoing policy making was limited because many groups lacked access to the GIS environment (Obermeyer, 1998). This has started to change over the last 5 years mainly due to availability of browser-based "Internet GIS" technology described in Chapter 6 (Internet Applications). However, the challenge remains to provide citizens with GIS applications that until recently were only available to professionals.

If bottom–up decision making is to succeed, all stakeholders must have access to GIS technology in various forms (Miller et al., 2004).

GIS Applications Are Numerous

GIS is as ubiquitous as it is today because it is a very effective tool for an incredible number of applications (Zimmer, 2001). Today's GIS is limitless! The number of GIS applications is limited only by our own imagination and the availability of data. For instance, the municipal applications alone are numerous, including:

- Water, wastewater, and stormwater operations
- Comprehensive planning

- Vulnerability assessment and security planning
- Permitting and code enforcement
- Building inspections
- Zoning
- Parcel mapping
- Pavement management
- Tracking customer complaints
- Grant applications
- Routing

The magnitude of GIS applications can be appreciated by the fact that this entire book is devoted to one application from the preceding list.

A recent survey conducted by Geospatial Information and Technology Association (GITA) indicates the following top ten applications in the utility industry (Engelhardt, 2001; GITA, 2001):

- Landbase model
- Work management
- Facility model analysis/planning
- Operations and maintenance
- Document management
- Customer information systems
- Workforce automation
- Regulatory reporting
- Environmental testing
- Marketing

The same survey lists the following technologies driving GIS implementation for utilities:

- Nonproprietary programming
- Pen/mobile computing/field data capture
- Internet/intranet
- Data exchange/open GIS
- Document management/workflow
- GPS
- Digital orthophotography
- Satellite imagery

The water industry was found to be focusing on the following applications:

- Work management
- Facility model analysis/planning
- Pen/mobile computing

Water utilities were found to be implementing more pen/mobile systems than any other market segment. It was also found that water utilities were seeking higher landbase accuracy of 5 ft compared with other utilities, for instance, the 50-ft accuracy sought by gas companies.

The major GIS applications for the water industry are summarized in the following list:

- GIS provides the ideal means of describing water and sewer infrastructure facilities, identifying problems and recommending solutions, scheduling and recording maintenance activities, and supporting technical analysis (e.g., hydraulic modeling) of the facilities. For example, GIS can be used for mapping the water mains and identifying water main breaks in terms of location, pressure, soil type, pipe size, pipe material, or pipe age. Such applications are described in Chapter 9 (Mapping Applications), Chapter 10 (Monitoring Applications), and Chapter 15 (Maintenance Applications).
- Various spatial data layers can be combined and manipulated in a GIS to address planning, operation, and management issues. For example, water and sewer line information can be combined with population statistics and ground elevation data to assess the adequacy of water and sewer utilities. The Metropolitan District of Greater Cincinnati (Ohio) uses GIS to locate storm sewer system problem areas. The trouble spots are identified and targeted for preventive maintenance by mapping the relationship between customer complaints and amount of rainfall (Mitchell, 1997). Such applications are described in Chapter 9 (Mapping Applications), and Chapter 15 (Maintenance Applications).
- GIS topology provides information about how the network elements are connected with each other and what is the direction of flow. This capability makes GIS ideally suitable for identifying customers of a utility network affected by service interruption, such as water main leaks and breaks. For instance, the Cherokee Metropolitan District (Colorado) uses a "Water/Wastewater" option on the district's GIS menu to display, plot, and identify the valves to be shut off to repair water system leaks and identify and notify the customers who will be out of water due to the valve closure (Mitchell, 1997). Such applications are described in Chapter 9 (Mapping Applications), and Chapter 15 (Maintenance Applications).
- GIS can be used to satisfy regulatory requirements that are increasingly reliant on computer-generated data and maps. For example, GIS can be used to develop water/sewer system inventory reports and watershed protection/management plans. Such applications are described in Chapter 10 (Monitoring Applications).
- GIS can be used to develop hydrologic and hydraulic (H&H) computer models for water and sewer systems, watersheds, and floodplains. These applications are described in Chapter 11 (Modeling Applications), Chapter 12 (Water Models), and Chapter 13 (Sewer Models).
- GIS can be used to design efficient meter-reading routes. This can be accomplished by linking the customer account database to the streets GIS layer. This application is described in Chapter 9 (Mapping Applications).
- GIS topology can help us to simulate the route of materials along a linear network. For example, we can assign direction and speed to a streams layer to simulate the fate of an accidental contaminant release by a factory through the stream network.
- GIS can be integrated with automated mapping/facilities management (AM/FM) systems to automate inspection, maintenance, and monitoring of water and sewer systems. Sample applications include:
  - Preparing work orders for inspection and maintenance activities
  - Scheduling TV inspection and cleaning of sewers
  - Identifying the valves that must be closed to repair a broken water pipe
  - Keeping track of leak detection survey for a water system

- Creating a map of customer complaints, pipe breaks, and basement flooding and identifying the reasons
- Improving management of labor resources through more efficient deployment of field crews. AM/FM applications are described in Chapter 14 (AM/FM/GIS Applications), and Chapter 15 (Maintenance Applications).
- GIS analyses can be used to develop a decision support system for the efficient operation and management of stormwater, best management practices (BMPs), floodplains, combined sewer overflows (CSOs), and sanitary sewer overflows (SSOs). Such applications are described in various chapters of this book.
- GIS can be used to integrate related technologies, such as relational database management systems (RDBMSs), the Internet, wireless communications, CAD, GPS, and remote sensing (satellite imagery). The integrated platform provides the best of all worlds. Such applications are described in various chapters of this book.

Disadvantages

As with any new technology, GIS has some drawbacks. The first issue is the substantial time and cost required to compile and analyze the necessary data. High initial costs are generally incurred in purchasing the necessary hardware, software, and for ongoing maintenance. Though the advantages of GIS applications are dramatic, the failure to effectively implement GIS can lead to disappointment and disillusionment with the technology. Improperly designed and planned GIS applications can result in costly and time-consuming efforts. GIS applications are disadvantageous when one fails to define a vision, understand the vision requirements, define the tools needed to attain the vision, and select appropriate technology to integrate those tools. Only when a GIS is fully understood with proper training and education should one expect its applications to be limitless. Needs analysis, described in Chapter 2, can be used to avoid this pitfall.

Another common pitfall in GIS application development is capturing more data than required by the application. This approach is called a "data-driven" or "bottom–up" approach. For example, a nonpoint source modeling project spent substantial time and effort to capture detailed soils series survey data from the U.S. Natural Resources Conservation Service (EPA, 2000). However, the final application ended up using the Universal Soil Loss Equation for erosion modeling, which required the simpler soil associations data instead of the detailed soils series data. Considerable money could have been saved if an appropriate model and its data requirements had been identified before starting the data conversion process. The use of inappropriate data in GIS applications may lead to misleading results. The data must be of appropriate scale and resolution and highly documented to be useful in GIS applications. Users must be extremely conscious of the nature of the source information to avoid abusive extrapolations and generalizations.

The GIS learning curve, privacy issues, and periodic shortage of skilled personnel are some other challenges of GIS implementation. However, as the *San Diego Union-Tribune* (1998) reports, "Those who overcome such hurdles soon find GIS applications breathtaking in scope and far-reaching in the potential to affect, if not shape and change, everyday lives."

SUCCESS STORIES

Some compelling examples demonstrating the benefits of GIS applications are described in the following subsections.

San Diego

The City of San Diego was an early convert to GIS technology and is considered a leader in GIS implementation. Having the motto, "We have San Diego covered," SanGIS is a joint agency of the City and the County of San Diego, responsible for maintenance of and access to regional geographic databases for one of the nation's largest county jurisdictions covering more than 4200 mi^2. SanGIS spent approximately $12 million during a 14-year period from 1984 to 1998 to collect GIS data. The conventional surveying approach would have cost them about $50 million (the *San Diego Union-Tribune*, 1998). The GIS/GPS approach has saved the City and the County of San Diego millions of dollars.

If we define a "savings factor" as the ratio of conventional approach (non-GIS) cost (or time) to GIS approach cost (or time), San Diego's success story resulted in a savings factor of 4.2 (50/12) or 420% cost savings.

Boston

The Massachusetts Water Resources Authority (MWRA) provides water and wastewater services to 2.5 million people in 60 municipalities of the Greater Boston area. The MWRA service area spans more than 800 mi^2 that contains several treatment plants, 780 mi of large-size pipelines, and dozens of pumping stations and tunnels. The MWRA recognized the potential for GIS to save ratepayers' money and initiated a GIS program in 1989. As the GIS data have been used repeatedly by individual communities to protect their water resources, the investment continues to pay off years later. For example, their geologic database "gBase" contains information on deep rock borings for the many tunnel and dam projects designed over the past 100 years. Ready access to this information guides current geologic exploration and enables MWRA to better locate new borings, which can cost $20,000 to $50,000 each (Estes-Smargiassi, 1998).

Cincinnati

Faced with a $10-billion network of aging infrastructure that included a water system with a capacity of 50 billion gallons a year, the City of Cincinnati, Ohio, conducted an in-depth feasibility study. The study showed that a GIS would save the City $11 million over a 15-year period, with payback anticipated within 8 years of implementation (American City & Country, 1993).

Knoxville

A $5.5-million GIS project that was started in 1986 jointly by the City of Knoxville, Knox County, and the Knoxville Utility Board (Tennessee) expected to pay for itself within 8 years (Dorris, 1989). In 1992, General Waterworks (King of Prussia, Pennsylvania) owned and managed 34 small- and medium-size water utility companies. The company's GIS experience indicated that a GIS at a medium-size water company can pay for itself in as little as 3 years (Goubert and Newton, 1992).

Dover

The planners in the City of Dover (New Hampshire) used GIS from Intergraph to determine where and how much solid waste needed to be collected and removed. Without GIS, collecting this information would have taken approximately 6 months. Using GIS, it only took a few weeks. This information was provided to bidders to help them base their bids on facts rather than assumptions. This application resulted in lower-priced bids (Thompson, 1991). *This success story resulted in a savings factor of 12 or 1200% cost savings.*

Charlotte

The City of Charlotte and Mecklenburg County of North Carolina, which are among the fastest growing metropolitan areas of the U.S., have developed a Watershed Information System called WISE that integrates data management, GIS, and standard hydrologic and hydraulic (H&H) programs such as HEC-1, HEC2, HEC-HMS, and HEC-RAS. Using this information technology (IT)–based integration method, the existing H&H models can be updated at a fraction (less than $100,000) of the cost of developing a new model (more than $1 million) (Edelman et al., 2001). *This success story resulted in a savings factor of 10 or 1000% cost savings.*

Elsewhere, an ArcInfo GIS software interface with H&H modeling packages HEC-1 and HEC-2 was found to offer substantial cost savings. It was determined that if the H&H models were not linked to the GIS, manual calculations would be required that would take at least five times longer than using the GIS linkage (Phipps, 1995). *This success story resulted in a savings factor of 5 or 500% cost savings.*

Thanks to a GIS approach, planning tasks previously requiring months to complete now take only days in Albany County, Wyoming.

Albany County

Albany County, located in southeastern Wyoming, covers 4,400 mi², a student-based population of 30,000, and 1,600 mi of roads. For rural communities like Albany County, building a GIS from scratch can be an expensive endeavor due to a lack of resources. For example, the cost associated with updating the county's existing digital aerial photography was estimated to exceed $100,000. A geographic-imaging approach consisting of GIS and satellite imagery, on the other hand, allowed the county to have high-resolution and up-to-date views of the entire county for $32,000. Thanks to the GIS approach, planning tasks previously requiring months to complete now take only days (Frank, 2001). *This success story resulted in a savings factor of over 3 or 300% cost savings.*

GIS Applications Around the World

Although most of the GIS application examples in this book are from the U.S., GIS applications for water and wastewater systems are growing throughout the world. In the early 1990s, the United Nations Institute for Training and Research (UNITAR) started to provide GIS training to specialists from developing countries with the broader goal that each participant would eventually establish a foundation for GIS technology in his or her own country. Established in 1992, the Center for Environment and Development for the Arab Region and Europe (CEDARE) recognized that GIS was a key tool for efficient collection, management, and analysis of environmental data in the Middle East. Recognized as one of the best GIS implementations in the world, the Environment Department of the State of Qatar developed a GIS in the mid-1990s and integrated it with their environment management tasks. Fueled by high-resolution satellite imagery from the Indian Remote Sensing (IRS) satellite IRS-1C, the GIS/GPS technology saw a boom in the late 1990s.

The Sistema de Agua Potabley Alcantarillado de León (SAPAL) in León, Mexico, began its GIS implementation in 1993 and built a complete inventory of their water system, including fittings, pipes, and pumps, using ESRI's ArcView and ArcInfo GIS software packages. The Center for Preparation and Implementation of International Projects, Moscow, implemented several environmental management GISs in the Yaroslavl, Vologda, Kostroma, and Ivanovo regions of Russia. The South Australian Water Corporation (SA Water) developed the Digitized Facilities Information System (DFIS) to extract digital data for conceptual design of more than 32,000 km of water and wastewater mains. SA Water's DFIS was used to develop a Water Master Plan for the Indonesian province of West Java for the design, construction, and operation of wastewater treatment, water supply, water storage, and flood control facilities. South Africa's Working for Water organization developed a prototype ArcView GIS-based project information management system to track more than 250 water resources management projects across the country (ArcNews, 2001).

EVOLVING GIS APPLICATIONS AND TRENDS

GIS technology is changing rapidly. New GIS applications are evolving frequently mainly due to the successful marriage of GIS and the Internet. GIS applications are being fueled by the recent advances in wireless, the Internet, networking, and satellite technologies. The cost of spatial data is falling rapidly due to competition in data acquisition, processing, and distribution. More intuitive and simpler interfaces are taking GIS beyond the world of the computer geeks, techno wizards, and GIS gurus. User interfaces are becoming friendlier, wizards are replacing obscure command lines, and the use of GIS by semiskilled end users is growing (Limp, 2001). These factors are resulting in the evolution of new GIS applications at unprecedented speed. Major evolving GIS applications and trends are listed in the following list:

- Storing both the geographic and the attribute data in one database
- Representing real-world features as objects rather than as geometric entities (e.g., points, lines, and polygons)
- Convergence of image processing and GIS into the large umbrella of information technologies (Limp, 2001)
- Using high-resolution submeter satellite imagery as base maps
- Capturing high-resolution images suitable for photogrammetric analysis using inexpensive and powerful digital cameras
- Using GPS technology for better locations of mobile sensor platforms and ground control points
- Accessing geographic data through wireless application protocol (WAP)–enabled devices, such as cellular phones, personal digital assistants (PDAs), mobile terminals, and pocket personal computers
- Using GIS and hydraulic modeling for vulnerability assessment and security planning to safeguard the water industry against sabotage and potential terrorist attacks
- Integration of computer, video camera, and GPS referred to as "video mapping" to document smoke tests and TV inspection of pipes
- Creation of an accurate virtual reality representation of landscape and infrastructure with the help of stereo imagery and automatic extraction of 3D information
- Rapid emergence of Extensible Markup Language (XML) as the *lingua franca* for data encoding and wireless Web applications
- Evolution of Geography Markup Language (GML) from XML for accurate and effective representation of GIS data in Web browsers (Waters, 2001)
- Evolution of a spatial Network Query Language (NQL) from the structured query language (SQL) for serving GIS data over the Internet, wireless, and PDAs (Waters, 2001)
- Migration from Unix workstations to Microsoft Windows desktop environment

FUTURE APPLICATIONS AND TRENDS

It is a rainy day on May 1, 2010, and it is Chrysa's first day on her new job. Chrysa is a newly hired sewer system maintenance person for a large city. She has a few years of prior sewer maintenance experience, but she is not thoroughly familiar with the sewer system of this city. As she is arranging her modular work space, the customer service department forwards her a basement flooding complaint. She talks to the customer, hangs up the phone, and slips on a headset as she leaves the office. The headset provides an enhanced reality system that combines glasses, earphones, and a tiny microphone, but weighs a little more than a pair of current sunglasses (Turner, 2001). When Chrysa reaches the subject property, she issues a simple voice command to view the buildings and sewer lines superimposed on a 3D wire-frame display of the landscape around her. As she walks around the property, her virtual reality display changes to show what is in front of her. From the 3D displays Chrysa notices that the basement of the subject property is only slightly higher than the sanitary sewer serving the property. She then uses her wireless and GPS-enabled PDA to connect to and query the City's centralized enterprise database. The GPS automatically identifies the Lot ID, and the database tells her that the subject property's service lateral does not have a backflow prevention valve.

Chrysa's display also appears simultaneously on the office computer and PDA of her supervisor, Brian. Brian sees a shopping center upstream of the property and clicks on it. The computer indicates that the shopping center construction was completed last month. Brian concludes that the new flow from the shopping center is overloading the sanitary sewer and causing the basement flooding. Brian also authorizes Chrysa to make arrangements for the installation of a backflow prevention valve. Chrysa clicks on the service lateral of the property and creates a work order with a few clicks on her PDA. Before leaving the property, she e-mails the work order to the City's sewer maintenance contractor and copies the same to Mr. Jones, the owner of the property.

Such scenarios may seem like science fiction, but much of the technology to support them already is under development or available in prototype form, including high-speed wireless, GPS, and mobile GIS.

Experts believe that in the near future most water industry professionals will be using GIS in the same way they have used a word processor or spreadsheet.

Around the mid-1990s, not too many people were too optimistic about the future of GIS applications. Today, the future of GIS looks bright. GIS is experiencing rapid growth as an information management tool for local and regional governments and utilities because of its powerful productivity and communication capabilities. Many current and evolving GIS applications have been described in the preceding text. Other potential applications will be discovered in the future, as GIS technology and regulatory requirements continue to evolve. All sectors of business and government that depend on accurate geographical information will continue to benefit from GIS applications (Robertson, 2001). In 1999, core worldwide GIS business (hardware, software, services) revenue grew by 10.6% to an estimated $1.5 billion (Daratech, 2000). In 2001, this revenue jumped to $7.3 billion. The world will be very different a decade from now. Advances already underway in IT, communications infrastructure, microelectronics, and related technologies will provide unprecedented opportunities for information discovery and management (Turner, 2001). Looking beyond 2005, GIS-related technologies (e.g., wireless, Internet, networking, GPS, and remote sensing) are expected to do even better. Integration of photogrammetry with other geospatial technologies is opening new doors for exploring and developing alternative spatial applications. The commercial remote sensing market is flourishing and is poised for revenues of $2.7 billion within 3 to 4 years. Location-based wireless services revenues in North America alone are expected to reach $3.9 billion by 2004 (Geospatial Solutions, 2001). A growing GIS market has resulted in lower GIS implementation costs, increased data accuracy, and continued improvement in GIS hardware and software. These developments will help us to develop broader GIS applications (Robertson, 2001).

Future prognosis is a popular hobby of GIS gurus. Most GIS experts believe that the future of GIS looks bright thanks to recent advances in GIS-related technologies. Due to GIS-related innovations, future GIS applications are expected to be centered around commercial remote sensing, location-based services (LBS), mobile GIS, and the Internet. The early 21st century will bring Web-based client/server solutions, interoperability, on-demand data, and software download via the Internet on a pay-per-use basis, and an abundance of online data that can be processed in the field using wireless Web-savvy devices (Waters, 2001). Instead of purchasing GIS data permanently, future GIS users will be able to subscribe to pay-per-view type plans and pay based on the extent and duration of their usage. The technology is moving so fast that you may be using some of these applications by the time you read this book.

Though GIS has come far, big changes are still in the works (Lanfear, 2000).

To some it may appear that the future of GIS is already here; but it is not. Though GIS has come far, big changes are still in the works (Lanfear, 2000). Based on the predictions of some GIS industry experts (Lanfear, 2000; Geospatial, 2001; Engelbrecht, 2001), the following trends are expected in future GIS applications:

- Spatial data will become more widely accepted, unit costs of spatial data will fall, and GIS applications will become more affordable.
- The explosion of spatially enabled consumer products will benefit the entire GIS applications industry. Soon GPS will reside inside our wrist watches. Location information via GPS will become as common as timekeeping on a wrist watch. The new generation of low-cost consumer GPS products and spatially enabled mobile phones and PDAs will provide consistent accuracies of 10 m or better — sufficient to navigate people to their destination.
- As the bandwidth of the Internet improves, it will enable us to deliver information-rich geospatial content directly to end users.
- Wireless Internet use will surpass wired use. By 2005, over 60% of world population will have wireless Internet connections.
- Billions of people will use a Web-enabled mobile device, and millions of those devices will be equipped with LBS capabilities, either through GPS, a wireless network, or a hybrid solution (Barnes, 2001). LBS will combine GIS applications with user-friendly mobile devices to provide needed information at any time or place. LBS devices paired with wireless technologies will provide on-demand geospatial information.
- High-resolution local GIS data will be merged with federal data inventories.
- In addition to GIS input (raw data), the World Wide Web clearinghouses will share the GIS output (processed data) and the processing tools themselves.
- The new generation of raster imaging tools will allow processing of satellite imagery inside spreadsheets and word processors.
- Commercial remote sensing companies will be selling more than satellite images; they will be offering subscription-based monitoring services for change detection and vegetation indices, etc.
- Maps of dubious accuracy will not be digitized into a GIS. Contemporary imagery will be obtained for each unique application or solution.

- There will be a digital database of land-cover imagery and vectors that will include the major landmasses of the entire world.
- Laser imaging detection and ranging (LIDAR) will become a true GIS tool and produce new varieties of GIS data products. Enhanced interpretation of the strength of airborne laser signals will enable automatic creation of low-grade digital orthophotos.
- GIS databases will be connected with real time sensor data inputs, which will provide new opportunities for mapmakers and resellers of GIS products and services. New markets will be developed to expand our selection of data sets.
- Last but not least, the future may create a new set of degree, licensing, and certification programs in the field of GIS.

GIS APPLICATION DEVELOPMENT PROCEDURE

As shown in Figure 1.1, developing GIS applications generally requires six typical steps:

1. Needs analysis (strategic planning)
2. Specifications (system design)
3. Application programming
4. Testing (pilot project)
5. Installation (hardware, software, and data)
6. Ongoing operation and maintenance (includes training)

Most of these activities are generally conducted in a needs analysis study as described in Chapter 2 (Needs Analysis). Step No. 3 requires software development as described in the following text. Step No. 4 is highly recommended because it provides an opportunity to test the system design for a small pilot project area and fine-tune the design and computer programs, if necessary. Pilot testing allows system designers and programmers to tweak the design to meet user needs. Pilot testing can even save money because the data or software can be changed before full

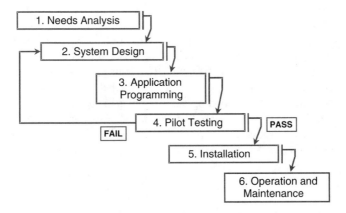

Figure 1.1   Six typical steps for developing GIS applications.

systemwide implementation, at which time it might be too expensive to make major changes.

APPLICATION PROGRAMMING

Application programming means writing the computer programs for GIS applications. Creative application developers strive to develop GIS applications to streamline the operations of their water industry business. They can find something useful in a GIS package and turn it into an innovative application. There are two methods of application programming:

1. GIS-based approach
2. Application-based approach

GIS-Based Approach

The easiest way to develop GIS applications is by extending the core capabilities of a GIS software. In this method, application functions are incorporated in a GIS. Application programs can be developed in a GIS or called from a GIS. This approach provides more GIS capability than the application-based approach. Application capability depends on the GIS software and the method of developing and integrating the application. The flexibility, adaptability, and openness of a GIS software are key to developing applications for that GIS platform. Depending on the complexity of the application, the GIS-based approach can be implemented using one or more of the following four methods:

1. GIS customization
2. Scripting
3. Extensions
4. External programs

GIS Customization

When supported by a GIS package, this method is implemented by customizing the GIS software's graphical user interface (GUI). Examples include adding or deleting a button or pull-down menu. For example, the buttons or menus that are irrelevant to a particular application can be deleted to simplify the learning process and prevent unwanted edits. New buttons or menus can be added to display helpful tips and messages. This method is more suitable for supporting applications rather than developing them. For example, it can be used to add a button that runs another program or script. This method is not very useful when used alone; it is therefore used in conjunction with the other methods described below. When used alone, it does not require computer programming.

Scripting

Scripting is suitable for simple applications such as creating a link to an H&H model. This method requires programming using a scripting language. Scripts are small computer programs written in a scripting language. A scripting language is a programming language that is (usually) embedded in another product, such as Microsoft's Visual Basic or Autodesk's AutoLISP. ESRI's ArcGIS software (Version 8 and higher) allows programming using Visual Basic for Applications (VBA). Older versions of ESRI's ArcInfo and ArcView software allow programming using Arc Macro Language (AML) and Avenue scripts, respectively. Avenue is ArcView's native scripting language built into ArcView. Avenue's full integration with ArcView benefits the user in two ways: first, by eliminating the need to learn a new interface; second, by letting the user work with Avenue without exiting ArcView (ESRI, 1995). A sample Avenue script for displaying sewer TV inspection video movies in ArcView GIS is given in the following text.

```
theVal = SELF
if (not (theVal.IsNull)) then
if (File.Exists(theVal.AsFileName)) then
  System.Execute("C:\Program Files\Windows Media
Player\MPLAYER2.EXE "+theVal)
else
  MsgBox.Warning("File "+theVal+" not found.,""Hot Link")
end
end
```

This "Movie&ImageHotLink" script enables users to click on a sewer line in the GIS map to play the digital movie of the sewer defect. It receives the movie filename, checks to see if the file exists, and then opens it with a user-specified movie player such as Windows Media Player. The reason for the small size of the script is the fact that it takes advantage of an existing program (i.e., it uses Media Player). Figure 1.2 shows an ArcView screenshot displaying the results of running this script. As the user click on a collapsed pipe, its video file starts playing in the Media Player window.

Figure 1.3 shows the integrated VBA programming interface of ArcGIS 8.3. The left window shows VBA code to automatically assign polygon feature IDs to point features (Lundeen, 2003). For example, this script can be used to create parcel IDs for water meters or to create basin names for manholes. The right window shows the water system layers and parcels in ArcMap. ESRI's ArcScripts Web site is a good place to find downloadable scripts for ESRI GIS software.

Extensions

Extensions are suitable for developing complex applications such as adding raster GIS capability to support hydrologic modeling. Scripts should be compiled and linked to GIS software before they can be executed, which is cumbersome. A set of scripts can be converted to an extension for faster and user-friendly installation and execution. Extensions are special scripts that are loaded into the GIS software to

Figure 1.2   Sample application that plays sewer TV inspection videos in ArcView GIS.

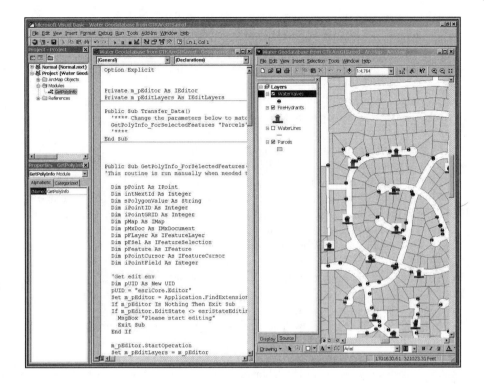

Figure 1.3   Integrated application development environment of ArcGIS (Version 8.3).

increase its capabilities. For example, ArcView 3.2 extensions can be loaded (or unloaded) by selecting them from the File\Extensions pull-down menu. Small basic extensions, such as the "hydro" (hydrology) extension are included with the ArcView software. Hundreds of free basic extensions written by the ESRI software user community can be downloaded from the ArcScripts Web site. Complex special-purpose extensions, such as Spatial Analyst Extension (adds raster GIS capability to ArcGIS) or 3D Analyst Extension (adds 3D capability to ArcGIS) should be purchased separately.

External Programs

Some scripting languages are not suitable for programming mathematically complex algorithms. For many applications (e.g., H&H modeling), the computer code is written and maintained by government agencies. Converting an existing code to a script might be error prone and cumbersome. In these situations, external programs can be called in from inside a GIS. The Movie&ImageHotLink script described earlier uses this method when it uses the external Media Player program. ArcGIS allows developing cross-platform applications with C++ or Java languages.

A dynamic link library (DLL) is an executable module that contains functions or procedures that can be called from an external program. For example, each major component of the Windows operating system is a DLL. Custom DLLs can be created using languages such as C and C++. Many GIS packages such as ArcView 3.x support DLLs that can be used to provide efficient interfacing with third-party external programs. Functions in DLLs can be called directly from Avenue, and data can be passed back and forth.

Application-Based Approach

In this method, GIS (mapping) functions are incorporated in existing applications. GIS functions are developed in, or are called from, an existing application. This method offers limited GIS and complete application functions. For instance, ESRI MapObjects and ArcObjects software provide programmable objects that can be plugged into VBA applications. RJN Group's CASSView module is an example of this approach. CASSView integrates an infrastructure maintenance management system in MapObjects environment. It requires the RJN CASS WORKS system and a minimum of a geo-coded street centerline to operate. CASSView provides CASS WORKS users with access to maintenance information, as-built records, maps, and GIS data. The end result is a database-driven desktop mapping application that provides visual data on demand.

In-house development of applications requires computer programming skills. It is often difficult to find people skilled in GIS-specific programming languages such as Avenue and AML. For advanced applications, the purchase price of commercial packages is usually less than the labor cost of in-house software development. A large variety of third-party applications from facilities management to H&H modeling is available for purchase from commercial vendors. Most commercially available software is listed in the "Business Partner" page of the GIS software Web site or advertised in the periodicals of the GIS software company.

USEFUL WEB SITES

ESRI user conference proceedings	www.esri.com/library/userconf/archive.html
ESRI water and wastewater industry	www.esri.com/industries/water/water.html
ESRI ArcScripts	arcscripts.esri.com
RJN Group (CASSView)	www.rjn.com
San Diego GIS home page	www.sangis.org

CHAPTER SUMMARY

This chapter showed that with GIS the possibilities to manage your water, wastewater, and stormwater systems are almost endless. It provides an ideal platform to integrate various business operations and technologies. GIS applications increase worker productivity and save time and money, and are critical to garnering financial support for the GIS departments of an organization. They equip us with better communication and decision-making tools. GIS applications are so numerous that the possibilities can be limited only by our own imagination. The following chapters will describe the procedures for and examples of developing GIS applications for the typical needs in the water industry. In particular, we will focus on the four areas of applications: mapping, monitoring, modeling, and maintenance. Referred to as the 4M applications, these are most critical to the effective operation and management of water, wastewater, and stormwater systems. Due to recent advances in Internet and wireless technologies, new GIS applications are evolving at an unprecedented rate. It is expected that future GIS applications will be even more exciting than those that have been listed in this chapter. The effect of these new applications will be profound. They will be inexpensive, user friendly, and ubiquitous, and will support the integration of GIS and related technologies in ways unlike anything current water industry professionals have ever envisioned.

CHAPTER QUESTIONS

1. What are GIS applications and how are they developed?
2. What are the pros and cons of GIS applications? How can you avoid the potential pitfalls of GIS applications?
3. What technologies are fueling the popularity of GIS applications?
4. List your ten favorite GIS applications.
5. How will your current GIS needs benefit from potential future GIS applications?

CHAPTER 2

Needs Analysis

A careful needs analysis is critical to successful GIS implementation and should be the first task of any GIS project.

Needs analysis requires a thorough analysis of existing data sources.

LEARNING OBJECTIVE

The learning objective of this chapter is to understand the value and methodology of needs analysis for GIS application projects.

MAJOR TOPICS

- Definition of needs analysis
- Needs analysis steps
- Needs analysis case studies

LIST OF CHAPTER ACRONYMS

COTS Commercial Off-the-Shelf
DBMS Database Management System
GUI Graphical User Interface
IT Information Technology
RDBMS Relational Database Management System

OCEAN COUNTY'S STRATEGIC PLAN

Ocean County Utilities Authority, located in central New Jersey, serves 36 municipalities through three wastewater treatment plants with a combined capacity of 80 MGD. The sewage collection system consists of 200 mi of sewer lines and 40 pumping stations. The Authority developed a GIS program as an integral part of its information technology (IT) strategy. Instead of treating the GIS as a stand-alone system, the Authority envisioned using it as integrated part of its other enterprise systems such as the SAP Maintenance Management System. Authority's IT-based approach included the following steps (Stupar et al., 2002):

- Needs analysis: This step identified goals of the GIS as they related to the Authority's business needs.
- Spatial database design: The objective of this step was to treat GIS data as an enterprise asset. ESRI's geodatabase data model was used to design the database.
- Data conversion: This step involved collecting the needed data and converting it to a geodatabase.
- Application development: This step designed and implemented applications for the GIS.

INTRODUCTION

Needs analysis (or needs assessment) identifies and quantifies the GIS needs of an organization and its stakeholders. Performing a needs analysis is the crucial first step to a successful GIS project. Needs analysis is analogous to strategic planning; it is a blueprint for funding, implementing, and managing a GIS. Like strategic planning, a careful needs analysis is critical to a successful GIS implementation and

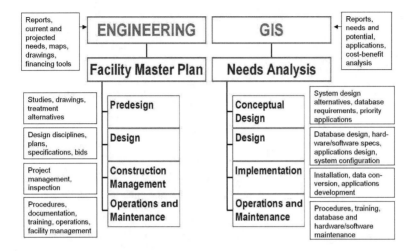

Figure 2.1   Comparison of needs-analysis study and facility master plan.

should be the first task of any GIS project. Needs analysis clarifies the project's specific needs and defines how a GIS will benefit an organization by relating specific organizational resources and needs to specific GIS capabilities (Wells, 1991). There are three major goals of needs analysis:

1. To define current work processes, data, and IT resources in place
2. To determine how the organization hopes to use GIS to streamline its operations
3. To make recommendations defining the path to meet these hopes

Needs analysis identifies potential applications that can be performed more efficiently using GIS technology. The applications are identified through stakeholder interviews and through an inventory of potential data resources.

The needs analysis findings are presented in a needs analysis study or needs assessment report. Depending on the size of the city or the organization, these studies can be generally completed in 6 to 12 months for $25,000 to $100,000. In certain ways, a GIS needs analysis study is analogous to a facility master plan in engineering. Figure 2.1 shows a comparison of the two documents. A facility plan provides information on the design, construction, and operations and maintenance of a structure. A needs analysis study provides the same information for a GIS application.

Table 2.1 lists the potential applications for the Borough of Ramsey, NJ, identified in a 1994 needs analysis study completed by Chester Engineers (Pittsburgh, Pennsylvania). This needs analysis study was completed in 1 year at a cost of $40,000.

NEEDS ANALYSIS STEPS

There are eight typical needs analysis steps:

1. Identify stakeholders
2. Talk to stakeholders

Table 2.1   Potential Applications of GIS Technology

Department	Representative Applications
Public works (responsible for operation and maintenance of the sewer system)	Emergency repair Utility markout Asset management Maintenance tracking and scheduling One-call support Critical water user location Wellhead protection
Engineering and planning	Generation of 200-ft notification lists Demographic analysis Planning review Permit approval Green space analysis Floodplain mapping Drainage management (storm sewer maps)
Tax office	Parcel location and data retrieval Tax-map production Property photo storage and retrieval Automated public access
Public safety	Fire hazard locations Hazardous-materials inventory Hydrant flow data retrieval Preferred hydrant location Crime location analysis

3. Inventory resources
4. Establish need priorities
5. Create system design
6. Conduct a pilot project
7. Prepare an implementation plan
8. Conduct the final presentation

Step 1. Stakeholder Identification

Stakeholders include all the people who can affect or can be affected by a GIS project, such as users, executives, and GIS team members. A broader classification can include users, managers, IT personnel, customers, clients, policymakers, elected officials, politicians, regulators, partners, and funding experts.

Step 2. Stakeholder Communication

This step requires talking with and listening to stakeholders to identify their GIS needs. Some or all of the following formal communication methods can be employed:

1. Introductory seminar
2. Work sessions
3. Focus groups
4. Interviews

The first three methods are optional and can be used depending on the size, scope, and budget of the project. The last method is almost always necessary to perform a comprehensive needs analysis study.

Introductory Seminar

An introductory seminar is warranted if most of the stakeholders are new to GIS. It should focus on three critical questions:

1. What is GIS?
2. Why use GIS?
3. Who else is using GIS?

The first item defines GIS for those new to it. This part should be done in layperson's language. Avoid technical terms and acronyms typically found in a college lecture on GIS-101. The second item answers the question of why GIS is important. It should emphasize that GIS benefits people, it makes things easier to do, and it increases productivity and reduces the cost of business functions. The third item presents compelling examples and success stories from other communities and organizations, such as:

- By using GIS, Charlotte, North Carolina, completed an H&H modeling project initially estimated at $1 million for less than $100,000.
- By using GIS and LIDAR elevation data, Chatham County, Georgia, saved $7 million in construction cost.
- By using GIS, Cincinnati, Ohio, expects to save $11 million in 15 years on a water infrastructure improvement project.

Success stories and associated savings factors presented in Chapter 1 (GIS Applications) can be used to answer questions 2 and 3.

Work Sessions and Focus Groups

These 1- to 2-h management level focus group sessions are conducted to gain a top–down view of the issues confronting the organization. They also allow capture of the business processes that can be enhanced using GIS applications. The discussions lead to establishing goals that can be forged into benchmarks for GIS implementation, such as:

- Creation of a map showing sewer pipes and manholes on an aerial photograph
- Creation of a map showing water mains, fire hydrants, and valves on city streets
- An application running on field (mobile) computers capable of identifying valves that should be closed to isolate a broken water main for repair

Interviews

This is often considered the single most important step in the development of a GIS plan (Henstridge, 1998). A number of selected stakeholders should be interviewed to identify the needs, uses, problems, and requirements the organization has for GIS and its applications. The interviews should attempt to quantify the cost and effort that can be reduced by implementing GIS. People with good communication skills should conduct this step. You should listen, learn, and encourage audience participation. Prepare a list of questions but avoid distributing a long questionnaire. Do not organize the interviews like a pop-up quiz. The typical interview duration is 1 to 2 h. The key questions that should be explored are:

- Do they want a GIS?
- Do they need a GIS?
- What is their wish list?
- What functions (operations) need to be improved?
- Do these functions need GIS?
- Can GIS improve these functions?
- What are the need priorities?

A GIS vision can evolve during the above communication steps. For large GIS projects, articulating a vision statement is highly recommended because, in the famous words of Yogi Berra, "If you don't know where you're going, you might not get there." The vision should reflect how users envision the business processes working in an ideal way. Most vision statements reflect easy access to accurate information (Zimmer, 2001).

When communicating with stakeholders, remember the famous metaphor, "to a man with a hammer, every problem looks like a nail." GIS cannot and will not solve every problem in an organization.

We sometimes set ourselves up for failure by creating unrealistic expectations of what GIS will do for us. Before the commencement of a needs analysis, some unscrupulous or unknowledgeable salespeople might make unsuspecting users believe that all applications will function flawlessly regardless of the quality or even existence of data. A GIS project's success is measured against expectations. Managing these expectations helps to achieve project goals and claim success. A mismatch between expectations and resources is often a source of GIS failure. Some GIS experts believe that an ill-defined scope and unrealistic expectations are the most common causes of GIS project failures (Somers, 1999; Davis, 2000; Meyers, 2002). Stakeholder communications provide an excellent opportunity to clearly define the scope of a GIS project and manage end-user expectations.

Step 3. Resource Inventory

This step requires inventorying GIS-related resources, such as maps and drawings, spatial and tabular data, databases, IT resources (hardware, software, and network), and the staff skills and interests. This step describes the as-is condition of the organization.

Lehigh County Authority located in Allentown, Pennsylvania, provides water and wastewater services to 12,000 customers in a service area of approximately 50 mi^2. In 1999, the Authority hired a consultant to complete a combined needs assessment and GIS strategic implementation plan. The needs analysis study included a complete inventory of the Authority's digital and hard-copy data sources and compiled the following list (Babbitt, 2002):

- A 1-in. = 800-ft schematic map
- A series of valve maps for each development
- Some CAD files
- Approximately 900 as-built drawings
- A valve and hydrant database in Microsoft Access
- An in-house customer information system in the ADMINS language

Step 4. Need Priorities

Because it is not feasible to implement a large number of applications simultaneously, the needs identified in Step 1 should be prioritized. This can be done by assigning a priority ranking to needs, such as mission critical, most urgent, very high, high, medium, and low. For example, Lehigh County Authority's needs analysis study identified 34 potential applications. Of these, five were considered high priority based on organization-wide use and business process improvements (Babbitt, 2002).

Remember that the people who are paying for your GIS project would be anxious to see the return on their investment. Applications that can be developed quickly (e.g., digital orthophotos) should, therefore, be given a higher priority to help show early progress during the implementation phase. The applications that are least costly or those that give the "biggest bang for the buck" (e.g., downloadable digital elevation models) can also be assigned a higher priority.

A common approach to implementing this step is to summarize and correlate the interview results into a matrix with priorities assigned to each need. Table 2.2 shows a popular method to quantify the needs by preparing a matrix of needs (applications) and the departments that need those applications. The last column shows the total score of each identified need as the sum of departments requiring that need.

Step 5. System Design

This step recommends a GIS to support the applications identified in the previous steps. It should evaluate the as-is state of the organization relative to key business functions and identify the required elements needed to create the desired improvements. System design includes determining specifications for the following components:

- Data conversion (mapping)
- Database
- Software
- Hardware
- User interface

Data Conversion (Mapping)

Approximately 75% of typical GIS costs are related to data conversion and creation. This component includes data conversion methods (scanning, digitization, etc.)

Table 2.2   Sample Departmental Needs Matrix

Needs (Applications)	Administration	Plants	Customer Service	Lab	Engineering	Collection System	Shop/Purchasing	Pretreatment	Solids	Total Score
Public access/general GIS data viewing	☑	☑	☑	☑	☑	☑	☑	☑	☑	9
Emergency management response	☑	☑	☑	—	☑	—	—	—	—	4
Special notification	☑	☑	☑	—	☑	—	—	—	—	4
Customer location and service verification	—	—	☑	—	☑	—	—	—	—	2
Customer service/complaint tracking	☑	—	☑	—	☑	☑	—	—	—	4
Sample point monitoring and tracking	—	☑	—	☑	☑	—	—	—	—	3
As-built indexing	—	—	—	—	☑	☑	—	—	—	2
Work order management	☑	☑	☑	—	☑	☑	☑	☑	☑	8
Project management tracking	☑	—	—	—	☑	—	—	—	—	2
Link TV inspections to GIS/work order management	—	—	—	—	☑	☑	—	—	—	2
Pretreatment tracking	—	—	☑	—	☑	☑	—	☑	—	4
Biosolids tracking	—	—	☑	—	☑	—	—	—	☑	3

and the source, resolution, and scale of the maps or map layers. Needs-driven design is particularly critical for data conversion and database components. The design and content of the mapping and database components determine the fundamental capabilities of the entire GIS. These components can be considered to be the foundation of the system. Once actual implementation has begun, extensive changes to either the mapping or the database design can be very costly. Additional information for this component is provided in Chapter 8 (Mapping).

Database

A GIS database stores descriptive information about map features as attributes. For example, a water system database includes attributes for pipes, valves, meters, hydrants, and so on; and a sewer system database contains attributes for pipes, manholes, catch basins, outfalls, and so on. The creation of an appropriate GIS database is the most difficult and expensive part of developing GIS applications. Successful GIS applications require a database that provides appropriate information in a useful and accessible form. The design of the database is, therefore, driven by application needs.

For data storage and manipulation, a database management system (DBMS) uses a data model, such as a hierarchical, network, or relational data model. Conventional GIS databases consist of graphic features with links or pointers (usually facility identification numbers, or IDs) to related attribute or tabular data. In the late 1990s, a new object-oriented (rule-based) data model was introduced that stores attributes as an intrinsic part of the graphics feature. These modern databases do not require links between features and attributes because they can store both the graphic features and attributes inside a single relational DBMS (RDBMS).

Database design involves three steps: (1) conceptual design, (2) logical design, and (3) physical design (Shamsi, 2002). Conceptual design does not depend on hardware or software. Logical design depends on software only. Physical design depends on hardware and provides a detailed definition of the structure and content of the database. Database design is presented with the help of a data dictionary, which documents the logical and physical structure of the layers of the GIS. Table 2.3 shows a sample data dictionary for a 2000 sewer mapping project in the city of Monongahela, Pennsylvania.

Table 2.3   Sample Data Dictionary

Layer	Name	Class	Attribute	Description
Study area	STUDY	Lines	none	Defines general study area and clips other layers
City boundary	BOUNDARY	Lines	none	Municipal boundary
Roads	ROADS	Lines	none	City streets from CAD file
Railroads	RAILS	Lines	none	Railroads from TIGER file
Surface water	RIVERS	Lines	none	River and creeks from CAD file
Subarea	SUBAREA	Lines	none	Digitized sewershed boundaries
		Polygons	Subarea_id	Subarea identification (ID) number
			Acres	Area in acres
			Mn_percimp	Mean percentage imperviousness
			Pop	Population
			Mn_popden	Mean population density
			Houses	Number of houses
			Mn_famsize	Mean family size
			Mn_mktval	Mean market value
			Mn_medinc	Mean median family income
Sewers	SEWERS	Lines	None	Collection system sewers
Interceptor	INTERCEP	Lines	None	Interceptor sewers
Treatment plant	TREATMNT	Lines	None	Treatment plant footprint
CSO discharge point	CSO	Points	Cso_id	ID Number of combined sewer overflow (CSO) discharge points

For detailed information on water, wastewater, and wastewater system database design, please refer to the author's companion book *GIS Tools for Water, Wastewater, and Stormwater Systems* (Shamsi, 2002).

Software Selection

GIS software represents less than 10% of the total GIS cost in most cases, yet people spend a lot of time in selecting the best software for their GIS applications. Many users are so proficient (and sometimes loyal) in using a given software that they will lose interest in GIS if they are not allowed to use their favorite GIS software. Because user needs are key to the success of a GIS project, their preferences should be taken seriously. Thus, despite its relatively lower cost, GIS software can have a significant impact on the success of a GIS project.

GIS software should have an "open" architecture. An open GIS allows for sharing of geographic data and integration among various GIS, GIS-related (e.g., remote sensing), and even non-GIS (e.g., hydraulic modeling) technologies. An open GIS can operate on different operating systems, platforms, and database management systems. It can be scaled to support a wide range of application needs, from an office engineer using GIS on a desktop, to a mobile field technician using a handheld device, to hundreds of users working on an enterprise system across multiple departments.

Today, GIS development software used for data conversion work is almost always a commercial off-the-shelf (COTS) program because custom development would be simply too expensive. Many commercial packages are available for implementing common GIS applications, such as work order management. Thus, no software development is necessary for data conversion or application software because reinventing the wheel is "penny-wise and pound-foolish." Some U.S. government agencies that invested in developing their own home-grown GIS packages did not see a return on their investment (Henstridge, 1998a). Custom development is warranted if COTS packages are not available for an application or if they are too expensive. As described in the following text, creation of a custom user interface also requires software development.

Custom software can be developed in-house if computer programmers skilled in the GIS-specific programming languages are available. If this is not possible, consultants can be hired. Development and (often overlooked) maintenance costs are the major disadvantages of in-house development. For some organizations (e.g., consultants) an advantage of the in-house software is the competitive edge of proprietary packages.

A needs analysis must carefully select an appropriate GIS software for the GIS needs of an organization. Fortunately, the field has narrowed and there are only a few major vendors today (Henstridge, 1998a). Some needs analysis studies utilize a decision matrix of user needs (applications) and available software capabilities to help select the most appropriate software. For example, Table 2.4 shows a sample decision matrix for various applications of a typical water utility that uses an ESRI GIS platform. This table indicates that ArcView, ArcInfo, and Spatial Database Engine (SDE) should be given high priority because they are required to implement all the GIS applications of the organization.

Table 2.4 Software Selection Matrix Example

Application	AV	SA	NA	3DA	AI	SDE	AO
Planning and engineering	✓	—	—	✓	✓	✓	✓
Operations and maintenance	✓	—	✓	—	✓	✓	✓
Construction	✓	—	✓	—	✓	✓	✓
Infrastructure management	✓	—	✓	—	✓	✓	✓
Water resources and hydrology	✓	✓	—	—	✓	✓	—
Finance and administration	✓	—	—	—	✓	✓	—

Notes: AV = ArcView; SA = Spatial Analyst; NA = Network Analyst; 3DA = 3D Analyst; AI = ArcInfo; SDE = Spatial Database Engine; and AO = ArcObjects.

Hardware Selection

The selection of appropriate GIS hardware depends on the scope of the GIS application project and the available resources. GIS hardware can be set up as a stand-alone workstation or in a network configuration. Typical hardware requirements are listed in the following text:

- Computer with fast processor, large memory, and extensive disk space
- DVD drive
- Large-screen, high-resolution, color monitor (21 in. minimum)
- Backup and storage device, such as a CD or DVD burner
- Color printer
- Large-format (wide-format) color plotter
- Digitizer
- Large-format scanner
- High-speed Internet connection (T1, cable, or DSL)
- GPS receiver
- Computer server (for networked configurations)

Hardware can be expensive, but like the GIS software, the hardware for a typical municipal GIS accounts for less than 10% of the total GIS investment. Thus, cost alone should not be used in hardware selection. The following additional factors should also be considered: service and support, capacity for growth (storage and memory), connectivity with peripheral devices (plotters, scanners, digitizers, GPS equipment, and field computers), and connectivity with Internet and networks (intranets and extranets). When comparing prices and hardware features, remember the golden rule: You get what you pay for. Another valuable advice for hardware procurement is: Try before you buy.

Today's laptops are just as powerful as desktops. For certain users, like a GIS manager who must commute between different offices, a laptop computer might be more suitable than a desktop computer. The laptops cost more but are worth the extra money for mobile users. As of 2003, the minimum requirements for a laptop to be able to run GIS software are: a 1.7 GHz processor; 512 MB memory; 60 GB hard drive; 1280 x 1024 resolution, 32-bit color XGA screen; DVD R/RW drive; Internal 56K modem; and an Ethernet or wireless Local Area Network (LAN) (Thrall, 2003).

Client/server architecture is a common computer system configuration in which one main server stores and distributes all data. Workstation users, called "clients," access and manipulate server data using client software installed on the workstation computer. A network-based design is necessary if more than one person will need to access the same GIS data, i.e., when a team rather than an individual is working on a project. For an individual researcher doing a thesis or term project, a personal workstation is sufficient. However, in real-world projects the entire project team must share the data, which requires a client/server architecture. Server maintenance is absolutely necessary. Without a dedicated technician, the client/server architecture is difficult to maintain.

Wide- or large-format plotters allow printing on 24- to 60-in. rolls. Today's wide-format plotters are available in both ink and laser technology. The cost depends on resolution, color, and speed. Common inkjet plotters range from $2,500 to $25,000. Common laser plotters range from $15,000 to $175,000. When budgeting for hardware cost, do not forget to include the maintenance cost; for example, supplies for printer and scanner such as toner, ink cartridge, paper rolls, etc. Compare the up-front cost of a laser printer to the savings of not replacing cartridges (approximately $40 each). The maintenance cost can range $100 to $500 per month based on usage. Thus, when pricing a plotter, do not forget the cost of media and ink. Compare the plotters using a cost-per-page figure for different media types. Make sure that the plotter you choose can handle a large number of plots per day at the quality you prefer. Request for a demonstration and plot your own maps before making the final selection. Some manufacturers like Hewlett Packard allow trading in the old plotter toward the purchase of a new one and have lease options starting at less than $100 per month.

As of 2000, ESRI software runs on many platforms including Digital Tru64 UNIX and Alpha Windows NT, IBM, AIX, HP-UX, Microsoft Windows 95/98/NT/2000, NEC EWS-UX, SGI IRIX, and Sun Solaris. Linux is being evaluated and tested as a viable platform (ESRI, 2000).

Orange County Geomatics/Land Information Office started to provide enterprise-wide access to GIS data in 1997 using hardware and software from Intergraph. Their Oracle database containing 7 GB of attribute information resided on a 200 MHz Quad Pentium Pro with 256 MB of RAM. The data could be viewed using Intergraph's MGE product suite (Goldstein, 1997).

Table 2.5 provides sample hardware and software specifications for a single-user installation specified in a 1994 needs analysis study for the Borough of Ramsey, New Jersey. The 1994 cost for this setup was approximately $3000. Table 2.6 provides hardware and software specifications and the approximate cost specified in 2000 for a four-seat, server-based GIS laboratory. The 2000 cost for this setup was approximately $145,000. A comparison of Table 2.5 and Table 2.6 indicates that computer hardware has changed substantially within the period of 6 years from 1994 to 2000.

User Interface

A user interface is a computer program that acts as an interpreter between the user and a computer. It generally operates on top of a GIS package as a menu-driven shell to the software's commands. A graphical user interface (GUI) is a popular

Table 2.5   Single-User Hardware and Software Specifications (Year 1994 Example)

Component	Recommended Specifications
CPU	Intel Pentium, 256 KB RAM cache
Video card	VL-bus Windows accelerator capable of 72 Hz vertical refresh rate at 1024 x 768 resolution with 256 simultaneous colors, noninterlaced
Monitor	17 in. size, performance to match video card, 0.28 mm dot pitch
RAM	16 MB, expandable to 32 MB, 36-bit SIMMs
Hard drive	500 MB IDE
Floppy drives	One 3.5 in. (1.44 MB), one 5.25 in (1.2 MB)
Expansion slots	Eight ISA slots, including two VL-bus slots
Drive bays	Total of five half-height bays, three with front access
Fax modem	14,400 bps Hayes compatible, internal
Printer/plotter	HP 650C DesignJet
Software	DOS 6.0, Windows 3.1, ArcView

type of user interface that replaces difficult-to-remember text commands by interactive computer graphics consisting of menus, dialogue boxes, input and output windows, and icons (Shamsi, 1997). For example, Internet Explorer™ is a GUI for the World Wide Web.

Creation of a custom GUI is optional. It is recommended when the GIS functions are very complex and/or the end users are complete novices. A custom GUI provides access (links) to all the GIS functions of an organization from a single program (screen) that bears the organization name and corporate logo. A custom GUI can also be used for user authentication using passwords.

Table 2.6   Hardware and Software Specifications, and Cost of a Four-User GIS Laboratory (Year 2000 Example)

Item	Year 2000 Cost ($)
One 400 MHz Pentium II NT server with 256 MB 100 MHz SDRAM, 3 x 4 GB hot swap hard drives, and Uninterruptible Power Supply (UPS)	9,000
Four Pentium II personal workstations (400 MHz, 128 MB RAM, 4.3 GB hard disk, writeable CD-ROM) with 21 in. monitors ($4,000 per workstation)	16,000
UPS backup for all personal workstations ($300 each)	1,200
Large-format color scanner with floor stand Calcomp ScanPlus III 800C (A–E size, 800 dpi max resolution)	20,000
Flatbed desktop scanner	100
Digitizer (42 x 60 in.) with floor stand. Must be WinTab compliant. Calcomp Summagrid V or GTCO Accutab.	10,000
Color plotter. Must be E size (36 in. wide). HP DesignJet 750C Plus (36-in.-wide sheet roll)	7,500
B&W LaserJet printer (HP LaserJet 5 or better)	500
Color inkjet desktop printer (HP DeskJet 1600C or better)	500
Iomega 2 GB Jaz drive for backup	500
ARC/INFO NT bundle 4 seats (includes ARC/INFO, ArcView, and ArcView extensions) ($20,000 each)	80,000
Total Cost	**145,300**

Some GIS packages allow customization of their default interfaces with little or no programming to meet user-specific needs. When available, this feature can be used to create simple GUIs. Custom GUIs for complex applications require computer programming or scripting. Custom GUI development needs will be gradually reduced and eventually eliminated as GIS packages complete their migration from a command-based to an interface-based environment.

Step 6. Pilot Project

This is a very important step. Inclusion of a pilot project in a GIS implementation plan allows the project team to test an initial strategy before committing to a specific course of action. For example, database design determines the fundamental capabilities of the entire GIS. Once actual implementation has begun, extensive changes to database design are very costly. Thus, before a systemwide implementation, the initial database design and data conversion methodology should be evaluated for a small portion of the project area (pilot area). The pilot project allows you to test the system design, evaluate the results, and modify the system design if necessary. It provides an opportunity to refine the proposed approach to full-scale data conversion and implementation. This step should also produce a working demonstration of the intended applications for the pilot area.

The Washington Suburban Sanitary Authority (WSSC) is among the ten largest water and wastewater utilities in the U.S. According to the 2003 records, it serves 420,000 customer accounts, provides water and sewer services to 1.6 million residents in a nearly 1000 mi^2 area, and employs more than 1500 people. In the early 1990s, WSSC undertook GIS implementation as a cooperative effort with five other regional agencies. The interagency organization called Geographic Information System for Montgomery and Prince George's Counties (GeoMaP) proceeded with studies to assess needs, analyze requirements, and evaluate alternatives. A detailed GIS system design was then completed for specifications of database, data conversion, hardware, software, and communications network. A pilot project was completed to evaluate the system design prior to full-scale implementation. The pilot was intended to be the first phase of GIS implementation rather than a separate experimental effort or prototype. The pilot project developed all components of the proposed database design for sample areas and designed and developed a sample set of GIS applications. The water and sewer facility layers were structured as ARC/INFO network coverages to support network analysis applications as well as to provide standard water and sewer reference maps. The sewer layer was found much easier to develop than the water layer (Cannistra et al., 1992).

Mohawk Valley Water (MVW) is a water utility serving more than 125,000 people in approximately 150 mi^2 area in central New York. MVW completed a 6-month pilot project in 2002 to develop a water distribution system geodatabase using ESRI's ArcGIS, ArcIMS, and ArcSDE software and a hydraulic model using Haestad Method's WaterCAD software. The pilot project covered approximately 5% of the representative distribution system area including three storage tanks and two pumping stations. The database design used ESRI's Water Distribution Data Model. A major goal of the pilot project was to estimate the time and money required to implement GIS and hydraulic modeling in the entire system (DeGironimo and Schoenberg, 2002). Another pilot project example, for the Borough of Ramsey, is described later in this chapter.

Step 7. Implementation Plan

An implementation plan identifies specific actions and resources required to implement a GIS. It provides information on schedule, cost, staffing, training, and operation and maintenance. It is used by a project or program manager to build the GIS, so it should be as detailed as possible. Phased implementation is preferable. The phases should be designed to distribute the cost evenly and to allow a comfortable pace over the project duration. Schedule success into your project schedule. Create your implementation plan with a few small but early successes to help you sustain the momentum over potential rough spots. Plan for early success by first implementing quick and high-impact applications that can be put to use right away. GIS integration with other information systems and applications (e.g., hydraulic modeling) is beneficial but technically complex. Because clearly identifiable benefits should be achieved in the earlier phases of an implementation plan, integration tasks should be scheduled for later phases where possible.

A successful implementation plan should be based on a holistic approach that achieves balance among stakeholders, technology, time, and budget.

Define key success factors before starting. Treat the success factors — such as acquiring hardware, installing and testing software, conducting training and workshops, populating the database, testing the database, and developing applications — as milestones. Project milestones provide a ruler against which success can be measured. Hitting these milestones will show the success of your GIS in clearly demonstrable ways and will allow you to measure success in small steps along the way. The milestones should specify the data sets, applications, and user access. A milestone like "Hydraulic Modeling by January 2004" leaves too much room for interpretation and, thus, dissatisfaction. A more useful milestone would be "Linking the water layer to EPANET model for the Engineering Department by January 2004."

User involvement is a key factor in the success of a GIS project. They should be incorporated in all stages of GIS development, rather than only in the needs analysis interviews. User training is paramount to the success of a GIS application project after it has been implemented.

The implementation plan should identify GIS staffing needs. This requires understanding the composition of the user community. When considering the best approach to GIS staffing and expertise investments, four factors should be evaluated: GIS expertise requirements, GIS project situation (e.g., early or later project phases), internal environment (e.g., hiring policies and procedures), and external environment (e.g., job market) (Somers, 1999a). All alternatives to fulfill GIS staffing needs including training existing staff, hiring new personnel, and using contractors and consultants should be evaluated. In-house projects do not require capital funds, and internal staff is more familiar with source documents. Contractors can usually complete the projects faster because they have specialized experience, more workers, and higher productivity.

The long-term viability of a GIS requires ongoing maintenance. Maintenance activities include updating the GIS data layers to reflect changes such as new streets, subdivisions, or sewer lines. Such routine maintenance activities can be contracted out on an hourly basis. The typical data maintenance cost for medium-size communities is $3000 to $5000 per year. The implementation plan should provide the maintenance procedure and responsibility for each layer in the GIS. For example, standards for site plans from developers and engineers should be developed so that changes can be easily incorporated into the GIS. Today, most plans are available in AutoCAD DXF or DWG file format and might require a data conversion effort for incorporating them into a GIS. For ESRI users, the Shapefile will be a more suitable format for accepting the site plans. If required, multiple-user editing of GIS layers should be provided by selecting appropriate GIS software. Separate operations and maintenance (O&M) manuals should be developed for large installations.

Procedures for obtaining ongoing operations support should be recommended. Today, annual software support can be easily purchased with the software. On-site

Table 2.7   Implementation Priorities and Project Phases

Needs (Applications)	Total Score	Cost to Implement	Difficulty to Implement	Mission Critical	Project Phase	Data Type
Public access/general GIS data viewing	9	L	L	VH	1	Core, facility, department
Work order management	8	M	L	VH	1	Core, facility
Emergency management response	4	M	L	H	1	Core, facility
Special notification	4	L	L	M	1	Core, facility
Customer service/complaint tracking	4	M/L	L	H	1	Core, facility
Pretreatment tracking	4	L	L	M	2	Core, facility, department
Sample point monitoring and tracking	3	L	L	L	2	Core, facility, department
Biosolids tracking	3	L	L	L	2	Core, facility, department
Customer location and service verification	2	M/L	M	H	1	Core, facility
As-built indexing	2	M	M	M	1	Core, facility
Project management tracking	2	M	M/L	M/L	2	Core, facility, department
Link TV inspections to GIS/work order management	2	M/H	M/H	M	1	Core, facility, department

Notes: L = low; M = medium; H = high; VH = very high; M/L = medium/low; and M/H = medium/high.

troubleshooting support can be obtained from consultants on an hourly or incident basis.

Finally, perceptions and expectations and hence the needs of people often change with time. A GIS implementation plan, especially if it is a long one, should be flexible to accommodate changes in organizations, people, and technology.

Table 2.7 shows a sample implementation plan summary. The first two columns show the needs and their scores from Table 2.2. Columns 3 to 5 assign rankings based on cost, implementation difficulty, and mission criticality, respectively. The sixth column assigns a project phase based on columns 2 to 5. The last column shows the data type required for each identified need.

Step 8. Final Presentation

This presentation should be conducted for the same audience that attended the introductory seminar. It should summarize the findings of the needs analysis study. The presentation should also include a live demonstration of the pilot project.

NEEDS ANALYSIS EXAMPLES

Some needs analysis examples are given in the following subsections that demonstrate the results of conducting various needs analysis steps described in the preceding text.

Pittsburgh, Pennsylvania

The city of Pittsburgh performed an in-house needs analysis in 1984 that spanned a period of 4 years and took 18 months of staff time (Wells, 1991). The heart of this needs analysis was a comprehensive inventory of existing files and maps. The needs analysis listed major functions and responsibilities of each city department. The major GIS applications included:

- Maintaining infrastructure maps and inventories
- Allowing network tracing and analysis for water, sewer, and stormwater systems
- A water and sewer facilities management system
- Computing shortest routes for emergency response and work crews
- Permit processing
- Address matching
- Producing maps
- Enhancing property and business tax administration
- Automating acquisition, maintenance, and disposal of tax-delinquent properties
- Integrating census and municipal data for community planning and budget analysis
- Facilitating production of routine reports, tables, form letters, and mailing lists
- Creating and maintaining disaster preparedness map layers (hazardous-material sites, hospitals, group care facilities, etc.)

Once the needs analysis was completed, the city hired a consultant (PlanGraphics, Inc.) to procure a GIS implementation project. The consultant translated the city's GIS needs into the following technical specifications:

- Database structure
- Data entry
- Graphic data manipulation
- Database query
- Data output and display
- User-directed software development
- System operation
- Applications
- System configuration and hardware

The consultant prepared a Request for Qualifications (RFQ) and Request for Proposals (RFP). The RFQ was sent to 80 firms and the RFP was sent to 11 firms. The proposals were submitted by five vendors.

Borough of Ramsey, New Jersey

The Ramsey Board of Public Works is responsible for the operation and maintenance of water distribution and sanitary sewer collection systems throughout the Borough of Ramsey. These operations necessitate the daily use of a variety of map products and associated geographically referenced (or "georeferenced") information resources, which describe or are related to a specific location, such as a land parcel, manhole, sewer segment, or building. In an effort to more efficiently manage its geographically referenced data, the board started to explore the benefits and applications of GIS technology and computer mapping in early 1993. As a first step, with the assistance of Chester Engineers, Inc. (Pittsburgh, Pennsylvania), they started a GIS pilot project (Shamsi et al., 1996). The goals of the pilot project were to:

- Thoroughly evaluate the benefits and costs of a GIS
- Develop specifications for GIS implementation
- Confirm the suitability of tasks selected for GIS automation
- Demonstrate GIS functional capabilities
- Firmly quantify unit costs of GIS implementation
- Demonstrate GIS benefits to the departments not participating in the pilot project
- Identify any technical problems
- Provide immediate tangible benefits
- Assess the quality of existing records and procedures

The GIS pilot project produced a GIS Needs Assessment Report, a GIS Implementation Plan, and a functioning GIS demonstration system for a selected portion of the borough. With the knowledge and experience gained from the pilot project, the board was prepared to pursue a broader implementation of GIS technology. The geographic limits of the pilot project area were selected on the basis of two criteria:

- Size: large enough to allow realistic evaluation, small enough to be affordable
- Location: encompassing a portion of the borough where activities supported by priority applications are likely to occur

The selected pilot area measuring approximately 2600 ft × 2600 ft was located in the central portion of Ramsey. The pilot project was completed in 1 year for approximately $40,000 (in 1994 dollars).

From the applications listed in Table 2.1, the following two priority applications were selected for pilot testing:

1. Public Works Department: Management of infrastructure asset location information to support emergency repair and utility markout functions. The GIS will automate the storage and retrieval of location and maintenance data for water and sewer infrastructure. Expected benefits include more rapid emergency repair response, more efficient utility markouts, and more efficient scheduling of routine maintenance.
2. Planning Department: Generation of 200-ft notification lists as part of plan review process. GIS automation will generate a 200-ft buffer, identify affected parcels, retrieve lot and block numbers, retrieve property owner names and addresses, and print a notification list. GIS benefits are expected to dramatically reduce the time required to complete these tasks.

The City of Bloomington, Indiana

The City of Bloomington (with a population of approximately 60,000) started their GIS strategy with a needs analysis in 1989. They completed their base map by the end of 1993, and got the system fully functional in the next year. They stored the GIS map and Oracle RDBMS data on workstations located at Bloomington's Utilities Service Center and distributed them through a fiber-optic network to city hall and to police and fire departments. The GIS software was loaded on ten Unix workstation host machines, most of which supported a number of X-terminal and P users across the network (Goldstein, 1997).

San Mateo County, California

San Mateo County, also referred to as the "Gateway to Silicon Valley," has approximately 200,000 property parcels and covers 552 mi². In 2001, the county started development of an enterprise-wide GIS base map. A success factor for the project was that all aspects of its development were planned and agreed on by all the stakeholders before the project started. A GIS Steering Committee, in consultation with a larger group of county departments, hired a consultant (GIS Consultants, Inc.) to conduct a needs analysis, which resulted in a strategic plan. Thereafter, a specific implementation plan was designed for base-map construction, followed by detailed and complete specifications and preliminary cost estimates. Formal needs analysis efforts were credited for the timely completion of base-map creation, hardware and software purchase, and training (Joffe et al., 2001).

CHAPTER SUMMARY

This chapter described the importance and benefits of a needs analysis prior to implementing a full-scale GIS project. Various steps involved in needs analysis were explained. When carefully done, needs analysis clarifies the project's specific needs, provides a process for arriving at interdepartmental understandings about the objectives and limits of a project, and establishes a detailed basis for developing GIS maps and applications.

CHAPTER QUESTIONS

1. What is needs analysis? How does it compare to strategic planning and facility master planning?
2. What steps are recommended for needs analysis? Which step is most critical?
3. What is a pilot project? What are the benefits of conducting a pilot project?
4. How would you prepare a needs analysis study for a water system mapping project?
5. You are tasked with integrating a large city's H&H model with the city's GIS. What will be your first step and why?

CHAPTER 3

Remote Sensing Applications

Can a satellite 400 miles above the ground surface help you locate a leaking pipe? Read this chapter to find out.

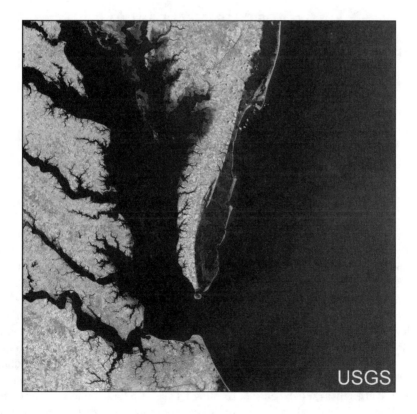

The Landsat 7 Enhanced Thematic Mapper (ETM+) scene of the lower Chesapeake Bay region acquired on July 5, 1999 (Image courtesy of USGS).

LEARNING OBJECTIVE

The learning objective of this chapter is to comprehend the applications of remote sensing technology in the water industry.

MAJOR TOPICS

- Remote sensing satellites
- Applications of satellite imagery
- Types of remote sensing data
- Digital orthophotos
- Using remote sensing for land-use classification
- Image processing software
- Anticipated future trends

LIST OF CHAPTER ACRONYMS

DEM Digital Elevation Model
DOP Digital Orthophoto
DOQ Digital Orthophoto Quadrangle
DOQQ Digital Orthophoto Quarter Quadrangle
LIDAR Laser Imaging Detection and Ranging
LULC Land Use/Land Cover
TM Thematic Mapper (onboard Landsat satellite)
USGS United States Geological Survey

ALBANY COUNTY'S REMOTE SENSING APPLICATION

Public-domain digital aerial photography data, such as USGS digital orthophoto quadrangles (DOQs) and digital orthophoto quarter quadrangles (DOQQs), usually become outdated in rapidly developing areas. For such areas, high-resolution satellite imagery may be a cost-effective source of more recent overhead images.

Albany County, located in southeastern Wyoming, covers 4,400 mi^2, has a student-based population of 30,000, and has 1,600 mi of roads. For rural communities such as Albany County, building a GIS from scratch can be an expensive endeavor due to lack of resources. The County's day-to-day mapping functions required a data layer of imagery for the entire county. Various data options were reviewed, including aerial flights, existing DOQs, and satellite imagery. New aerial imagery was eliminated because it was too expensive. Existing DOQs were not suitable because they were 7 years old and did not reflect recent county growth trends. In addition, costs associated with updating the County's existing digital aerial photography exceeded $100,000. High-resolution satellite imagery, on the other hand, allowed the County to have high-resolution up-to-date views of the entire county for $32,000. For 85 mi^2 of populated areas, the County selected 1-m pan-sharpened IKONOS satellite imagery (described later in this chapter). For the rest of the county, 90 quads of

CARTERRA DOQ 5-m black and white (B&W) imagery was selected. Both products were produced by Space Imaging (Thornton, Colorado). Thanks to this geographic-imaging approach, planning tasks previously requiring months to complete took only days after the County implemented this project (Frank, 2001).

In the Albany County of Wyoming, addition of high-resolution up-to-date imagery to GIS data reduced the completion of typical planning tasks from months to a few days.

INTRODUCTION

The technologies that are commonly used in conjunction with GIS are commonly referred to as GIS-related technologies. Examples include remote sensing, global positioning system (GPS) surveying, the Internet, and wireless technologies. This chapter will focus on remote sensing, one of the most successful GIS-related-technologies. Other related technologies are described elsewhere in the book.

Remote sensing allows obtaining data of a process from a location far away from the user. Remote sensing can, therefore, be defined as a data collection method that does not require direct observation by people. Remote sensing is the process of detection, identification, and analysis of objects through the use of sensors located remotely from the object. Three types of remote sensing systems are useful in the water industry:

1. Aerial photographs
2. Satellite imagery
3. Radar imagery

The data from these systems are commonly referred to as remote sensing or remotely sensed data. Sometimes, remote sensing data are incorrectly confused with supervisory control and data acquisition (SCADA) data used to operate water and wastewater treatment plants. Remote sensing data collected using airplanes are called aerial photographs or aerial photos. Digital remote sensing data collected from satellites are called satellite imagery or images. Digital pictures of the Earth are taken by satellites from 400 to 500 mi above the ground compared with aerial photographs that are taken by aircraft from 1 mi above the ground (for low-altitude photography) to 7 to 8 mi above the ground (for high-altitude photography). The chart in Figure 3.1 shows the altitude difference between the aircraft- and satellite-type remote sensing systems. Radar imagery or images are another type of remote sensing data but their usage is not widespread in the water industry. Although the definition of remote sensing includes aerial photos and radar data, remote sensing is often considered synonymous with satellite imagery.

The American Society for Photogrammetry and Remote Sensing (ASPRS) values the U.S. remote sensing industry at about $1.3 billion (as of 2001) and forecasts 13% annual growth, giving values of $3.4 billion by 2005 and $6 billion by 2010. The industry currently consists of about 220 core companies employing about 200,000 employees in the areas of remote sensing, photogrammetry, and GIS imaging. A 2001

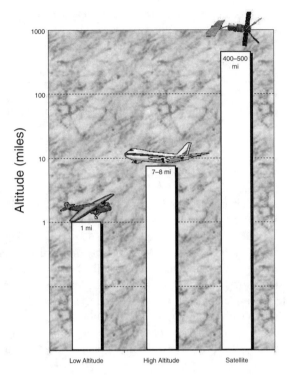

Figure 3.1   Altitude difference in aerial photography and satellite imagery.

ASPRS study concludes that utilities are one of the greatest untapped potential markets and that a shortage of trained workers is one of the greatest challenges to the growth of the remote sensing industry (Barnes, 2001a).

Although vector GIS data are still an important and vital tool for many water industry applications, the newer raster GIS applications of satellite imagery are beginning to make a major move into the GIS and mapping market. The benefits of satellite imagery are (Schultz, 1988):

1. They enable aerial measurements in place of point measurements.
2. They offer high spatial and/or temporal resolution.
3. All information is collected and stored at one place.
4. Data are available in digital form.
5. Data acquisition does not interfere with data observation.
6. Data can be gathered for remote areas that are otherwise inaccessible.
7. Once the remote sensing networks are installed, data measurement is relatively inexpensive.

Satellite imagery is stored in a pixel (raster) format that makes it ideally suited for incorporation into a GIS (Engman, 1993). Thus, satellite imagery can be treated as raster-type GIS data. Image processing equipment and methods can be used to extract useful information from hard copy and digital images and combine it with other data layers in a GIS. Image data sources including scanned paper maps, aerial photographs, and satellite imagery can be used in a GIS when reprojected as image maps. Projected images can be used as a background or as a base map upon which other vector layers are overlaid.

Casual GIS users can easily import remote sensing imagery into their GIS projects as an image theme (or layer). However, advanced remote sensing applications and image analyses require formal remote sensing training and digital image processing skills. The incorporation of remote sensing data in a GIS requires a digital image processing software such as ERDAS IMAGINE, Geomatica, ER Mapper, or ENVI, or a raster GIS software with image processing capability, such as ArcGRID or IDRISI. Such programs are described later in this chapter.

These are exciting times both for the GIS and the remote sensing industries, thanks to dramatic price and performance breakthroughs in GIS hardware and software. The increasing use of GIS is contributing to a renewed interest in satellite imagery by nongeographers, such as civil and environmental engineers. Although GIS technology is promoting the use of satellite imagery, satellite imagery is also in turn advancing the use of GIS. Although non-GIS stand-alone image processing software can be used for exploring satellite imagery, those with GIS capabilities are more suitable because they can combine imagery with additional information, such as demographic and topographic data (Corbley, 2000).

REMOTE SENSING APPLICATIONS

Satellite imagery is not restricted to the visible (0.4 to 0.7 μm wavelength) part of the electromagnetic spectrum. Satellite sensors can record Earth images at wavelengths not visible to the human eye, such as near-infrared and thermal-infrared bands. Different satellite bands provide information about different objects and conditions of the Earth. For example, thermal-infrared band (10.4 to 12.5 μm wavelength) data are useful for soil–moisture discrimination. These bands of satellite data can be used as different data layers in a GIS for further analysis.

Remote sensing applications in the water industry are as diverse and numerous as the GIS applications themselves. Typical examples are listed below:

1. Satellite remote sensing has contributed to water resources applications and research for three decades (Jackson, 2000). Remote sensing data are especially useful in watershed hydrologic modeling. Satellite imagery can be used to estimate input parameters for both the lumped-parameter and distributed-type hydrologic models.
2. Satellite imagery can be used for delineating watersheds and streams. For example, SPOT satellite's stereographic capability can generate topographic data. Terra satellite can provide digital elevation models (DEMs) from stereo images. (These and other satellites are discussed later in this chapter.) Topographic and DEM data collected by satellites can be processed in GIS for automatic delineation of watershed boundaries and streams.
3. Remote sensing data are used for land-use classification. GIS can help to refine or verify the imagery-based land-use classes.

4. Satellite and radar data can be used to estimate the area and intensity of rainfall.
5. Remote sensing can produce surface temperature data through thermal-infrared images.
6. Microwave remote sensing can produce soil-moisture data.
7. Remotely sensed temperature and moisture data can be combined to estimate evaporation and evapotranspiration rates.
8. Remote sensing data are used to estimate vegetation indices and the leaf area index. These parameters can be combined to delineate areas where a subsurface supply of water is available for vegetation.
9. Remote sensing can be used in real time flood forecasting with a distributed hydrologic model into which radar rainfall data can be input.
10. Other applications (Singh, 1995; ASCE, 1999) are:
    • Utility routing
    • Weather forecasting
    • Environmental impact assessment of large water resources projects
    • Snow and ice conditions (microwave region)
    • Forecasting seasonal and short-term snowmelt runoff
    • Evaluation of watershed management strategies for conservation planning
    • Inventory surface water, such as rivers, lakes, reservoirs, swamps, and flooded areas
    • Water quality parameters such as algae, chlorophyll, and aquatic life
    • Thermal and chemical pollution and oil spills
    • Drought assessment and forecasting
    • Geologic and geomorphologic information
    • Groundwater mapping

REMOTE SENSING SATELLITES

Satellite data became available to water industry professionals in 1972 when the U.S. government launched the first Landsat satellite, which was specifically designed to provide imagery of the Earth (Miotto, 2000). In the late 1970s and early 1980s, a second generation of Landsat satellites was developed. Landsats 4 and 5 were launched in July 1982 and March 1984, respectively. They were equipped with two instruments:

    • Multispectral scanner (MSS) having 80-m resolution and 4 spectral bands
    • Thematic mapper (TM) having 30-m resolution and 7 spectral bands

MSS sensors capture imagery at different wavelengths of light to produce color images. Landsat 4 was retired in 1991, and Landsat 5's MSS sensor failed in October 1993. The successor satellite, Landsat 6, failed to achieve orbit in 1993. To keep the imagery flowing, Landsat 7 was launched on April 15, 1999.

Popular satellite-based sensors and platforms include Landsat MSS and TM, AVHRR, AVIRIS, SPOT XS, GOES, SEASAT, SIR, RADARSAT, SRTM, TOPSAT, ERS-1 and 2, and JERS-1 (Luce, 2001; Lunetta and Elvidge, 1998). The remote sensors that provide hydrologically useful data include aerial photographs, scanning radiometers, spectrometers, and microwave radars. The satellites that provide hydro-logically useful data are the NOAA series, TIROS N, SPOT, Landsat, and the geostationary satellites GOES, GMS, and Meteosat. Satellites can capture imagery in areas where conventional aircraft cannot fly. However, bad weather, especially cloud cover, can prevent satellites from capturing imagery (Robertson, 2001).

SPATIAL RESOLUTION

The spatial resolution of an image is defined as the size of the smallest feature that can be discerned on the image. Spatial coverage is defined as the area of the Earth's surface captured by the image. In general, the higher the spatial resolution, the smaller is the spatial coverage. For example, NASA's Terra Satellite MODIS sensor has 36 spectral channels at 250 m, 500 m, and 1 km. A standard MODIS image covers 1200 km × 1200 km, whereas a standard IKONOS satellite image covers 11 km × 11 km. At such a large spatial coverage, MODIS spatial resolution is more than 50 times coarser than the IKONOS imagery (Space Imaging, 2001). In 2001, the approximate number of 30-m (or better) resolution satellites in the world was 30, and the number of 10-m (or better) resolution satellites was 14 (Limp, 2001).

Figure 3.2 provides a comparison of image resolution. It shows five images at various resolutions for the same geographic area (Gish, 2001). The top-left image, with the highest resolution, is a 0.15-m (0.5-ft) B&W orthophoto taken in 1993. The top-right image is a 0.6-m (2-ft) 1998 B&W orthophoto. The center-left image is a 1-m (3.28-ft) 1999 color-infrared orthophoto taken in invisible light in the infrared bands. The center-right image is a simulated B&W SPOT image with a 10-m (32.8-ft) resolution. Finally, the bottom image has the lowest resolution of 30 m (98.4 ft) and consists of Landsat 7 TM color imagery taken in 2000.

In remote sensing, B&W or gray-scale imagery is called *panchromatic* and color imagery is called *multispectral*. Panchromatic satellite-imagery resolution varies from 15 m (49 ft) for the Landsat 7 satellite, 10 m (33 ft) for the French SPOT satellite series, 5 m (16 ft) for the Indian Remote Sensing series, 1 m (3.2 ft) for the IKONOS satellite (Gilbrook, 1999), to 60 cm (2 ft) for the QuickBird-22 satellite. Until recently, satellite images tended to have very low resolutions. In January 2000, IKONOS high-resolution satellite imagery became available in the commercial marketplace for the first time.

Based on their spatial resolution, remote sensing data can be divided into three categories:

1. Low-resolution data corresponding to imagery with a resolution greater than 30 m
2. Medium-resolution data corresponding to imagery with a resolution between 5 and 30 m
3. High-resolution data corresponding to imagery with a resolution less than 5 m

Low-Resolution Satellite Data

The United States Earth Observing System (EOS) satellites are an excellent source of low- and medium-resolution satellite data. There are four EOS satellites currently in orbit: Landsat 7, QuickSAT, ACRIMSAT, and Terra. Terra, launched by NASA in December 1999, has three remote sensing instruments that could be

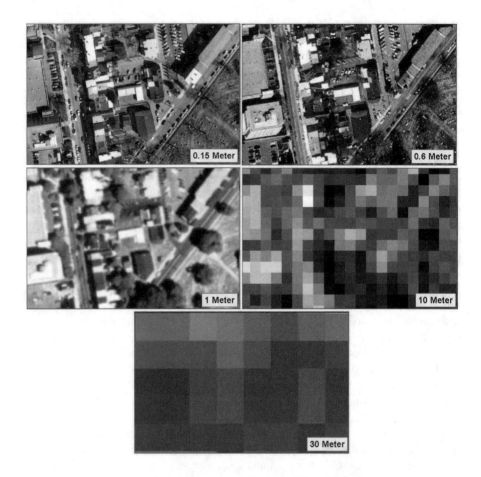

Figure 3.2   Image resolution comparison. Top left: 0.15-m B&W orthophoto (1993); top right: 0.6-m B&W orthophoto (1998); center left: 1-m color infrared orthophoto (1999); center right: 10-m simulated SPOT; bottom: 30-m Landsat TM (2000).

useful for certain water resources applications: Advanced Spaceborne Thermal Emission and Reflection Radiometer (ASTER), Moderate Resolution Imaging Spectroradiometer (MODIS), and Multiangle Imaging Spectroradiometer (MISR). ASTER provides digital elevation maps prepared from stereo images. MODIS provides data on cloud- and snow-cover characteristics. MISR data can distinguish different types of clouds, land cover, and vegetation canopy. Although low-resolution satellite imagery works well for regional level studies, it is not very useful in water industry applications.

Medium-Resolution Satellite Data

Table 3.1 provides a summary of medium-resolution satellites. Landsat 7 is the most recent satellite in the Landsat series. By May 2001, Landsat 7 had captured more than 200,000 15-m scenes throughout the world. The Enhanced Thematic

Table 3.1 List of Major Medium-Resolution Satellites

Feature	Landsat 7	SPOT 4	IRS-1C
Company	USGS	French Government	Indian Government
Launch date	April 15, 1999	1998	December 28, 1995
B&W resolution	15	10	5
Color resolution	30	20	23
Swath width (km)	185	120	70
Global cover repeat days	16	26	24

Mapper Plus (ETM+) sensor onboard Landsat 7 provides 15-m panchromatic and 30-m multispectral resolutions. Landsat 7 offers imagery of the highest resolution and lowest price of any Landsat. The USGS ground-receiving station in Sioux Falls, South Dakota, records 250 Landsat scenes a day that are available online within 24 hours. Landsat 7 is expected to have a design life of 5 years. Each Landsat image covers about 10,000 mi^2. Landsat 7 is very useful in water resources applications.

Figure 3.3 shows a modified Landsat TM image for southwestern Pennsylvania, which can be used in a GIS to consistently map land use/land cover (LULC) throughout the state. These images, called Terrabyte, are extracted from the 30-m resolution TM data using an extractive process based on a research trust at Penn State University under cooperation between the Office for Remote Sensing of Earth Resources in the Environmental Resources Research Institute and the Center for

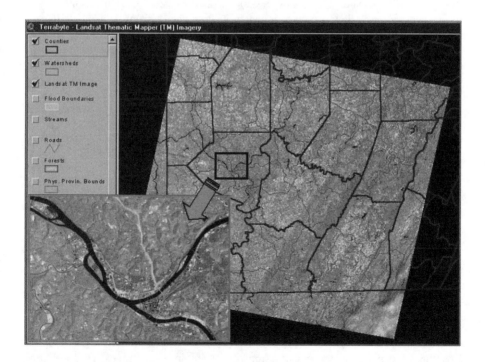

Figure 3.3 Terrabyte Landsat Thematic Mapper image for southwestern Pennsylvania.

Statistical Ecology and Environmental Statistics in the Department of Statistics, with sponsorship from the National Science Foundation and the Environmental Protection Agency. The Terrabyte images are not intended to provide fine detail such as individual buildings at the site level, but rather to convey a sense of landscape organization. Each pixel record occupies one byte; hence the name Terrabyte. Terrabyte condensations for ten satellite scenes will fit on one CD-ROM, whereas two scenes of original satellite data would more than fill one CD-ROM. Terrabyte CD-ROMs of Pennsylvania data have been distributed by Pennsylvania Mapping and Geographic Information Consortium (PaMAGIC) (www.pamagic.org).

In addition to the EOS satellites, France's SPOT 4 satellite provides 10-m panchromatic and 20-m color imagery. In the U.S., 60 × 60 km SPOT scenes cost $750 (pre-1998) to $1500 (post-1998). India's Indian Remote Sensing (IRS) satellite IRS-1C provides 5-m panchromatic and 23-m or 188-m color imagery. Commercial companies and government agencies around the world had plans to launch more than 25 medium-resolution (30 m or better) satellites by the end of 2003.

High-Resolution Satellite Data

High-resolution data correspond to imagery whose resolution is less than 5 m. Traditionally, water industry professionals have purchased aerial photography services on an as-needed basis, which is costly and time-consuming. Now, thousands of square miles of GIS-ready seamless imagery is available in various formats with the promise to bring remote sensing data to any desktop (Robertson, 2001). Until recently, some water industry professionals used 5-m panchromatic imagery from India's IRS-1C satellite or 10-m panchromatic imagery from France's SPOT 4 satellite for their high-resolution data needs. The recent launches of IKONOS (1-m) and QuickBird-2 (60-cm) satellites have changed this by starting to provide high-resolution panchromatic imagery, which meets the U.S. National Map Accuracy Standards for 1:5000-scale maps. 1-m imagery represents an accuracy level commensurate with 1:2400 mapping, which is more than adequate for many planning and H&H modeling applications.

GIS applications are poised to bring the recently available high-resolution satellite imagery directly to the dispatch office of a water, wastewater, or stormwater utility.

High-Resolution Satellites

There are three major satellites that are providing high-resolution satellite imagery today: IKONOS, OrbView, and QuickBird. Table 3.2 provides more information about high-resolution satellites.

High-resolution imagery shows detailed features, such as houses, schools, street centerlines, rights-of-way, trees, parks, highways, and building facilities. They can be used for base-map and land-registry updates, infrastructure mapping analysis and management, natural resource inventories, ecological assessments, transportation mapping, and planning the construction of new highways, bridges, and buildings (Murphy, 2000).

Table 3.2 List of Major High-Resolution Satellites

Feature	QuickBird-2	IKONOS	OrbView-3
Company	DigitalGlobe	Space Imaging	ORBIMAGE
Launch date	October 18, 2001	September 24, 1999	June 26, 2003
B&W resolution (m)	0.61	1	1
Color resolution (m)	2.5	4	4
Swath width (km)	16.5	12	8
Global cover repeat days	148	247	3
Standard scene size	40 km × 40 km	13 km ×13 km	User defined
Web site	digitalglobe.com	spaceimaging.com	orbimage.com

DigitalGlobe's (formerly Earth Watch) QuickBird-1, designed for 1-m panchromatic and 4-m color resolution, failed to achieve its proper orbit after being launched from Plesetsk, Russia, on November 20, 2000. QuickBird-1's unfortunate failure is a good example of a "beneficial loss." QuickBird-2 was launched on October 18, 2001, on a Boeing Delta II rocket from Vanderberg Air Force Base in California. At the time of this writing, QuickBird-2 provides the only commercial satellite imagery of resolution less than 1 m. Figure 3.4 shows the sample QuickBird imagery taken in 2002 and a photograph of the San Diego Convention Center area in California that hosts the world's largest GIS conference (ESRI Annual User Conference) every year. Note that the boats at the marina and the north and south towers of the Marriot

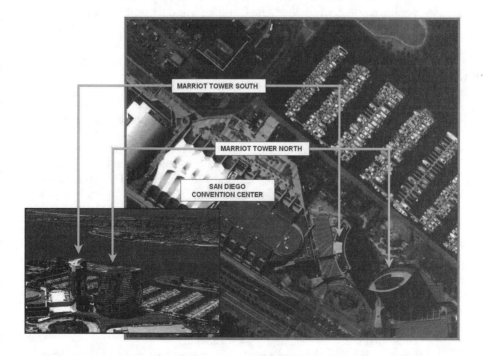

Figure 3.4 Sample QuickBird imagery for San Diego, California. (Image courtesy of Digital-Globe.)

Hotel adjacent to the convention center are clearly visible on the imagery. Digital-Globe is planning to launch another satellite dubbed WorldView in 2006, which will be capable of collecting 50-cm panchromatic and 2-m multispectral imagery. World-View's 800-km-high orbit will allow the satellite to visit imagery collection sites more frequently, letting users repeat their image acquisition about once a day.

IKONOS provides 1-m panchromatic and 4-m multispectral (color) imagery. The satellite weighs 1600 lb and orbits 438 mi above the ground surface. IKONOS products are available under the CARTERRA™ brand name in TIFF and GeoTIFF format. CARTERRA also provides DOQ — B&W, color, or false color IR 5-m imagery, cut into a convenient 7.5-min USGS quadrangle format. Orthorectified CARTERRA DOQs provide an image map suitable for water resources management, urban and rural planning, change detection, and map creation and revision.

High-Resolution Imagery Applications

GIS applications are poised to bring the recently available high-resolution satellite imagery directly to the dispatch office of a water, wastewater, or stormwater utility. For years, aerial photography has been used in many utility GISs, and that use will continue because of its submeter resolution. High-resolution satellite imagery is now available commercially at a reasonable cost. Therefore, when a 1-m resolution is sufficient, satellite imagery can be used as a base map instead of orthorectified digital orthophotos. High-resolution satellite imagery provides digital data at a fraction of the cost people pay for aerial photographs of the same level of accuracy. There is no question that the launch of high-resolution satellites marks a new era in the remote sensing industry.

Typical applications of high-resolution satellite imagery for the water industry are described in the following list:

- High-resolution satellite imagery can enable the water and wastewater system utilities to gather information quickly and inexpensively, allowing them to perform daily operations more efficiently.
- Multispectral imagery can detect vegetation stress before it is visible to the naked eye. Watermain leaks and manhole overflows can impact the soil and vegetation conditions around them. These potential indicators may be used to identify leaks and overflows. *Thus, although satellites cannot directly locate leaking pipes, they can provide the surrogate data that can lead to locating them.*
- High-resolution imagery is especially useful in remote areas of the world where there are no governments and commercial archives and where cost and regulatory hurdles preclude aerial missions.
- Medium-resolution imagery cannot capture some landscape characteristics, such as distribution of shrubs vs. bare ground or gaps in the forest crown (Space Imaging, 2001). High-resolution imagery bridges the gap between field measurements and medium-resolution imagery, providing a continuum from point measurements to medium resolution. High-resolution imagery can also be used to "ground truth" the low- and medium-resolution imagery.
- High-resolution imagery can be used to study urban growth and detailed urban land-use mapping. It can be used to identify growth trends in order to develop the necessary infrastructure in advance.

- Damage from natural disasters (earthquakes, hurricanes, fires, and floods) can be analyzed and response plans prepared. High-resolution imagery can be used to prepare more accurate flood-prediction computer models, monitor stormwater runoff, and study erosion and sedimentation.
- Right-of-way encroachments can be identified by periodically running automated change-detection routines on new imagery.
- Inexpensive pipeline siting and corridor selection can be performed using least-cost path analysis.
- Ecological assessments can be conducted.

Data Sources

TerraServer was started as a joint research project by Aerial Images, Microsoft, USGS, and Compaq. It is considered one of the world's largest online atlases of high-resolution satellite imagery and aerial photography (Thoen, 2001).

In 2001, USGS, PCI Geomatics, Oracle, and Sun Microsystems teamed up to provide a new data delivery service called Real-time Acquisition and Processing of Imagery Data (RAPID). It provides same day service for conversion of TM imagery into easy-to-use data that can be downloaded using an Internet connection. Using RAPID, users can have georeferenced, GIS-ready processed imagery within 10 min of receipt from USGS.

Some consumer-oriented companies are also selling high-resolution satellite imagery. For example, Eastman Kodak Company's CITIPIX imagery database consists of 95 major North American metropolitan areas, including 7000 cities and towns and 600 U.S. and Canadian counties. This ready-to-use "Earth Imaging Products" consist of orthorectified imagery in 6-in., 1-ft, 2-ft, and 1-m resolutions. Kodak's 24-bit color images exceed National Map Accuracy Standards' accuracy requirement at 1:1200. These products are intended for applications in architecture, engineering, construction, telecommunication, utilities, insurance, and real-estate industries as well as local, state, and provincial governments.

The cost of spatial data is falling rapidly due to competition in data acquisition, processing, and distribution. As satellite imagery has become more widely accepted, its unit cost has started to decline. For example, Landsat-4 and -5 imagery used to cost $4400 per scene; now the same scene costs $600. After the launch of QuickBird-2 and OrbView-3 satellites, the price of IKONOS imagery has come down from $62/km^2 to $29/km^2 (with a 100 km^2 minimum order) — a decrease of over 50%.

DIGITAL ORTHOPHOTOS

Digital orthophotos are a special type of high-resolution remote sensing imagery. Traditional aerial photos contain image displacements caused by camera lens distortion, camera tip and tilt, terrain (topographic) relief, and scale (Michael, 1994). Because of these problems, an aerial photograph does not have a uniform scale, and therefore, it is not a map. The distortions are removed through a rectification process to create a computer file referred to as a digital orthophoto (DOP). The image rectification is done with the help of geodetic surveying and photogrammetry. A DOP

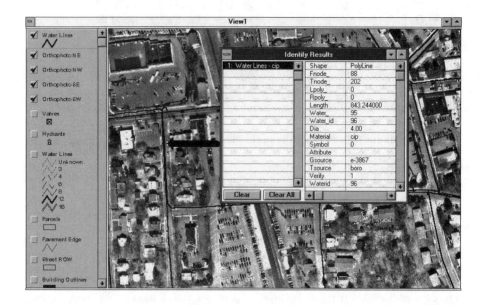

Figure 3.5   Digital orthophoto base map for water system mapping.

is a uniform-scale photographic image and can be considered a photographic map. As the name implies, orthophotos are orthographic photographs or simply photo maps.

DOPs are very detailed, can be easily interpreted, and provide excellent accuracy that can be easily quantified and verified. DOPs are a valuable source for developing an accurate landbase in a GIS mapping project. Because it is a photographic image, the DOP displays features that may be omitted or generalized on other cartographic maps. This makes the digital orthophoto valuable as a base map layer in a GIS. Figure 3.5 shows a DOP base map for a water distribution system mapping project. The DOP has been processed to remove scale distortion and create an accurate and true-to-scale base map with an accuracy of ±1.25 ft. Waterlines have also been plotted on the digital orthophoto.

USGS Digital Orthophotos

A digital orthophoto quadrangle (DOQ) is an orthorectified raster image of a low-altitude USGS aerial photograph in Universal Transverse Mercator (UTM) projection in North American Datum of 1983 (NAD83). DOQs have the geometric properties of a map and meet the National Map Accuracy Standards. DOQs are based on the 1:80,000-scale National High Altitude Photography (NHAP) aerial photos with an altitude of 40,000 ft. They cover an area equal to 7.5 min (1:24,000) USGS quads (quadrangles); hence their name. They have a resolution of 2 m and an accuracy of 40 ft. They are available as 5-MB compressed JPEG files that can be purchased for $35.5 per county CD.

Digital orthophoto quarter quadrangles (DOQQs) are based on the 1:40,000-scale National Aerial Photography Program (NAPP) aerial photos with an altitude

Figure 3.6   USGS digital orthophoto quarter quadrangle (DOQQ) for Washington, D.C. (Image courtesy of USGS.)

of 20,000 ft. They cover an area equal to a quarter of a 7.5-min USGS quads (3.75 min of latitude by 3.75 min of longitude in geographic extent); hence their name. They have a resolution of 1 m and an accuracy of 33 ft. DOQQs also have a UTM map projection system and NAD83 datum. A sample DOQQ image for Washington, D.C., is shown in Figure 3.6.

DOQQs are available as 40 to 50 MB uncompressed files that can be purchased for $60 per file on CD-ROM from the USGS EarthExplorer Web site. State or regional mapping and spatial data clearinghouse Web sites are the most valuable source of free local spatial data. For example, Pennsylvania Spatial Data Access system (PASDA), Pennsylvania's official geospatial information clearinghouse and Pennsylvania's node on the National Spatial Data Infrastructure (NSDI), provides free downloads of DOQQ and other spatial data.

Because DOQQs are based on the NAD83 datum, most U.S. ArcView 3.x users who want to use DOQQs in their GIS projects will have to reproject their NAD27 vector themes into NAD83. NAD83 is an Earth-centered datum (GRS 80 ellipsoid) unlike NAD27, which is based on an arbitrary starting point in Meades Ranch (Kansas). The positions or points or features based on these reference datums will differ considerably. Though reprojected themes will line up reasonably well with features in DOQQs, a more precise alignment can be achieved by adjusting the false easting and northing values to "tweak" the theme (Miller, 2000).

Like DOPs, DOQs and DOQQs also can be used as GIS base maps to overlay other thematic data layers. DOQs can be used in land management, habitat analysis, evacuation planning, and in many other areas (Miller, 2000).

Case Study: Draping DOQQ Imagery on DEM Data

The city of Ventura, California, is located 70 mi north of Los Angeles, covers almost 21 mi², and is home to 105,000 people. In an effort to redefine its GIS, the City acquired 11 USGS DOQQs in 1999. The DOQQs were mosaicked and reprojected to California State Plane with ERDAS IMAGINE software and compressed with LizardTech's MrSID software to create a seamless image. They also created a panoramic view of the city using IMAGINE VirtualGIS software by draping the DOQQ mosaic over 10-m DEMs obtained from USGS and Intermap Technologies (Ottawa, Ontario, Canada). The image draping demonstrated the value of orthorectified imagery and helped the City to derive other GIS data layers, such as slope, aspect, and shaded relief (ERDAS, 2001a).

EXAMPLES OF REMOTE SENSING APPLICATIONS

A USGS project helping to restore the original water flow of Everglades is a good example of blending various types of remote sensing data. This project uses aerial videography, laser imaging detection and ranging (LIDAR)*-based topography, and Landsat-based vegetation to create hydrodynamic models of water flow (Miotto, 2000).

The Arizona Department of Water Resources (ADWR) uses satellite data (Landsat and SPOT) and GIS software (ERDAS IMAGINE and ArcInfo) to monitor water use, regulate water rights, and enforce limits on irrigated acreage expansion. ADWR purchases growing-season imagery at an annual cost of $15,000. An investigation is started if the images show crop growth that does not correspond to water rights (McKinnon and Souby, 1999).

A 1997 USGS study for the California Gulch Superfund Site in Leadville, Colorado, demonstrated an application of remote sensing in locating and pinpointing sources of acid mine drainage and a tremendous variety of other surface materials. Mineral maps produced by the study helped expedite the cleaning of hazardous material and reduced the cleanup cost. These maps were produced using a new tool called imaging spectroscopy, the latest in remote sensing science.

Satellite imagery can be used for irrigation monitoring and assessment, or to evaluate erosion potential or landslides. Other remote sensing applications for the water industry include land-use classification, soil moisture mapping, and estimation of meteorological data. These applications are described in the following subsections.

LULC Classification

The San Diego Association of Governments (SANDAG) deals with one of the nation's largest county jurisdictions covering more than 4200 mi². Before 1988, SANDAG used costly and time-consuming aerial-photography and photo-interpretation techniques to create LULC maps and updated them only once every 5 years. To meet

---

* LIDAR is a new remote sensing technology that measures ground surface elevation from an airplane to create DEM data. Additional information is provided in Chapter 4 (DEM Applications).

the special challenge of keeping track of this rapidly changing area in a cost-effective manner, SANDAG turned to GIS. It used raster GIS and image processing software ERDAS, vector GIS software ArcInfo from ESRI, color-infrared aerial photographs, and satellite imagery. Switching to satellite imagery and GIS as a land inventory tool allowed SANDAG to see the region in a new way and permitted rapid change detection. The GIS-based LULC-mapping approach provided SANDAG with current and verified LULC data for modeling transportation, infrastructure, and water needs (Kindleberger, 1992).

In 2003, a digital database of land-cover imagery and vectors was created that includes the major landmasses of the entire world. Called GeoCover LC (Land Cover), this was the first worldwide LULC database prepared using medium-resolution Landsat 7 TM imagery.

LULC classification is one of the most common applications of satellite imagery. Remote sensing data are a valuable source of information for land-use modeling. Derivation of LULC classes from low-level aerial photography is referred to as the *conventional* method compared with the remote sensing techniques that employ satellite imagery. Studies have shown that the remote sensing techniques for LULC classification are more cost-effective. The cost benefits have been estimated on the order of 6 to 1 in favor of the satellite imagery approach. Although remote sensing-based land-use statistics may not be as detailed as those derived using the conventional manual method, many computed parameters such as runoff curve numbers and discharges are nearly the same (Engman, 1993).

The choice of sensors for LULC applications is dictated by the time and space resolution needed for interpretation. The sensor chosen must permit a suitable classification of LULC. The quality of this classification will depend on the LULC types, the quantity of images available, and the dates on which they were taken as well as the classification technique used. Ambiguities in interpretation can be reduced by using field information (ground truthing) to improve classifications. The number of LULC categories depends on the intended application. For example, in watershed nutrient modeling this number must permit the estimation of total nutrient export from a subwatershed and to compare the possible effects of land-use changes. At a minimum, urban, agricultural, grasslands, and forested areas must be clearly distinguished. A distinction between types of farming will provide better nutrient estimates (Payraudeau et al., 2000).

When collecting LULC data, keep in mind the four "Cs": currency, construction, categories, and consistency. Currency determines how current your LULC data are. Construction pertains to new developments that can significantly alter land use and affect soil erosion, sedimentation, and even water quality. Categories represent the LULC classes (polygons). One may not want to have an "agricultural" class created by lumping together cropland, pasture, and dairy farms. And, finally, consistency determines how consistent in time, scale, resolution, and classification various sources of LULC data are (Slawecki et al., 2001).

Vegetation is an important part of a watershed's ecosystem. To preserve and maintain the watershed ecosystems, resource managers need high-quality vegetation maps to monitor temporal vegetative changes and to pinpoint habitats and species likely to be affected by management decisions. Based on the spectral return from the infrared imagery and using classification techniques of a remote sensing software package, people can identify trees and other vegetation growing in a watershed and even determine the total irrigated area. California Department of Forestry and Fire Protection (Sacramento, California) updates California's LULC mapping and monitoring program on a 5-year cycle. The program utilizes image classification and GIS modeling to generate data that describe the condition and extent of various land-cover types, as well as the magnitude and cause of such changes. Data sources include DOQs, aerial photographs, and satellite images from Landsat TM, Indian Remote Sensing, and SPOT. LULC layers are produced using automated image classification of satellite imagery. This approach efficiently and consistently maps large areas at a low cost. Next, the LULC layer is converted to vector format and hand-edited with high-resolution DOQ as a backdrop. GIS models are run in ESRI's ArcInfo GRID program to label each LULC polygon with a vegetation type. The updated LULC map is created by adding the new 5-year LULC change layer to the old LULC map (Rosenberg et al., 2001).

Dry- and wet-weather flows from sewersheds depend on land use. The hydrologic and environmental effects of land do not depend on the amount of land-use change alone. The location of land-use change is equally important. For example, a development adjacent to a water body might produce a greater impact on water quantity and quality. GIS and remote sensing are ideally suited to studying the temporal and spatial variation of land use. Remote sensing has been recently recognized as a tool that holds great promise for water quality monitoring activities. In 2003, three U.S. agencies, the Environmental Protection Agency (EPA), National Aeronautic and Space Administration (NASA), and the National Oceanic and Atmospheric Administration (NOAA) entered into a partnership to more fully explore how remote sensing may support state water quality management activities (Mehan, 2003).

Digital LULC data are used to derive several watershed parameters, such as the percent imperviousness and runoff curve number, as described in Chapter 11 (Modeling Applications) and Chapter 13 (Sewer Models). This runoff-coefficient estimation technique is a good example of maximizing the capabilities of remote sensing. The LULC layer can be developed from supervised (interactive) or unsupervised (automatic) classification of satellite imagery. GIS can be used to reclassify or aggregate the usually large number of imagery-based LULC classes into a small number of user-specified LULC classes. GIS can also help to refine or verify the imagery-based LULC classes. For example, population-density and unit-type (single or multiple family) attributes of census GIS data (e.g., census blocks) can be used to reclassify the typical residential land-use classes (e.g., low-, medium-, and high-density residential) into single-family residential and multifamily residential land-use classes.

LULC changes observed in remote sensing imagery are of interest to planners, ecologists, hydrologists, and atmospheric scientists. Remote sensing imagery taken at different times can be compared to detect changes in LULC using a process called *change detection*. Remote sensing change detection can also be used for near-real-time detection and mapping of fires, power outages, and floods. Remote sensing change detection techniques can also be adapted for assessing the differences between modeled and observed images for validation of distributed hydrologic and ecosystem models (Luce, 2001; Lunetta and Elvidge, 1998).

Standard multispectral data provide only a grid of numbers that literally reflect the amount of electromagnetic energy measured at a specified small square on the Earth's surface. The most common and pervasive problem in any spectral classification of these data is the presence of mixed cells called "mixels." A 10 m × 10 m location on the ground may be half lawn and shrub and half rooftop. Obviously, because the spectral reflectance value for this cell cannot show two half-values, the value becomes the average of the two surface types, statistically dissimilar from both vegetation and urban areas. Taking into consideration the fact that cells are not spectrally divided into only two distinct surface types but every possible combination, the difficulties in spectrally based classification procedures become apparent. A key challenge to using multispectral satellite imagery data is to "unmix" (usually urban) land-use areas that contain numerous mixed pixels due to a highly variable landscape. High-spectral (or hyperspectral) imagery that divides the electromagnetic spectrum into a large number of segments for a finer classification of the Earth's surface is an important evolving trend. The mixel problem can be eliminated by using hyperspectral data from new satellites, such as NASA's Earth Observing-1 (EO-1) and Orbimage's OrbView 3. One-and-a-half months after its launch aboard the EO-1 spacecraft Hyperion, NASA's first hyperspectral imager was transmitting 30-m resolution images of the Earth in 220 spectral bands from the visible to shortwave infrared. Hyperion captures 7.5 km × 180 km images with high radiometric accuracy. NASA's ASTER is capable of collecting 14 bands of data at 15- to 19-m resolutions. The next OrbView launch is expected to provide a 200-band hyperspectral capability with 8-m resolution. The hyperspectral analysis software packages such as ENVI use special techniques like spectral angle mapping or linear spectral unmixing to isolate different contributions to a single pixel. The IMAGINE software from ERDAS provides a subpixel classifier to handle the mixed cells (ERDAS, 2001).

Note that local governments are usually the best sources of LULC data because they are involved most directly in land-planning activities. Many cities maintain accurate, aerial photo-based parcel maps for tax assessment. Local LULC data can include outlines of buildings, driveways, and pavements, which are excellent sources of data for applications that require highly accurate parameters, such as estimation of percent imperviousness by LULC classes and urban runoff modeling (Slawecki et al., 2001). Before using GIS and remote sensing techniques to create LULC data, one should check with local governments and regional organizations in the study area to determine if existing LULC maps or data are available.

Soil Moisture Mapping

Recent developments have opened the doors for exploring remotely sensed data in the microwave region. The strength of the microwave signal is directly related to the amount of water present below the target surface. This feature makes microwave remote sensing particularly attractive in water resources studies. Microwave sensors can produce soil-moisture data that measure the surface dielectric properties. Daily soil-moisture maps can be composed to establish antecedent conditions for runoff modeling. NASA's Tropical Rainfall Mapping Mission (TRMM) Microwave Imager has been used to establish soil moisture over a 140 km × 280 km region in Oklahoma.

Estimating Meteorological Data

Many watersheds, especially those smaller than 1000 km², do not have recording rain gauges. These data gaps can be filled by the rainfall data provided by weather satellites and radars. Detailed applications of remote sensing data in rainfall measurement are provided in Chapter 10 (Monitoring Applications).

Evapotranspiration data are critical in water budget computations of a watershed. Evapotranspiration measurements require expensive instruments for an area no larger than a field. Estimation of evapotranspiration for large areas using computer models may be difficult and time consuming. Research hydrologists at the University of Nebraska-Lincoln have developed a way to measure watershed evapotranspiration using vegetative productivity (greenness) extracted from satellite imagery. Because producing green matter involves using a specific amount of water, evapotranspiration can be calculated for large areas. Evapotranspiration estimates are used to determine irrigation-water and groundwater recharge, two of the most difficult-to-estimate parameters in most water budget computations.

GEOGRAPHIC IMAGING AND IMAGE PROCESSING SOFTWARE

In recent days, geographic imaging has become an essential part of GIS applications. Without geographic imaging capability, a GIS software is like a car without a spare wheel, or a computer without a modem, or a pizza without toppings. Imagine tens or hundreds of image tiles that have unbalanced color, brightness, or contrast, which must be manually adjusted to plot nicely. Imagine a GIS project involving 6-in. resolution images for an entire county and the user waiting for the computer to redraw the screen as he or she zooms in and out. Geographic-imaging and image processing products solve these problems by helping users to visualize, manipulate, analyze, measure, and integrate geographic imagery and geospatial information. These programs can be classified in two categories:

1. Preprocessors: These programs (such as MrSID described in the following text), help to preprocess raster data for GIS applications. Some preprocessing programs can join multiple images to provide georeferenced color-balanced mosaics. They combine multiple frames of scanned film, digital photos, or satellite images into a single picture and apply mapping-coordinates information to the mosaicked image for spectral and spatial analysis.
2. GIS Extensions: These programs add image processing capability to a GIS software. For example, Image Analysis is an extension that adds image processing capability to ArcView GIS software.

Representative software are listed in Table 3.3 and described in the following subsections.

ERDAS Software Products

ERDAS, Inc. (Atlanta, Georgia), is a leading geographic imaging and image processing software company. The company's software products help organizations

Table 3.3 Geographic-Imaging and Image Processing Software

Software	Version	Vendor	Cost ($)	Web site
ERDAS IMAGINE	8.7	Leica Geosystems, Atlanta	>5000	gis.leica-geosystems.com www.erdas.com
ER Mapper	6.1	Earth Resources Mapping, West Perth, Australia	5000	www.ermapper.com
Image Analysis	1.1	ERDAS and ESRI	2500	
Geomatica EASI/PACE	— 7.0	PCI Geomatics, Ontario, Canada	N/A	www.pcigeomatics.com
Image Analyst for MicroStation	—	Z/I Imaging Corp.	N/A	www.ziimaging.com
Geographic Transformer	4.2	Blue Marble Geographics	800	www.bluemarblegeo.com
MrSID	1.4	LizardTech	1000–3500	www.lizardtech.com
ENVI	3.4	Research Systems, Inc.	4000	www.rsinc.com

visualize, manipulate, analyze, measure, and integrate geographic imagery and geospatial information into 2D and 3D environments. In July 2001, ERDAS's geographic imaging software division was acquired by Leica Geosystems (Heerbrugg, Switzerland) to form a new GIS and mapping division. In the GIS community, Leica Geosystems is known for its GPS and field data collection related equipment.

The ERDAS flagship product is called IMAGINE (previously known as ERDAS; the latest version is 8.7); it provides remote sensing capabilities and a broad range of geographic imaging tools. It contains tools to make data production faster and easier, such as on-the-fly reprojection, a batch wizard to automate routine procedures, the ability to create and edit ESRI Shapefiles, and faster and easier map production capabilities. It also provides enhanced mosaicking capabilities for creating seamless output images. Using specialized color-balancing procedures to remove "hot spots" from aerial photographs and satellite imagery, IMAGINE's simple but powerful mosaicking tool can run in an automated mode or allow users to intervene for quality assurance. For example, its cropping feature removes an image's rough edges, and the Exclude Areas tool can be used to define pixel groups likely to skew the image histogram and cause erroneous mosaicking artifacts.

The ERDAS IMAGINE product suite consists of three components that can be combined to create a scaleable solution for project-specific needs:

- IMAGINE Essentials: A mapping and visualization tool that allows different types of geographic data to be combined with imagery and organized for a mapping project.
- IMAGINE Advantage: This component builds upon the geographic imaging capabilities of IMAGINE Essentials by adding more precise mapping and image processing capabilities. It analyzes data from imagery via image mosaicking, surface interpolation, and advanced image interpretation and orthorectification tools.
- IMAGINE Professional: This is a suite of sophisticated tools for remote sensing and complex image analysis. It contains all the capabilities of IMAGINE Essentials and IMAGINE Advantage and adds radar analysis and advanced classification tools like the IMAGINE Expert Classifier. It also includes graphical spatial data modeling, an advanced feature for geographic data analysis.

ERDAS MapSheets is a mapping and geographic presentation software package. Because of it's compatibility with Microsoft Office, MapSheets allows using Object Linking and Embedding (OLE) technology to incorporate maps and images into reports, presentations, and spreadsheets. Reportedly, it is as easy to use as a word processor or a spreadsheet because it works directly with Microsoft Office software. It allows adding a map to a Microsoft Word report, using Excel to query the map attributes, using corporate data with Microsoft Access, and making presentations in PowerPoint. MapSheets allows reshaping images and drawings that have different projections. Its change-detection feature allows viewing changes from one image or drawing to another.

ERDAS Software Application Example

With a population of over 500,000 people, Colorado Springs, Colorado, is the U.S.' 18th-fastest growing city. In 1999, the Water Resources Department of Colorado

Springs Utilities (CSU) embarked on a pipeline-mapping project using high-resolution aerial photographs. CSU used 6-in. resolution DOPs of the Rockies to accurately, quickly, and inexpensively create a base map of two water pipelines that stretch approximately 130 mi. The resulting data were combined with GPS and ArcInfo GIS data to create an accurate set of detailed pipeline maps at a scale of 1 in. = 200 ft. CSU utilized ERDAS' Stereo Analyst, IMAGINE Virtual GIS, and IMAGINE Ortho-BASE geographic imaging software products. The compatibility between ERDAS and ESRI products allowed CSU to leverage its existing ArcInfo data to analyze pipeline accessibility, maintenance history and schedules, and future water needs. CSU also used high-resolution aerial photographs to create a detailed map of its water treatment plant in about 2 weeks. The ability to process more high-resolution imagery in-house in less time helped CSU to cost-effectively and efficiently maintain its GIS and save an estimated $1.8 million in mapping cost (ArcNews, 2001a).

ArcView Image Analysis Extension

ArcView Image Analysis Extension was developed as a collaborative effort between ESRI and ERDAS. It provides a direct path from IMAGINE to ArcView for users with complex geographic imaging and processing needs and provides readily available image and remote sensing data. It allows georeferencing imagery to Shapefiles, coverages, global positioning system points, or reference images; image enhancement; automatic mapping of feature boundaries; change detection for continuous and thematic imagery; multispectral categorizations for LULC mapping and data extraction; vegetation greenness mapping; and mosaicking imagery from different sources and different resolutions. One of the most useful tools in Image Analysis Extension is the Image Align tool designed to coregister image data to vector layers. The intuitive point-marking scheme is designed to make image recti-fication simple for novices. In addition to the standard rectification process from user-specified control points and GPS-collected points, the software also displays selected satellite data types in proper map position automatically. This capability is called image calibration, and is based on positional information (ephemeris) provided in commercial data sources. Image Analysis' spectral-categorization capabilities include unsupervised image classification with ISODATA classifier and finding like areas for single-class identification. These capabilities can be used for automatic LULC classification from satellite imagery.

ERDAS Stereo Analyst is another ArcView extension. It allows users to collect and visualize spatial data in true stereo and to roam with real-time pan and zoom. Features include the ability to collect and edit 3D Shapefiles and visualization of terrain information, tree stands, and watersheds.

MrSID

Water system and sewer system GIS data generally have DOP base maps that are stored in extremely large files. For example, the City of Loveland, Colorado, had four aerial photo images of 1.3 gigabytes (GB) each (Murphy, 2000a). The recent explosion of high-resolution imagery has dramatically increased the size of raster data. Raster images are storage-hungry but compressed images lose resolution because there is a trade-off between image size and resolution. Large image files take a long time to display on the computer. Fortunately, the data-reading capacity of image processing software has increased from a few gigabytes to the terabyte range. Image-compression techniques have played a critical role in storing large raster data sets in compressed file formats, such as Earth Resource Mapping's ECW format and LizardTech's MrSID format.

Multiresolution Seamless Image Database (MrSID) is a relatively new image file type (SID) from LizardTech. It encodes large, high-resolution images to a fraction of their original file size while maintaining the original image quality. Images become scalable and can be reduced, enlarged, zoomed, panned, or printed without compromising integrity. MrSID's selective decompression and bandwidth-optimization capabilities also increase the file transfer speed. MrSID provides the world's highest compression ratios that average at about 40 but can be as high as 100. For example, Mecklenburg County, North Carolina, had 708 sheets of 1-in. = 1000-ft B&W digital orthophotos for 538 mi^2 of the entire county. The scanned 9 in. × 9 in. films at a resolution of 1-ft to 1-pixel created approximately 23 MB georeferenced TIFF files. This procedure created 16 GB of imagery stored on 27 CD-ROMs. Delivering a compression ratio of 1:28, MrSID took 14 hours on a Pentium PC with 512 MB of RAM to compress 16 GB of imagery to a single 608 MB MrSID Portable Image Format that could be stored on a single 650 MB CD-ROM (Kuppe, 1999). Similarly, MrSID was able to compress 18 GB of Washington, D.C., DOPs onto one CD-ROM.

MrSID allows fast viewing of massive images. For instance, displaying a 50 MB TIFF image can take several minutes, whereas it requires only a few seconds to open the same image in the MrSID format. Reportedly, MrSID software can automatically mosaic hundreds of image tiles of virtually any size into a single, seamless image that is geometrically and geographically accurate with all georeferencing data intact. MrSID images can be viewed in most popular GIS software packages, such as ESRI's ArcInfo, ArcView, MapObjects, ArcIMS, ArcExplorer, and ArcPad; Autodesk, Intergraph, and GE Smallworld. MrSID ArcView GIS extension allows MrSID images to be instantly decompressed and displayed within ArcView GIS. It takes advantage of MrSID's image-compression and retrieval capability and offsets the problems of working with large images in ArcView GIS. MrSID's ArcView GIS extension gives users the ability to work with raster images of any size while providing instantaneous, seamless, multiresolution browsing of large raster images in ArcView.

PCI Geomatics

PCI Geomatics (Ontario, Canada) provides a suite of image processing, remote sensing, orthophoto, and GIS software tools. In 2001, PCI Geomatics software was used to produce "ImageMap USA," the first seamless color satellite image of the continental U.S. at 15-m spatial resolution. This mosaic is composed of approximately 450 individual Landsat 7 satellite scenes. PCI Geomatics' software products are described below.

Geomatica™ unites previously separate technologies that were dedicated to remote sensing, image processing, GIS (both vector and raster), cartography, and desktop photogrammetry into a single integrated environment. Geomatica represents the most aggressive movement toward the integration of GIS and image processing functions in one software package (Limp, 2001). Although Geomatica is available in several configurations, all have a consistent user interface and data structure.

EASI/PACE desktop remote sensing software allows working with Landsat, SPOT, air photos, or radar data in one image processing package. Users can use data from GIS databases and extract information for resource analysis, mapping and environmental applications. It can also be used to make image maps and update GIS data.

ImageWorks is a GUI-based software that incorporates the most frequently needed image-display, enhancement, and data management tools in a user-friendly interface. The software provides integrated raster and vector display capability.

OrthoEngine Suite provides capabilities for producing map-accurate imagery. It can orthorectify many types of image data, including aerial photos, digital camera frames, and Landsat, SPOT, IRS, IKONOS, RADARSAT, ERS, and JERS imagery. It allows DEM generations from SPOT, IRS, or RADARSAT imagery. It also provides manual and automatic mosaicking tools to create seamless mosaics of multiple images.

FLY! is a terrain-visualization tool that drapes imagery and vectors over DEM data to create 3D perspective scenes in near real-time. An intuitive point-and-click user interface enables users to control flight speed, direction, elevation, and perspective parameters interactively during flight.

Radar imagery from satellites and aircraft has become a significant tool for a wide range of remote sensing applications. Data from sources such as RADARSAT, ERS, and JERS provide timely and consistent sources of information, regardless of weather conditions or illumination. With the expanding use of radar-imaging systems comes the need for new processing tools to extract useful information. Imaging radar data requires special handling and analysis techniques. PCI Geomatics has developed RADARSOFT to meet these needs.

PAMAP is a full-featured GIS software that combines an easy-to-use interface with increased interactivity and dynamic links to many industry-standard external databases for efficient storage of attribute data. It uses PCI Geomatics' Generic Database (GDB) technology, which supports over 50 different raster and vector data types facilitating easy exchange of spatial and attribute data.

GeoGateway allows users to visualize large data files, reproject them into any of the over 25 supported projection systems (or define a new projection system), subset the files into smaller windows of data, and write the resulting data out to a different, more distributable format, or even to an enterprise relational database such as Oracle.

Blue Marble Geographics

Blue Marble Geographics provides a host of geographic imaging, image processing, and map-projection software tools and utilities. The Geographic Transformer allows users to establish an "image-to-world" relationship between image and map coordinates and reproject an image into a georeferenced image map. It is a simple software to georeference, reproject, tile, and mosaic images. The Geographic Transformer AVX is an ArcView extension that integrates these functions directly into the ArcView GIS environment. Geographic Calculator is a stand-alone utility than can convert coordinates from one system to another. It offers more than 12,000 predefined coordinate systems as well as user-defined systems, datums, and units. Geographic Calculator supports DXF, DWG, SHP, MIF, and TAB files.

FUTURE DIRECTIONS

Remote sensing has dramatically improved our ability to forecast weather, monitor pollution, and respond to natural disasters. As we continue to transition into the information age, we will uncover even more remote sensing applications. According to industry experts, QuickBird and LIDAR are only the tip of the iceberg. There are several new developments in the remote sensing field that will have a significantly positive impact on future applications in the water industry (Merry, 2000). These include the launch of the EOS series of satellites, the availability of satellite hyperspectral remote sensing data, and high-resolution commercial remote sensing data. Hyperspectral sensors examine hundreds of individual bands of the electromagnetic spectrum to reveal extremely subtle characteristics of the Earth's surface. The EOS time-series data is organized in an information system that will allow direct assimilation of remote sensing data into computer models for calibration purposes. New sensors, particularly in the microwave region, promise great potential for hydrologic applications. The U.S. Navy's scheduled launch of the Naval Earth Map Observer (NEMO) will provide data on shallow-water bathymetry, bottom type composition, currents, oil slicks, atmospheric visibility, beach characteristics, and near-shore soil and vegetation (Lillesand and Kiefer, 1999).

In the future, maps of dubious accuracy will not be digitized into a GIS. Contemporary imagery will be obtained for each unique application or solution. The new technology will enable improved analysis of reflected laser signals of LIDAR systems. Enhanced interpretation of the strength of the laser signals will enable automatic creation of low-grade DOPs. LIDAR technology one day will become so accurate that it will be possible to distinguish between different types of trees. As our ability to develop DEMs increases, the potential uses of LIDAR technology will multiply.

The human eye is accustomed to seeing objects such as tanks and ponds, whereas remote sensing data recognize nothing but pixels. This limitation creates the pixel problem described earlier in the chapter, which can be eliminated using the latest hyperspectral imagery. In addition to this solution, new remote sensing techniques are being developed to enable identification of objects. In such applications, the spatial and textural properties of the image are used along with the spectral properties of a single pixel to extract real-world objects from remotely sensed data (Limp, 2001). Geographic imaging software companies like ERDAS are developing hyperspectral analysis tools that will provide task-oriented, wizard-based capabilities to simplify and automate numerous preprocessing steps for analyzing hyperspectral imagery.

Autonomous aerial imaging combines the advantages of both aerial and satellite imaging. Autonomous imaging is already available for military missions, and it is anticipated that it will soon become available for commercial applications. On April 23, 2001, the U.S. Air Force's Global Hawk became the first autonomous powered aircraft to cross the Pacific Ocean. The result of a collaboration between the U.S. Air

Force and the Australian Defence Organization, Global Hawk can fly at altitudes of 65,000 ft and image 40,000 mi^2 in less than 24 hours. For commercial applications, the aircraft could be quickly deployed to an area of interest (or crisis) and data could be downloaded and processed in near real-time-like satellite imagery, but with a higher pixel resolution. This technology can reduce overhead cost on pilot and technician hours and make real-time high-resolution imagery affordable to water industry professionals.

USEFUL WEB SITES

IKONOS satellite	www.spaceimaging.com
Landsat 7 satellite	http://landsat7.usgs.gov
OrbView satellite	www.orbimage.com
Pennsylvania Spatial Data Access system (PASDA)	www.pasda.psu.edu
QuickBird satellite	www.digitalglobe.com
Terra satellite	http://terra.nasa.gov
TerraServer	www.terraserver.com
USGS DOQ	www-wmc.wr.usgs.gov/doq/
USGS EarthExplorer	http://earthexplorer.usgs.gov

CHAPTER SUMMARY

The increasing use of GIS is contributing to a renewed interest in remote sensing — the process of observing and mapping from a distance. GIS technology is promoting the use of remote sensing data such as aerial photographs, satellite imagery, and radar data. Remote sensing technology offers numerous applications in the water industry. Space-based satellite imagery in GIS-ready format can be used as cost-effective base maps for mapping water industry systems. It can be used for land-use classification, watershed delineation, and measuring rainfall. Remote sensing data are available in various resolutions; not all of which are suitable for water industry applications. The recent availability of high-resolution (0.5 to 5 m) satellite imagery is poised to bring remote sensing technology to a water utility near you. For additional information, the famous Lillesand and Kiefer (1999) textbook *Remote Sensing and Image Interpretation* is suggested.

CHAPTER QUESTIONS

1. What is remote sensing and how is it related to satellite imagery?
2. How is remote sensing related to GIS and how is it used in GIS applications?
3. What is a digital orthophoto and how is it used in GIS?
4. How are remote sensing data used for land-use classification?
5. What type of software is required for using remote sensing data?

CHAPTER **4**

DEM Applications

Can a laser device mounted in an airplane create a GIS-ready ground
surface elevation map of your study area or measure the elevation of
your manholes? Read this chapter to find out.

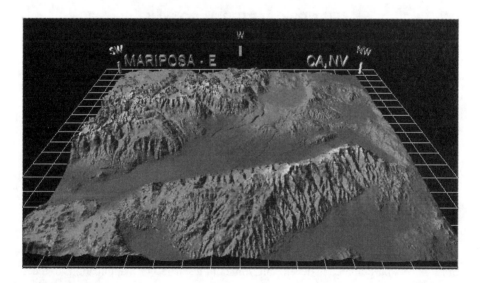

1:250,000 USGS DEM for Mariposa East, California (plotted using DEM3D viewer software
from USGS).

LEARNING OBJECTIVE

The learning objective of this chapter is to learn how to use digital elevation models (DEM) in GIS for water industry applications.

MAJOR TOPICS

- DEM basics
- DEM data resolution and accuracy
- USGS DEM data
- DEM data from remote sensing technology
- DEM data from LIDAR and IFSAR technologies
- DEM analysis techniques and software packages
- DEM application case studies and examples

LIST OF CHAPTER ACRONYMS

3-D Three-Dimensional
DEM Digital Elevation Model
DTM Digital Terrain Model
ERDAS Earth Resource Data Analysis System
IFSAR Interferometric Synthetic Aperture Radar
LIDAR Laser Imaging Detection and Ranging/Light Imaging Detection and Ranging
NED National Elevation Detection and Ranging
TIN Triangular Irregular Network

HYDROLOGIC MODELING OF THE BUFFALO BAYOU USING GIS AND DEM DATA

In the 1970s, the Hydrologic Engineering Center (HEC) of the U.S. Army Corps of Engineers participated in developing some of the earliest GIS applications to meet the H&H modeling needs in water resources. In the 1990s, HEC became aware of the phenomenal growth and advancement in GIS. The capability of obtaining spatial data from the Internet coupled with powerful algorithms in software and hardware made GIS an attractive tool for water resources projects. The Buffalo Bayou Watershed covers most of the Houston metropolitan area in Texas. The first recorded flood in 1929 in the watershed devastated the city of Houston. Since then, other flooding events of similar vigor and intensity have occurred. During 1998 to 1999, the hydrologic modeling of this watershed was conducted using the Hydrologic Modeling System (HMS) with inputs derived from GIS. The watersheds and streams were delineated from the USGS DEM data at 30-m cell resolution, stream data from USGS digital line graph (DLG), and EPA river reach file (RF1). When used separately, software packages such as ArcInfo, ArcView, and Data Storage System (DSS) were found to be time consuming, requiring the combined efforts of many people. HEC integrated these existing software tools with new programs developed in this project into a comprehensive GIS software package called HEC-GeoHMS. The low-relief terrain of the study area required human interpretation of drainage paths, urban drainage facilities, and man-made hydraulic structures (e.g., culverts and storm drains), which dictated flow patterns that could not be derived from DEM terrain representation. To resolve this issue, the project team took advantage of the flexibility in HMS to correct drainage patterns according to human interpretations and local knowledge (Doan, 1999).

DEM BASICS

Topography influences many processes associated with the geography of the Earth, such as temperature and precipitation. GIS application professionals must be able to represent the Earth's surface accurately because any inaccuracies can lead to poor decisions that may adversely impact the Earth's environment. A DEM is a numerical representation of terrain elevation. It stores terrain data in a grid format for coordinates and corresponding elevation values. DEM data files contain information for the digital representation of elevation values in a raster form. Cell-based raster data sets, or grids, are very suitable for representing geographic phenomena that vary continuously over space such as elevation, slope, precipitation, etc. Grids are also ideal for spatial modeling and analysis of data trends that can be represented by continuous surfaces, such as rainfall and stormwater runoff.

DEM data are generally stored using one of the following three data structures:

- Grid structures
- Triangular irregular network (TIN) structures
- Contour-based structures

Regardless of the underlying data structure, most DEMs can be defined in terms of (x,y,z) data values, where x and y represent the location coordinates and z represents the elevation values. Grid DEMs consist of a sampled array of elevations for a number of ground positions at regularly spaced intervals. This data structure creates a square grid matrix with the elevation of each grid square, called a pixel, stored in a matrix format. Figure 4.1 shows a 3D plot of grid-type DEM data. As shown in Figure 4.2, TINs represent a surface as a set of nonoverlapping contiguous triangular facets, of irregular size and shape. Digital terrain models (DTMs) and digital surface models (DSMs) are different varieties of DEM. The focus of this chapter is on grid-type DEMs.

Usually, some interpolation is required to determine the elevation value from a DEM for a given point. The DEM-based point elevations are most accurate in relatively flat areas with smooth slopes. DEMs produce low-accuracy point elevation values in areas with large and abrupt changes in elevation, such as cliffs and road cuts (Walski et al., 2001).

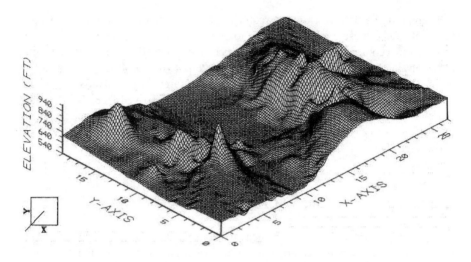

Figure 4.1   Grid-type DEM.

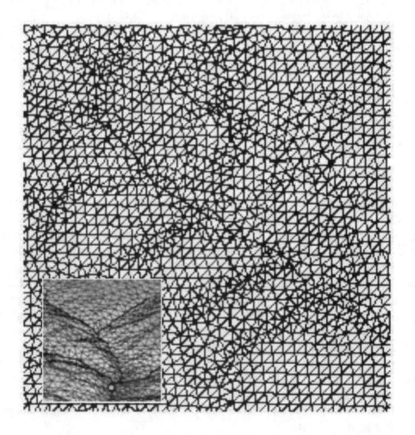

Figure 4.2   TIN-type DEM.

DEM APPLICATIONS

Major DEM applications include (USGS, 2000):

- Delineating watershed boundaries and streams
- Developing parameters for hydrologic models
- Modeling terrain gravity data for use in locating energy resources
- Determining the volume of proposed reservoirs
- Calculating the amount of material removed during strip mining
- Determining landslide probability

Jenson and Dominique (1988) demonstrated that drainage characteristics could be defined from a DEM. DEMs can be used for automatic delineation of watershed and sewershed boundaries. DEM data can be processed to calculate various watershed and sewershed characteristics that are used for H&H modeling of watersheds and sewersheds. DEMs can create shaded relief maps that can be used as base maps in a GIS for overlaying vector layers such as water and sewer lines. DEM files may be used in the generation of graphics such as isometric projections displaying slope, direction of slope (aspect), and terrain profiles between designated points. This aspect identifies the steepest downslope direction from each cell to its neighbors.

Raster GIS software packages can convert the DEMs into image maps for visual display as layers in a GIS. DEMs can be used as source data for digital orthophotos. They can be used to create digital orthophotos by orthorectification of aerial photos, as described in Chapter 3 (Remote Sensing Applications). DEMs can also serve as tools for many activities including volumetric analysis and site location of towers. DEM data may also be combined with other data types such as stream locations and weather data to assist in forest fire control, or they may be combined with remote sensing data to aid in the classification of vegetation.

Three-Dimensional (3D) Visualization

Over the past decade, 3D computer modeling has evolved in most of the engineering disciplines including, but not limited to, layout, design, and construction of industrial and commercial facilities; landscaping; highway, bridge, and embankment design; geotechnical engineering; earthquake analysis; site planning; hazardous-waste management; and digital terrain modeling. The 3D visualization can be used for landscape visualizing or fly-through animation movies of the project area. 3D animations are highly effective tools for public- and town-meeting presentations. GIS can be used to create accurate topographic elevation models and generate precise 3D data. A DEM is a powerful tool and is usually as close as most GISs get to 3D modeling.

3D graphics are commonly used as a visual communication tool to display a 3D view of an object on two-dimensional (2D) media (e.g., a paper map). Until the early 1980s, a large mainframe computer was needed to view, analyze, and print objects in 3D graphics format. Hardware and software are now available for 3D modeling of terrain and utility networks on personal computers. Although DEMs are raster images, they can be imported into 3D visualizations packages. Affordable and user-friendly software tools are bringing more users into the world of GIS.

These software tools and 3D data can be used to create accurate virtual reality representations of landscape and infrastructure with the help of stereo imagery and automatic extraction of 3D information. For example, Skyline Software System's (www.skylinesoft.com) TerraExplorer provides realistic, interactive, photo-based 3D maps of many locations and cities of the world on the Internet.

Satellite imagery is also driving new 3D GIS applications. GIS can be used to precisely identify a geographic location in 3D space and link that location and its attributes through the integration of photogrammetry, remote sensing, GIS, and 3D visualization. 3D geographic imaging is being used to create orthorectified imagery, DEMs, stereo models, and 3D features.

DEM RESOLUTION AND ACCURACY

The accuracy of a DEM is dependent upon its source and the spatial resolution (grid spacing). DEMs are classified by the method with which they were prepared and the corresponding accuracy standard. Accuracy is measured as the root mean square error (RMSE) of linearly interpolated elevations from the DEM, compared with known elevations. According to RMSE classification, there are three levels of DEM accuracy (Walski et al., 2001):

- Level 1: Based on high-altitude photography, these DEMs have the lowest accuracy. The vertical RMSE is 7 m and the maximum permitted RMSE is 15 m.
- Level 2: These are based on hypsographic and hydrographic digitization, followed by editing to remove obvious errors. These DEMs have medium accuracy. The maximum permitted RMSE is one half of the contour interval.
- Level 3: These are based on USGS digital line graph (DLGs) data (Shamsi, 2002). The maximum permitted RMSE is one third of the contour interval.

The vertical accuracy of 7.5-min DEMs is greater than or equal to 15 m. Thus, the 7.5-min DEMs are suitable for projects at 1:24,000 scale or smaller (Zimmer, 2001a). A minimum of 28 test points per DEM are required (20 interior points and 8 edge points). The accuracy of the 7.5-min DEM data, together with the data spacing, adequately support computer applications that analyze hypsographic features to a level of detail similar to manual interpretations of information as printed at map scales not larger than 1:24,000. Early DEMs derived from USGS quadrangles suffered from mismatches at boundaries (Lanfear, 2000).

DEM selection for a particular application is generally driven by data availability, judgment, experience, and test applications (ASCE, 1999). For example, because no firm guidelines are available for selection of DEM characteristics for hydrologic modeling, a hydrologic model might need 30-m resolution DEM data but might have to be run with 100-m data if that is the best available data for the study area. In the U.S., regional-scale models have been developed at scales of 1:250,000 to 1:2,000,000 (Laurent et al., 1998). Seybert (1996) concluded that modeled watershed runoff peak flow values are more sensitive to changes in spatial resolution than modeled runoff volumes. An overall subbasin area to grid–cell area ratio of 10^2 was found to produce reasonable model results.

Table 4.1   DEM Applications

DEM Resolution	Approximate Cell Size	Watershed Area (km²)	Typical Application
1 sec	30 m	5	Urban watersheds
3 sec	100 m	40	Rural watersheds
15 sec	500 m	1,000	River basins, States
30 sec	1 km	4,000	Nations
3 min	5 km	150,000	Continents
5 min	10 km	400,000	World

The grid size and time resolution used for developing distributed hydrologic models for large watersheds is a compromise between the required accuracy, available data accuracy, and computer run-time. Finer grid size requires more computing time, more extensive data, and more detailed boundary conditions. Chang et al. (2000) conducted numerical experiments to determine an adequate grid size for modeling large watersheds in Taiwan where 40 m × 40 m resolution DEM data are available. They investigated the effect of grid size on the relative error of peak discharge and computing time. Simulated outlet hydrographs showed higher peak discharge as the computational grid size was increased. In a study, for a watershed of 526 km² located in Taiwan, a grid resolution of 200 m × 200 m was determined to be adequate.

Table 4.1 shows suggested DEM resolutions for various applications (Maidment, 1998). Large (30-m) DEMs are recommended for water distribution modeling (Walski et al., 2001).

The size of a DEM file depends on the DEM resolution, i.e., the finer the DEM resolution, the smaller the grid, and the larger the DEM file. For example, if the grid size is reduced by one third, the file size will increase nine times. Plotting and analysis of high-resolution DEM files are slower because of their large file sizes.

USGS DEMS

In the U.S., the USGS provides DEM data for the entire country as part of the National Mapping Program. The National Mapping Division of USGS has scanned all its paper maps into digital files, and all 1:24,000-scale quadrangle maps now have DEMs (Limp, 2001).

USGS DEMs are the (x,y,z) triplets of terrain elevations at the grid nodes of the Universal Transverse Mercator (UTM) coordinate system referenced to the North American Datum of 1927 (NAD27) or 1983 (NAD83) (Shamsi, 1991). USGS DEMs provide distance in meters, and elevation values are given in meters or feet relative to the National Geodetic Vertical Datum (NGVD) of 1929. The USGS DEMs are available in 7.5-min, 15-min, 2-arc-sec (also known as 30-min), and 1° units. The 7.5- and 15-min DEMs are included in the large-scale category, whereas 2-arc-sec DEMs fall within the intermediate-scale category and 1° DEMs fall within the small-scale category. Table 4.2 summarizes the USGS DEM data types.

This chapter is mostly based on applications of 7.5-min USGS DEMs. The DEM data for 7.5-min units correspond to the USGS 1:24,000-scale topographic quadrangle map series for all of the U.S. and its territories. Thus, each 7.5-min

Table 4.2   USGS DEM Data Formats

DEM Type	Scale	Block Size	Grid Spacing
Large	1:24,000	7.5 ft × 7.5 ft	30 m
Intermediate	Between large and small	30 ft × 30 ft	2 sec
Small	1:250,000	1° × 1°	3 sec

by 7.5-min block provides the same coverage as the standard USGS 7.5-min map series. Each 7.5-min DEM is based on 30-m by 30-m data spacing; therefore, the raster grid for the 7.5-min USGS quads are 30 m by 30 m. That is, each 900 m² of land surface is represented by a single elevation value. USGS is now moving toward acquisition of 10-m accuracy (Murphy, 2000).

USGS DEM Formats

USGS DEMs are available in two formats:

1. DEM file format: This older file format stores DEM data as ASCII text, as shown in Figure 4.3. These files have a file extension of dem (e.g., lewisburg_PA.dem). These files have three types of records (Walski et al., 2001):
   • Type A: This record contains information about the DEM, including name, boundaries, and units of measurements.

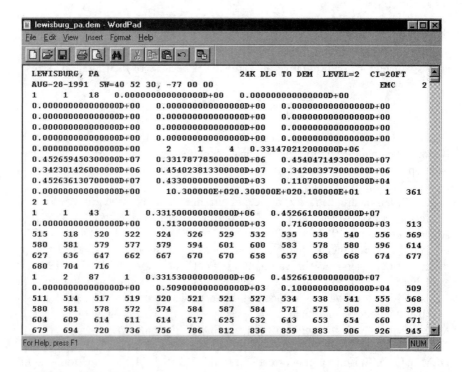

Figure 4.3   USGS DEM file.

- Type B: These records contain elevation data arranged in "profiles" from south to north, with the profiles organized from west to east. There is one Type-B record for each south–north profile.
- Type C: This record contains statistical information on the accuracy of DEM.
2. Spatial Data Transfer Standard (SDTS): This is the latest DEM file format that has compressed data for faster downloads. SDTS is a robust way of transferring georeferenced spatial data between dissimilar computer systems and has the potential for transfer with no information loss. It is a transfer standard that embraces the philosophy of self-contained transfers, i.e., spatial data, attribute, georeferencing, data quality report, data dictionary, and other supporting metadata; all are included in the transfer. SDTS DEM data are available as tar.gz compressed files. Each compressed file contains 18 ddf files and two readme text files. For further analysis, the compressed SDTS files should be unzipped (uncompressed). Standard zip programs, such as PKZIP, can be used for this purpose.

Some DEM analysis software may not read the new SDTS data. For such programs, the user should translate the SDTS data to a DEM file format. SDTS translator utilities, like SDTS2DEM or MicroDEM, are available from the GeoCommunity's SDTS Web site to convert the SDTS data to other file formats.

National Elevation Dataset (NED)

Early DEMs were derived from USGS quadrangles, and mismatches at boundaries continued to plague the use of derived drainage networks for larger areas (Lanfear, 2000). The NED produced by USGS in 1999 is the new generation of seamless DEM that largely eliminates problems of quadrangle boundaries and other artifacts. Users can now select DEM data for their area of interest.

The NED has been developed by merging the highest resolution, best-quality elevation data available across the U.S. into a seamless raster format. NED is designed to provide the U.S. with elevation data in a seamless form, with a consistent datum, elevation unit, and projection. Data corrections were made in the NED assembly process to minimize artifacts, perform edge matching, and fill sliver areas of missing data. NED is the result of the maturation of the USGS effort to provide 1:24,000-scale DEM data for the conterminous U.S. and 1:63,360-scale DEM data for Alaska. NED has a resolution of 1 arc-sec (approximately 30 m) for the conterminous U.S., Hawaii, and Puerto Rico and a resolution of 2 arc-sec for Alaska. Using a hill-shade technique, USGS has also derived a shaded relief coverage that can be used as a base map for vector themes. Other themes, such as land use or land cover, can be draped on the NED-shaded relief maps to enhance the topographic display of themes. The NED store offers seamless data for sale, by user-defined area, in a variety of formats.

DEM DATA AVAILABILITY

USGS DEMs can be downloaded for free from the USGS geographic data download Web site. DEM data on CD-ROM can also be purchased from the USGS EarthExplorer Web site for an entire county or state for a small fee to cover the shipping and handling cost. DEM data for other parts of the world are also available.

The 30 arc-sec DEMs (approximately 1 km² square cells) for the entire world have been developed by the USGS Earth Resources Observation Systems (EROS) Data Center and can be downloaded from the USGS Web site. More information can be found on the Web site of the USGS node of the National Geospatial Data Clearinghouse. State or regional mapping and spatial data clearinghouse Web sites are the most valuable source of free local spatial data. For example, the Pennsylvania Spatial Data Access system (PASDA), Pennsylvania's official geospatial information clearinghouse and its node on the National Spatial Data Infrastructure (NSDI), provides free downloads of DEM and other spatial data.

DEM DATA CREATION FROM REMOTE SENSING

In February 2000, NASA flew one of its most ambitious missions, using the space shuttle *Endeavor* to map the entire Earth from 60° north to 55° south of the equator. Mapping at a speed of 1747 km² every second, the equivalent of mapping the state of Florida in 97.5 sec, the Shuttle Radar Topography Mission (SRTM) provided 3D data of more than 80% of Earth's surface in about 10 days. The SRTM data will provide a 30-m DEM coverage for the entire world (Chien, 2000).

Topographic elevation information can be automatically extracted from remote sensing imagery to create highly accurate DEMs. There are two ways in which DEM data can be created using remote sensing methods: image processing and data collection.

Image Processing Method

The first method uses artificial intelligence techniques to automatically extract elevation information from the existing imagery. Digital image-matching methods commonly used for machine vision automatically identify and match image point locations of a ground point appearing on overlapping areas of a stereo pair (i.e., left- and right-overlapping images). Once the correct image positions are identified and matched, the ground point elevation is computed automatically. For example, the French satellite SPOT's stereographic capability can generate topographic data. USGS Earth Observing System's (EOS) Terra satellite can provide DEMs from stereo images.

Off-the-shelf image processing software products are available for automatic extraction of DEM data from remote sensing imagery. For instance, Leica Geosystems' IMAGINE OrthoBASE Pro software can be used to automatically extract DEMs from aerial photography, satellite imagery (IKONOS, SPOT, IRS-1C), and digital video and 35-mm camera imagery. It can also subset and mosaic 500 or more individual DEMs. The extracted DEM data can be saved as raster DEMs, TINs, ESRI 3D Shapefiles, or ASCII output (ERDAS, 2001b).

Data Collection Method

In this method, actual elevation data are collected directly using lasers. This method uses laser-based LIDAR and radar-based IFSAR systems described in the following text.

LIDAR

Unlike photogrammetric techniques, which can be time consuming and expensive for large areas, this method is a cost-effective alternative to conventional technologies. It can create DEMs with accuracy levels ranging from 20 to 100 cm, which are suitable for many engineering applications. This remote-sensing technology does not even involve an image. Laser imaging detection and ranging (LIDAR) is a new system for measuring ground surface elevation from an airplane. LIDAR can collect 3D digital data on the fly. LIDAR sensors provide some of the most accurate elevation data in the shortest time ever by bouncing laser beams off the ground. LIDAR technology, developed in the mid-1990s, combines global positioning system (GPS), precision aircraft guidance, laser range finding, and high-speed computer processing to collect ground elevation data. Mounted on an aircraft, a high-accuracy scanner sweeps the laser pulses across the flight path and collects reflected light. A laser range-finder measures the time between sending and receiving each laser pulse to determine the ground elevation below. The LIDAR system can survey up to 10,000 acres per day and provide horizontal and vertical accuracies up to 12 and 6 in., respectively. Chatham County, home of Savannah, Georgia, used the LIDAR approach to collect 1-ft interval contour data for the entire 250,000 acre county in less than a year. The cost of conventional topographic survey for this data would be over $20 million. The County saved $7 million in construction cost by using data from Airborne Laser Terrain Mapping (ALTM) technology, a LIDAR system manufactured by Optech, Canada. The new ALTM data were used to develop an accurate hydraulic model of the Hardin basin (Stones, 1999).

Chatham County, Georgia, saved $7 million in construction cost by using LIDAR data.

Boise-based Idaho Power Company spent $273,000 on LIDAR data for a 290 km stretch of the rugged Hell's Canyon, through which the Snake River runs. The cost of LIDAR data was found to be less than aerial data and expensive ground-surveying. The company used LIDAR data to define the channel geometry, combined it with bathymetry data, and created digital terrain files containing ten cross sections of the canyon per mile. The cross-section data were input to a hydraulic model that determined the effect of power plants' releases on vegetation and wildlife habitats (Miotto, 2000).

IFSAR

Interferometric Synthetic Aperture Radar (IFSAR) is an aircraft-mounted radar system for quick and accurate mapping of large areas in most weather conditions without ground control. Because it is an airborne radar, IFSAR collects elevation data on the first try in any weather (regardless of fog, clouds, or rain), day or night, significantly below the cost of satellite-derived DEM. The IFSAR process measures elevation data at a much denser grid than photogrammetric techniques, using over-lapping stereo images. A denser DEM provides a more detailed terrain surface in an image. IFSAR is efficient because it derives the DEM data by digital processing of a single radar image. This allows elevation product delivery within days of data collection. A DEM with a minimum vertical accuracy of 2 m is necessary to achieve the precision level orthorectification for IKONOS imagery. DEMs generated from the IFSAR data have been found to have the adequate vertical accuracy to orthorectify IKONOS imagery to the precision level (Corbley, 2001).

Intermap Technologies' (Englewood, Colorado, www.intermaptechnologies. com) Lear jet-mounted STAR-3i system, an airborne mapping system, has been reported to provide simultaneous high-accuracy DEMs and high-resolution orthorectified imagery without ground control. STAR-3i IFSAR system typically acquires elevation points at 5-m intervals, whereas photogrammetric sources use a spacing of 30 to 50 m. STAR-3i can provide DEMs with a vertical accuracy of 30 cm to 3 m and an orthoimage resolution of 2.5 m.

DEM ANALYSIS

Cell Threshold for Defining Streams

Before starting DEM analysis, users must define the minimum number of upstream cells contributing flow into a cell to classify that cell as the origin of a stream. This number, referred to as the cell "threshold," defines the minimum upstream drainage area necessary to start and maintain a stream. For example, if a stream definition value of ten cells is specified, then for a single grid location of the DEM to be in a stream, it must drain at least ten cells. It is assumed that there is flow in a stream if its upstream area exceeds the critical threshold value. In this case, the cell is considered to be a part of the stream. The threshold value can be estimated from existing topographic maps or from the hydrographic layer of the real stream network. Selection of an appropriate cell threshold size requires some user judgment. Users may start the analysis with an assumed or estimated value and adjust the initial value by comparing the delineation results with existing topographic maps or hydrographic layers. The cell threshold value directly affects the number of subbasins (subwatersheds or subareas). A smaller threshold results in smaller subbasin size, larger number of subbasins, and slower computation speed for the DEM analysis.

The D-8 Model

The 8-direction pour point model, also known as the D-8 model, is a commonly used algorithm for delineating a stream network from DEMs. As shown in Figure 4.4, it identifies the steepest downslope flow path between each cell and its eight neighboring cells. This path is the only flow path leaving the cell. Watershed area is accumulated downslope along the flow paths connecting adjacent cells. The drainage network is identified from the user-specified threshold area at the bottom of which a source channel originates and classifies all cells with a greater watershed area as part of the drainage network. Figure 4.4 shows stream delineation steps using the D-8 model with a cell threshold value of ten cells. Grid A shows the cell elevation

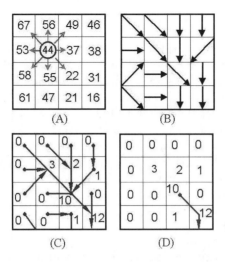

Figure 4.4   Figure 11-4. D-8 Model for DEM-based stream delineation (A) DEM elevation grid, (B) flow direction grid, (C) flow accumulation grid, and (D) delineated streams for cell threshold of ten.

values. Grid B shows flow direction arrows based on calculated cell slopes. Grid C shows the number of accumulated upstream cells draining to each cell. Grid D shows the delineated stream segment based on the cells with flow accumulation values greater than or equal to ten.

DEM Sinks

The D-8 and many other models do not work well in the presence of depressions, sinks, and flat areas. Some sinks are caused by the actual conditions, such as the Great Salt Lake in Utah where no watershed precipitation travels through a river network toward the ocean. The sinks are most often caused by data noise and errors in elevation data. The computation problems arise because cells in depressions, sinks, and flat areas do not have any neighboring cells at a lower elevation. Under these conditions, the flow might accumulate in a cell and the resulting flow network may not necessarily extend to the edge of the grid. Unwanted sinks must be removed prior to starting the stream or watershed delineation process by raising the elevation of the cells within the sink to the elevation of its lowest outlet. Most raster GIS software programs provide a FILL function for this purpose. For example, ArcInfo's GRID extension provides a FILL function that raises the elevation of the sink cells until the water is able to flow out of it.

The FILL approach assumes that all sinks are caused by underestimated elevation values. However, the sinks can also be created by overestimated elevation values, in which case breaching of the obstruction is more appropriate than filling the sink created by the obstruction. Obstruction breaching is particularly effective in flat or low-relief areas (ASCE, 1999).

Stream Burning

DEM-based stream or watershed delineations may not be accurate in flat areas or if the DEM resolution failed to capture important topographic information. This problem can be solved by "burning in" the streams using known stream locations from the existing stream layers. This process modifies the DEM grid so that the flow of water is forced into the known stream locations. The cell elevations are artificially lowered along the known stream locations or the entire DEM is raised except along known stream paths. The phrase *burning in* indicates that the streams have been forced, or "burned" into the DEM topography (Maidment, 2000). This method must be used with caution because it may produce flow paths that are not consistent with the digital topography (ASCE, 1999).

DEM Aggregation

Distributed hydrologic models based on high-resolution DEMs may require extensive computational and memory resources that may not be available. In this case, high-resolution DEMs can be aggregated into low-resolution DEMs. For example, it was found that the 30-m USGS DEM would create 80,000 cells for the 72.6 km^2 Goodwater Creek watershed located in central Missouri. Distributed modeling of 80,000 cells was considered time consuming and impractical (Wang et al., 2000). The 30 m $\times$ 30 m cells were, therefore, aggregated into 150 m x 150 m (2.25 ha) cells. In other words, 25 smaller cells were aggregated into one large cell, which reduced the number of cells from 80,000 to approximately 3,000. Best of all, the aggregated DEM produced the same drainage network as the original DEM. The aggregation method computes the flow directions of the coarse-resolution cells based on the flow paths defined by the fine-resolution cells. It uses three steps: (1) determine the flow direction of the fine-resolution DEM, (2) determine outlets of coarse-resolution DEM, and (3) approximate the flow direction of coarse-resolution DEM, based on the flow direction of the fine-resolution DEM.

Slope Calculations

Subbasin slope is an input parameter in many hydrologic models. Most raster GIS packages provide a SLOPE function for estimating slope from a DEM. For example, ERDAS IMAGINE software uses its SLOPE function to compute percent slope by fitting a plane to a pixel elevation and its eight neighboring pixel elevations. The difference in elevation between the low and the high points is divided by the horizontal distance and multiplied by 100 to compute percent slope for the pixel. Pixel slope values are averaged to compute the mean percent slope of each subbasin.

SOFTWARE TOOLS

The DEM analysis functions described in the preceding subsections require appropriate software. Representative DEM analysis software tools and utilities are listed in Table 4.3.

Table 4.3 Sample DEM Analysis Software Tools

Software	Vendor and Web site	Notes
Spatial Analyst and Hydro extension	ESRI, Redlands, California www.esri.com	ArcGIS 8.x and ArcView 3.x extension
ARC GRID extension		ArcInfo 7.x extension
Analyst		ArcGIS 8.x and ArcView 3.x extension
IDRISI	Clark University Worcester, Massachusetts www.clarklabs.org	
ERDAS IMAGINE	Leica Geosystems, Atlanta, Georgia gis.leica-geosystems.com www.erdas.com	Formerly, Earth Resource Data Analysis System (ERDAS) software
TOPAZ	U.S. Department of Agriculture, Agricultural Research Service, El Reno, Oklahoma grl.ars.usda.gov/topaz/TOPAZ1.HTM	
MicroDEM	U.S. Naval Academy www.usna.edu/Users/oceano/pguth/website/microdem.htm	Software developed by Peter Guth of the Oceanography Department
DEM3D viewer	USGS, Western Mapping Center, Menlo Park, California craterlake.wr.usgs.gov/dem3d.html	Free download, allows viewing of DEM files through a 3D perspective

Some programs such as Spatial Analyst provide both the DEM analysis and hydrologic modeling capabilities. ASCE (1999) has compiled a review of hydrologic modeling systems that use DEMs. Major DEM software programs are discussed in the following text.

Spatial Analyst and Hydro Extension

Spatial Analyst is an optional extension (separately purchased add-on program) for ESRI's ArcView 3.x and ArcGIS 8.x software packages. The Spatial Analyst Extension adds raster GIS capability to the ArcView and ArcGIS vector GIS software. Spatial Analyst allows for use of raster and vector data in an integrated environment and enables desktop GIS users to create, query, and analyze cell-based raster maps; derive new information from existing data; query information across multiple data layers; and integrate cell-based raster data with the traditional vector data sources. It can be used for slope and aspect mapping and for several other hydrologic analyses, such as delineating watershed boundaries, modeling stream flow, and investigating accumulation. Spatial Analyst for ArcView 3.x has most, but not all, of the functionality of the ARC GRID extension for ArcInfo 7.x software package described below.

Spatial Analyst for ArcView 3.x is supplied with a Hydro (or hydrology) extension that further extends the Spatial Analyst user interface for creating input data for hydrologic models. This extension provides functionality to create watersheds and stream networks from a DEM, calculate physical and geometric properties of the watersheds, and aggregate these properties into a single-attribute table that can be attached to a grid or Shapefile. Hydro extension requires that Spatial Analyst be already installed. Hydro automatically loads the Spatial Analyst if it is not loaded.

Depending upon the user needs, there are two approaches to using the Hydro extension:

1. Hydro pull-down menu options: If users only want to create watershed subbasins or the stream network, they should work directly with the Hydro pull-down menu options (Figure 4.5). Table 4.4 provides a brief description of each of these menu options. "Fill Sinks" works off an active elevation grid theme. "Flow Direction" works off an active elevation grid theme that has been filled. "Flow Accumulation" works off an active flow direction grid theme. "Flow Length" works off an active flow direction grid theme. "Watershed" works off an active flow accumulation grid theme and finds all basins in the data set based on a minimum number of cells in each basin. The following steps should be performed to create watersheds using the Hydro pull-down menu options, with the output grid from each step serving as the input grid for the next step:
   - Import the raw USGS DEM.
   - Fill the sinks using the "Fill Sinks" menu option (input = raw USGS DEM). This is a very important intermediate step. Though some sinks are valid, most are data errors and should be filled.
   - Compute flow directions using the "Flow Direction" menu option (input = filled DEM grid).
   - Compute flow accumulation using the "Flow Accumulation" menu option (input = flow directions grid).

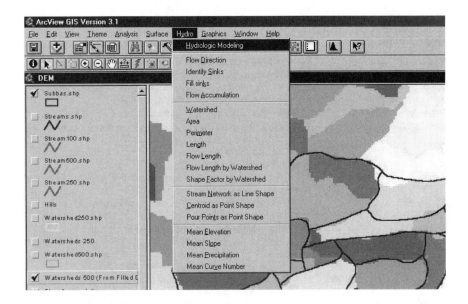

Figure 4.5   Hydro extension pull-down menu.

- Delineate streams using the "Stream Network" menu option (input = flow accumulation grid).
- Delineate watersheds using the "Watershed" menu option.

2. Hydrologic Modeling Dialogue: If users want to create subbasins and calculate many additional attributes for them, they should use the Hydrologic Modeling Dialogue (Figure 4.6), which is the first choice under the Hydro pull-down menu. The Hydrologic Modeling Dialogue is designed to be a quick one-step method for calculating and then aggregating a set of watershed attributes to a single file. This file can then be used in a hydrologic model, such as the Watershed Modeling System (WMS) (discussed in Chapter 11 [Modeling Applications]), or it can be reformatted for input into HEC's HMS model, or others. The following steps should be performed to create watersheds using the Hydrologic Modeling Dialogue:

- Choose "Delineate" from DEM and select an elevation surface.
- Fill the sinks when prompted.
- Specify the cell threshold value when prompted. This will create watersheds based on the number of cells or up-slope area defined by the user as the smallest watershed wanted.

Additional DEM analysis resources (tutorials, exercises, sample data, software utilities, reports, papers, etc.) are provided at the following Web sites:

- ESRI Web site at www.esri.com/arcuser/ (do a search for "Terrain Modeling")
- University of Texas at Austin (Center for Research in Water Resources) Web site at www.crwr.utexas.edu/archive.shtml

The last four Hydro options (Table 4.4) work with existing data layers. They do not create elevation, slope, precipitation, and runoff curve number layers. They

Table 4.4   Hydro Extension Menu Options

Hydro Menu Option	Function
Hydrologic modeling	Creates watersheds and calculates their attributes
Flow direction	Computes the direction of flow for each cell in a DEM
Identify sinks	Creates a grid showing the location of sinks or areas of internal drainage in a DEM
Fill sinks	Fills the sinks in a DEM, creating a new DEM
Flow accumulation	Calculates the accumulated flow or number of up-slope cells, based on a flow direction grid
Watershed	Creates watersheds based upon a user-specified flow accumulation threshold
Area	Calculates the area of each watershed in a watershed grid
Perimeter	Calculates the perimeter of each watershed in a watershed grid
Length	Calculates the straight-line distance from the pour point to the furthest perimeter point for each watershed
Flow length	Calculates the length of flow path for each cell to the pour point for each watershed
Flow length by watershed	Calculates the maximum distance along the flow path within each watershed
Shape factor by watershed	Calculates a shape factor (watershed length squared and then divided by watershed area) for each watershed
Stream network as line shape	Creates a vector stream network from a flow accumulation grid, based on a user-specified threshold
Centroid as point shape	Creates a point shape file of watershed centroids
Pour points as point shape	Creates a point shape file of watershed pour points
Mean elevation	Calculates the mean elevation within each watershed
Mean slope	Calculates the mean slope within each watershed
Mean precipitation	Calculates the mean precipitation in each watershed
Mean curve number	Calculates the mean curve number for each watershed

simply compute mean areal values of these four parameters for the subbasins, using the existing GIS layers of these parameters. Thus, the GIS layers of elevation, slope, precipitation, and runoff curve number must be available to use the mean functions of the Hydro extension.

Figure 4.7 shows Hydro's raindrop or pour point feature. Using this capability, the user can trace the flow path from a specified point to the watershed outlet. Hydro also calculates a flow length as the maximum distance along the flow path within each watershed. The flow path can be divided by the measured or estimated velocity to estimate the time of concentration or travel time that are used to estimate runoff hydrographs. Travel time can also be used to estimate the time taken by a hazardous waste spill to reach a sensitive area or water body of the watershed. Laurent et al. (1998) used this approach to estimate travel time between any point of a watershed and a water resource (river or well). This information was further used to create a map of water resources vulnerability to dissolved pollution in an area in Massif Central, France. Subbasin area can be divided by flow length to estimate the overland flow width for input to a rainfall-runoff model such as EPA's Storm Water Management Model (SWMM).

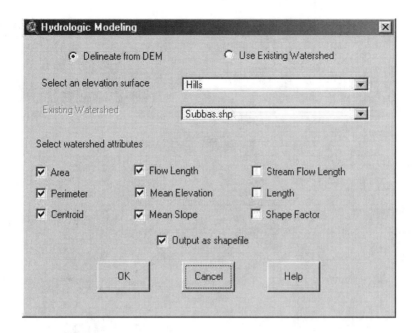

Figure 4.6 Hydro extension Hydrologic Modeling Dialogue.

ARC GRID Extension

ARC GRID is an optional extension for ESRI's ArcInfo® 7.x GIS software package. GRID adds raster geoprocessing and hydrologic modeling capability to the vector-based ArcInfo GIS. For hydrologic modeling, the extension offers a Hydrologic Tool System and several hydrologically relevant functions for watershed and stream network delineation.

The FLOWDIRECTION function creates a grid of flow directions from each cell to the steepest downslope neighbor. The results of FLOWDIRECTION are used in many subsequent functions such as stream delineation. The FLOWACCUMULATION function calculates upstream area or cell-weighted flow draining into each cell. The WATERSHED function delineates upstream tributary area at any user-specified point, channel junction, or basin outlet cell. This function requires step-by-step calculations. Arc Macro Language (AML) programs can be written to automate this function for delineating subbasins at all the stream nodes.

GRID can find upstream or downstream flow paths from any cell and determine their lengths. GRID can perform stream ordering and assign unique identifiers to the links of a stream network delineated by GRID. Spatial intersection between streams and subbasins can define the links between the subbasins and streams. This method relates areal attributes such as subbasin nutrient load to linear objects such as streams. The NETWORK function can then compute the upstream accumulated nutrient load for each stream reach (Payraudeau et al., 2000). This approach is also useful in DEM-based runoff quality modeling.

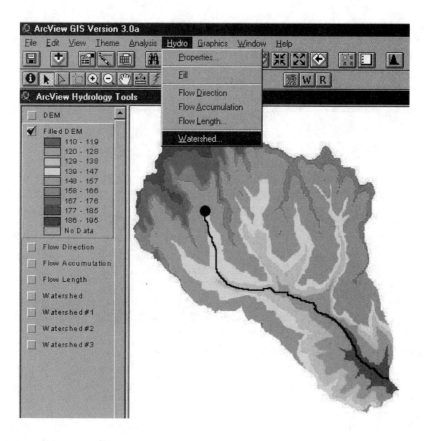

Figure 4.7   Hydro extension's pour point feature.

IDRISI

IDRISI is not an acronym; it is named after a cartographer born in 1099 A.D. in Morocco, North Africa. IDRISI was developed by the Graduate School of Geography at Clark University. IDRISI provides GIS and remote sensing software functions, from database query through spatial modeling to image enhancement and classification. Special facilities are included for environmental monitoring and natural resource management, including change and time-series analysis, multicriteria and multiobjective decision support, uncertainty analysis (including Bayesian and Fuzzy Set analysis), and simulation modeling (including force modeling and anisotropic friction analysis). TIN interpolation, Kriging, and conditional simulation are also offered.

IDRISI is basically a raster GIS. IDRISI includes tools for manipulating DEM data to extract streams and watershed boundaries. IDRISI GIS data has an open format and can be manipulated by external computer programs written by users. This capability makes IDRISI a suitable tool for developing hydrologic modeling applications. For example, Quimpo and Al-Medeij (1998) developed a FORTRAN program to model surface runoff using IDRISI. Their approach consisted of delineating watershed subbasins from DEM data and estimating subbasin runoff curve numbers from soils and land-use data.

Figure 4.8 shows IDRISI's DEM analysis capabilities. The upper-left window shows a TIN model created from digital contour data. The upper-right window shows a DEM created from the TIN with original contours overlayed. The lower-right window shows an illuminated DEM emphasizing relief. The lower-left window shows a false color composite image (Landsat TM bands 2, 3, and 4) draped over the DEM (IDRISI, 2000).

TOPAZ

TOPAZ is a software system for automated analysis of landscape topography from DEMs (Topaz, 2000). The primary objective of TOPAZ is the systematic identification and quantification of topographic features in support of investigations related to land-surface processes, H&H modeling, assessment of land resources, and management of watersheds and ecosystems. Typical examples of topographic features that are evaluated by TOPAZ include terrain slope and aspect, drainage patterns and divides, channel network, watershed segmentation, subcatchment identification, geometric and topologic properties of channel links, drainage distances, representative subcatchment properties, and channel network analysis (Garbrecht and Martz, 2000). The FILL Function of TOPAZ recognizes depressions created by embankments and provides outlets for these without filling, a better approach than the fill-only approach in other programs (e.g., IDRISI or Spatial Analyst).

CASE STUDIES AND EXAMPLES

Representative applications of using DEM data in GIS are described in this section.

Watershed Delineation

A concern with streams extracted from DEMs is the precise location of streams. Comparisons with actual maps or aerial photos often show discrepancies, especially in low-relief landscapes (ASCE, 1999). A drainage network obtained from a DEM must be comparable to the actual hydrologic network. Thus, it is worthwhile to check the accuracy of DEM-based delineations. This can be done by comparing the DEM delineations with manual delineations. Jenson (1991) found approximately 97% similarity between automatic and manual delineations from 1:50,000-scale topographic maps.

The objective of this case study was to test the efficacy of DEM-based automatic delineation of watershed subbasins and streams. It was assumed that manual delineations are more accurate than DEM delineations. Thus, a comparison of manual and DEM delineations was made to test the accuracy of DEM delineations.

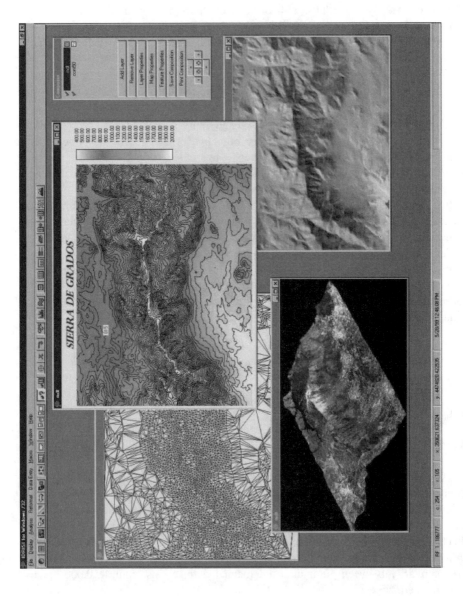

Figure 4.8 IDRISI's DEM analysis features.

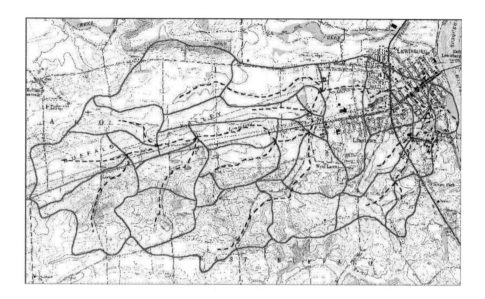

Figure 4.9 Bull Run Watershed showing manual subbasins and streams.

The case study watershed is the Bull Run Watershed located in Union County in north-central Pennsylvania (Shamsi, 1996). This watershed was selected because of its small size so that readable report-size GIS maps can be printed. The proposed technique has also been successfully applied to large watersheds with areas of several hundred square miles. Bull Run Watershed's 8.4 mi² (21.8 km²) drainage area is tributary to the West Branch Susquehanna River at the eastern boundary of Lewisburg Borough. The 7.5-min USGS topographic map of the watershed is shown in Figure 4.9. The predominant land use in the watershed is open space and agricultural. Only 20% of the watershed has residential, commercial, and industrial land uses.

Manual watershed subdivision was the first step of the case study. The 7.5-min USGS topographic map of the study area was used for manual subbasin delineation, which resulted in the 28 subbasins shown in Figure 4.9. This figure also shows the manually delineated streams (dashed lines).

Next, ArcView Spatial Analyst and Hydro extension were used to delineate subbasins and streams using the 7.5-min USGS DEM data. Many cell threshold values (50, 100, 150, …, 1000) were used repeatedly to determine which DEM delineations agreed with manual delineations. Figure 4.10, Figure 4.11, and Figure 4.12 show the DEM subbasins for cell thresholds of 100, 250, and 500. These figures also show the manual subbasins for comparison. It can be seen that the 100 threshold creates too many subbasins. The 500 threshold provides the best agreement between manual and DEM delineations.

Figure 4.13, Figure 4.14, and Figure 4.15 show the DEM streams for cell threshold values of 100, 250, and 500. These figures also show the manually delineated streams for comparison purposes. It can be seen that the 100 threshold creates too many streams (Figure 4.13); the 500 threshold looks best (Figure 4.15) and

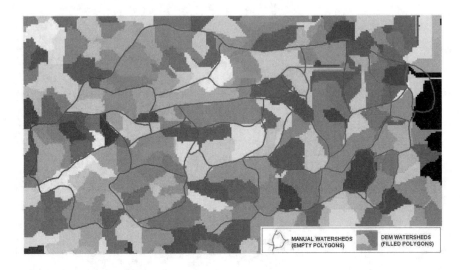

Figure 4.10 Manual vs. DEM subbasins for cell threshold of 100 (too many subbasins).

provides the best agreement between the manual and DEM streams. The upper-right boundary of the watershed in Figure 4.15 shows that one of the DEM streams crosses the watershed boundary. This problem is referred to as the boundary "cross-over" problem, which is not resolved by altering threshold values. It must be corrected by manual editing of DEM subbasins or using DEM preprocessing methods such as the stream burning method described earlier.

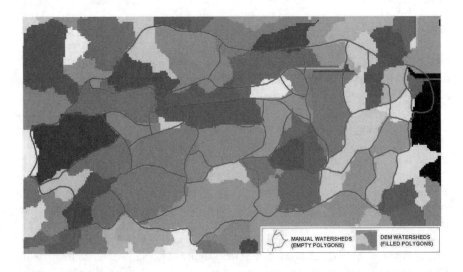

Figure 4.11 Manual vs. DEM subbasins for cell threshold of 250 (better).

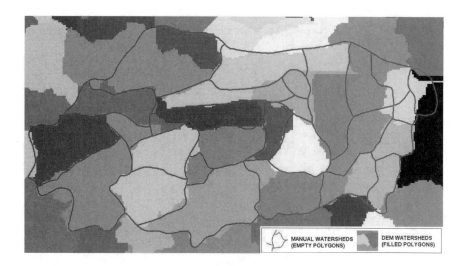

Figure 4.12   Manual vs. DEM subbasins for cell threshold of 500 (best).

Figure 4.16 shows DEM-derived subbasin and stream maps for a portion of the very large Monongahela River Basin located in south western Pennsylvania, using the 30-m USGS DEM data and a cell threshold value of 500 cells.

From the Bull Run watershed case study, it can be concluded that for rural and moderately hilly watersheds, 30-m resolution DEMs are appropriate for automatic delineation of watershed subbasins and streams. The 30-m DEMs work well for the mountainous watersheds like those located in Pennsylvania where subbasin boundaries

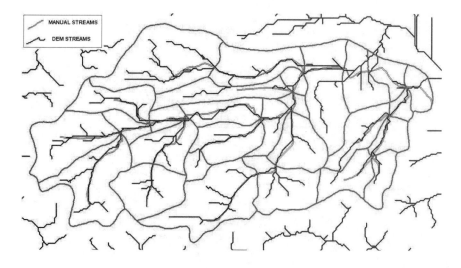

Figure 4.13   Manual vs. DEM streams for cell threshold of 100 (too many streams).

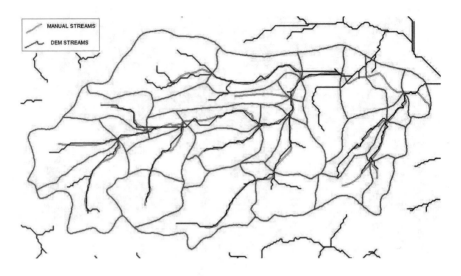

Figure 4.14   Manual vs. DEM streams for cell threshold of 250 (better).

are well defined and distinct due to highly variable topography. For 30-m USGS DEMs, a cell threshold of 500 is appropriate. Stream networks generated from the 30-m USGS DEMs at a cell threshold value of 500 were satisfactory, with a minor watershed boundary crossover problem. Correction of boundary crossover problems might require some manual intervention to ensure the efficacy of the DEM-based automatic

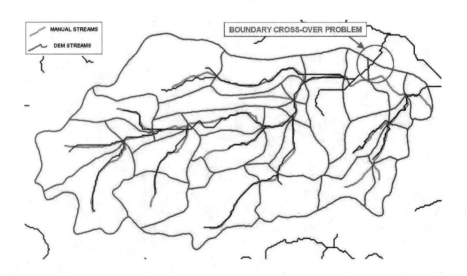

Figure 4.15   Manual vs. DEM streams for cell threshold of 500 (best).

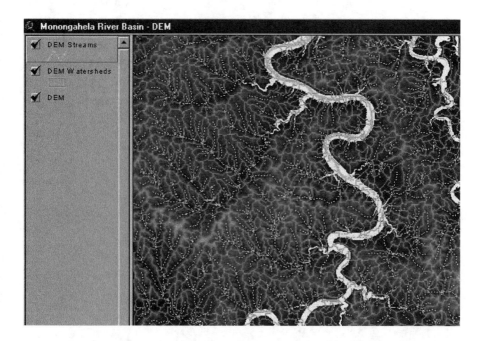

Figure 4.16 DEM-derived watersheds and streams for a large watershed.

watershed delineation approach. Compared with the manual watershed delineation approach, the DEM approach offers significant time savings, eliminates user subjectivity, and produces consistent results.

Sewershed Delineation

Watersheds define the drainage areas of natural streams and rivers. Sewersheds define the drainage areas of man-made sewers and storm drains. The conventional manual method of sewershed delineation is cumbersome, and an automated method is desirable. Can we use DEMs for automatic delineation of sewersheds? Unfortunately, the current research indicates that DEMs are generally applicable to natural landscapes and may not apply to urbanized settings (ASCE, 1999). This limitation is expected to be more pronounced in low-relief or flat areas. DEM delineations are based on the premise that water flows in the downslope direction, i.e., that all the sewers are gravity pipes. Therefore, DEM-based sewershed delineations will not be accurate for a pumped system with force mains and will require manual editing of the sewershed boundaries.

Figure 4.17 shows a comparison of DEM watersheds (filled polygons) and manually delineated sewersheds for the gravity sewers area of the Borough of Charleroi located near Pittsburgh, Pennsylvania. The watersheds were created from the 30-m resolution USGS DEM of the study area using a cell threshold of 100. The mean land slope of the study area is approximately 15%. The comparison shows that some sewersheds (for example, Sewershed Nos. 3, 4, 5, and 6) are reasonably

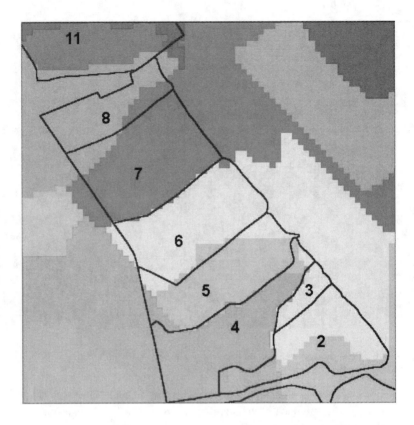

Figure 4.17 Comparison of DEM watersheds (filled polygons) and manually drawn sewer-sheds.

close to the watershed boundaries and some sewersheds (for example, Sewershed No. 2) are quite different from the watershed boundaries.

The Massachusetts Water Resources Authority (MWRA) recently determined that DEMs were not too helpful in delineating sewershed boundaries (Herrlin, 2000). MWRA, instead, used a "Euclidean Distance" approach consisting of the following steps:

1. All the sewers tributary to a pour point (point where a trunk sewer connects an interceptor) are identified using a custom network-tracing utility (upstream trace). The group of pipes identified like this defines a tributary area.
2. A new attribute with the name of the tributary area is manually added to all the sewers. This step is quite laborious but needs to be done only once.
3. Vector sewer layers are converted to a grid format. This is necessary because the Euclidean distance approach works only on raster layers.
4. The Euclidean distance allocation function is applied to all the sewer cells. The Euclidean distance is calculated in the same manner as finding the hypotenuse of a triangle. This generates a sewershed boundary around sewers.
5. The raster Euclidean distance boundaries are converted back to vector (polygon) format.

Some techniques that may be used for improving the accuracy of DEM-based sewershed delineations are listed below:

1. Use large-scale, high-resolution DEMs derived from digital orthophotos. These relatively dense DEMs have break lines at breaks in grades, which may provide valuable information for delineating sewersheds in flat urban areas. LIDAR technology can also provide highly accurate and cost-effective DEMs and 3D break lines that can be used for delineating sewersheds.
2. Burning in the DEMs with streams improves the accuracy of delineated watersheds in flat areas. Burning in the DEMs with sewers may improve the delineated sewersheds.
3. Adding manhole rim elevations to DEMs may be helpful in delineating sewersheds. Alternatively, DEMs may be burned in with manhole rim elevations.

Water Distribution System Modeling

The performance of a water distribution system depends on variations in ground surface elevation. For example, high-elevation areas may experience inadequate pressure if appropriate pumping is not provided. Thus, all distribution system models require input data for the elevation of each model node. DEMs can be used for 3D representation of water systems and estimating elevation at the nodes of a water distribution network model.

Shamsi (1991) showed applications of personal computer-based 3D graphics in network and reliability modeling of water distribution systems. Optional 3D extensions of GIS software, such as ESRI's 3D Analyst, can create 3D maps of terrain, water demand, and modeled pressures. Figure 4.18 shows three stacked surfaces: ground surface elevations, demands, and modeled average daily pressures for the Borough of White Haven, Pennsylvania. The land area of the borough is 773 acres, the population is 1,091 persons, and the average daily demand is 197,000 gal/d. The distribution system consists of water mains ranging in size from 2 to 8 in. It is interesting to note the high-pressure areas coinciding with the low-elevation areas, and vice versa. It also quickly becomes apparent that the high-demand areas predominate in the lower elevations. A display such as Figure 4.18 is an effective visual aid to displaying the hydraulic model results. Most importantly, such graphics can be easily understood by system operators without the extensive training that is usually required to understand the model output in tabular format.

Node elevation is a critical input parameter for all water distribution system models. Traditionally, the node elevations have been determined manually by overlaying the water distribution network over a topographic map of the study area. Unfortunately, due to its tedious nature, this approach is cumbersome. GIS is now being used to automatically extract the node elevations from a DEM of the study area. This approach requires a vector layer of the network nodes, a DEM grid, and a user script. The script appends a new "Elevation" field to the "Node" layer table, queries the DEM grid at node locations, and writes the elevation values in the Elevation field.

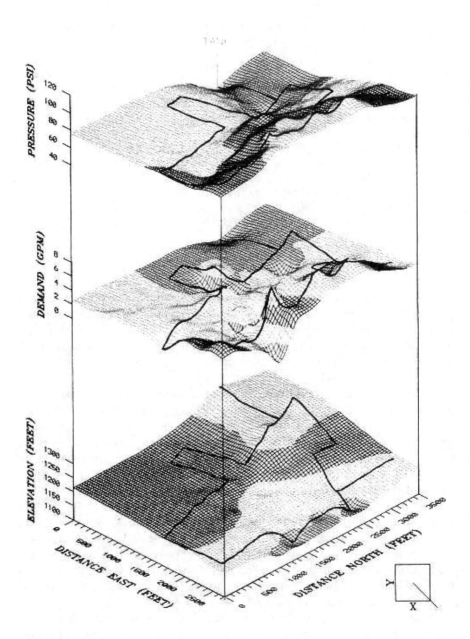

Figure 4.18 Hydraulic modeling results as 3D surfaces.

WaterCAD Example

If node locations are available in an ArcView Shapefile, ArcView 3D Analyst extension (Version 3.x) can be used to calculate node elevation from a DEM grid. This extension uses bilinear interpolation to calculate the node elevation from the grid cell elevation data. The elevation data are pulled from the DEM overlay, and the node file is converted to a 3D file. The steps for this procedure for Haestad Method's WaterCAD modeling software are listed below (Walski et al., 2001):

1. Start ArcView and load the 3D Analyst extension.
2. Add both the DEM grid and the node point themes to the same view.
3. Select "Theme\Convert to 3D Shapefile." Select "Surface" for the elevation file type and enter the name of the DEM grid (input). Also provide a name for the 3D Shapefile (output).
4. Use an Avenue script to extract elevation data from the 3D Shapefile and place them in the node theme table. A sample script called addxyz.ave (Hsu, 1999; Shamsi, 2002) can be downloaded from the ESRI ArcScripts Web site (www.esri.com/arcscripts/scripts.cfm).
5. Select "File\Export" from the ArcView pull-down menus and enter the export filename for a dBase IV (DBF) or delimited text (TXT) file.
6. Select "File\Synchronize\database Connections" from the WaterCAD's pull-down menus and set up a link between the export file (created in the previous step) and WaterCAD.
7. Select "Elevation" as the WaterCAD field into which the elevation data should be imported and the name of the corresponding field in the export file (e.g., "Z-coord").
8. Select the link in the Database Connection Manager and click the "Synchronize In" button.
9. Finally, validate the success of elevation data import process by checking the elevations at a number of nodes against known elevations.

USEFUL WEB SITES

Center for Research in Water Resources, University of Texas at Austin	www.crwr.utexas.edu/archive.shtml
DEM translators	software.geocomm.com/translators/sdts/
Pennsylvania Spatial Data Access system (PASDA)	www.pasda.psu.edu
ESRI DEM analysis software	www.esri.com
Leica geosystems software	gis.leica-geosystems.com
USGS DEM download	edc.usgs.gov/geodata/
USGS EartExplorer site for buying DEM data	earthexplorer.usgs.gov
USGS NED home page	edcnts12.cr.usgs.gov/ned/default.asp
USGS node of the National Geospatial Data Clearinghouse	nsdi.usgs.gov

CHAPTER SUMMARY

GIS provides an integrated platform for using DEMs. DEM data can be used for automatic watershed and stream delineation and computation of watershed hydrologic parameters. DEMs can be used as source data for digital orthophotos and as base maps in a GIS. DEM data can be created from topographic maps and remote sensing data. DEMs are not intended to replace elevation data obtained from surveys, high-accuracy GPS, or even well-calibrated altimeters. Rather, DEMs should be used as a labor-saving process to relieve the user of the tedium and human error involved in the conventional manual interpolation method from paper maps. Public-domain DEMs generally do not have adequate resolution and accuracy for engineering design. They are suitable for applications where a high degree of accuracy is not required, such as H&H modeling.

CHAPTER QUESTIONS

1. What are DEMs and how are they used in GIS?
2. List at least ten applications of DEMs?
3. What are the limitations of DEM data? How are they overcome?
4. How are DEMs used for delineating (1) streams, (2) watersheds, and (3) sewer-sheds?
5. How are DEMs created using the remote sensing technology?

CHAPTER 5

GPS Applications

GPS technology is no longer limited to determining coordinates. It has become an efficient and increasingly popular way for collecting GIS attribute data for water and wastewater infrastructure.

GIS data collection for sewer system manholes using GPS.

LEARNING OBJECTIVE

The learning objective of this chapter is to discover the applications of global positioning system (GPS) technology in water industry GIS projects.

MAJOR TOPICS

- GPS basics
- GPS applications in water industry
- GPS applications in GIS
- GPS survey steps
- GPS equipment
- GPS software

LIST OF CHAPTER ACRONYMS

DGPS Differential Global Positioning System
GPS Global Positioning System
GLONASS Global Navigation Satellite System (Russian System)
NAVSTAR Navigation System by Timing and Ranging
PDOP Positional Dilution of Precision
RTK Real Time Kinematic
SA Selective Availability
WGS84 World Geodetic System of 1984

STREAM MAPPING IN IOWA

Application	Stream traversing and mapping channel features using GPS
GPS software	Trimble's Pathfinder Office
GPS equipment	Trimble Pathfinder Pro XR
GPS accuracy	1 m for discrete data and 5 m for continuous data
GPS data	Streambank conditions, bottom sediment material and thickness, channel cross sections, debris dams, tile lines, tributary creeks, and cattle access points
Study area	Neal Smith National Wildlife Refuge, Jesper County, Iowa
Organization	Iowa Department of Natural Resources

In 2000, GIS and GPS were used together to map a 12-km portion of Walnut Creek located near Prairie City in Jesper County, Iowa (Schilling and Wolter, 2000). The objective of the mapping project was to locate channel features and identify spatial trends among alluvial system variables that could be used to identify and prioritize portions of the stream channel and watershed in need of further investigation or restoration. Using Trimble Pathfinder Pro XR GPS equipment, discrete locations (channel cross sections, debris dams, tile lines, tributary creeks, and cattle access points) were mapped to an accuracy of 1 m, whereas the continuous data (bank erosion rates, streambed materials, and thickness) were recorded to an accuracy of

5 m. To record continuous conditions, the GPS equipment was operated in continuous line mode with location recorded every 5 sec. GPS data were exported into a GIS format (ESRI Shapefile), using Pathfinder Office software. Field descriptions of the continuous line segments and discrete features were added to the GPS location information to create various GIS layers. Segment lengths varied from 10 to 50 m. Discrete channel features were located by pausing the continuous line mode of the GPS and taking points at feature locations. Stream survey data were used to model watershed conditions, identify water sampling points, and evaluate and select appropriate channel-rehabilitation measures.

This chapter is intended for professionals in the geographic "positioning" field. It presents applications of GIS and GPS for water industry infrastructure management. The use of GPS in collecting attributes data is discussed, and methods of data attribution are described. A review of GPS equipment and software is presented. GPS accuracy issues are also discussed.

GPS BASICS

GPS, also referred to as Navigation System by Timing and Ranging (NAVSTAR), is a satellite-based radio navigation system developed and operated by the U.S. Department of Defense. The Russian government operates a similar system called Global Navigation Satellite System (GLONASS). At the present time, GPS includes 29 active satellites located in 6 orbital planes.

GPS systems utilize a constellation of satellites orbiting the earth twice daily (i.e., passing over approximately the same world location every 12 hours) and transmitting precise time and position signals. GPS receivers read signals from orbiting satellites to calculate the exact spot of the receiver on Earth as geographic coordinates (latitude and longitude) referenced to the World Geodetic System of 1984 (WGS84) datum. The signals from at least four satellites should be available to determine the coordinates of a position. Physical obstructions such as mountains, trees, and buildings, and other factors such as satellite malfunction and rephasing operations can restrict GPS signals and degrade GPS accuracy. Ideal GPS operating conditions that provide the best accuracy are listed below (Lyman, 2001):

- Low positional dilution of precision (PDOP), a measure of best geometrical configuration of satellites
- Good signal strength
- Little or no multipath (reflection of GPS signals off distant reflective environment such as mountains and buildings)
- Little or no signal degradation because of geomagnetic storms and ionospheric or atmospheric effects

The accuracy of GPS coordinates can be increased by applying differential corrections. Differential corrections move user points closer to their "actual" location. This is done by comparing the user's new data on unknown locations with the data collected at the same time on a point with known coordinate values (Zimmer, 2001b). The GPS receivers that can receive and apply the corrections in real time are called real time kinematic (RTK) receivers. Non-RTK receivers require postprocessing of raw GPS data in the office. In the U.S., federal and state agencies are cooperating to make differential GPS readily available to all users.

GPS precision can vary from a few millimeters to hundreds of meters. The required precision depends on the project-specific requirements. The available precision varies with GPS mode (static, RTK, or kinematic), GPS equipment, time of occupation, and location (vegetation, reflection, and buildings). GPS survey cost increases with the accuracy requirements. The typical utility precision standard is 3 to 5 cm.

To prevent misuse by hostile forces, the U.S. Department of Defense had introduced an intentional error called the Selective Availability (SA) error in their GPS signals. The recent advances in GPS technology had reduced the effectiveness of the SA error. On midnight May 2, 2000, the SA error was removed 6 years ahead of schedule by a presidential order. This event marked an important day in the history of GPS because it increased the GPS accuracy up to ten times. The SA removal has improved the accuracy of inexpensive GPS receivers. SA removal has also increased the accuracy of GPS receivers operating in an autonomous (unassisted) mode without a base station. No significant impact has been noted in the performance of survey grade receivers (Murphy, 2001).

GPS technology is making major progress in improving the speed, reliability, and accuracy of the mathematical processes by which coordinates are calculated from the satellite beams (ASCE, 2001).

GPS APPLICATIONS IN THE WATER INDUSTRY

At the present time, the GPS revolution is well under way. For example, a Mercedes-Benz driver equipped with TeleAid system can press an SOS button to summon a tow truck, police, or ambulance. This button uses GPS technology to transmit the specific location, model, and color of the vehicle. The GPS applications for the water, wastewater, and stormwater systems, though not as dramatic as TeleAid, are revolutionizing the way these systems are designed, constructed, operated, and maintained. Representative GPS applications for the management of water, wastewater, and stormwater systems are:

1. GPS can be used to increase the accuracy of existing system maps by verifying and correcting locations of the system components. Frequent field changes often mean utility lines can be several feet off horizontally and/or vertically from where they appear on the plans. Thus, unless updated frequently, most utility plans, especially in growing cities, are outdated frequently. GPS data collection is no longer limited to collecting coordinates of point features. Now users can bike along a channel to map line features, or walk around a detention pond to map polygon features.
2. New water system or sewer system maps can be created if they do not exist.
3. Water system or sewer system attributes can be collected for populating the GIS database.

Surveying

Traditionally, people have used geodetic surveying to locate and map utility infrastructure components. Such transit survey required traversing between a known point to the point of interest, which often took several hours per point. GPS has been found to be up to 50% faster than the traditional methods (Anderson, 1998). GPS survey takes only a few minutes or seconds per point. For example, up to 300 points can be surveyed in 1 day, using a GPS survey. A two-man crew with bicycle-mounted equipment can survey up to 500 points per day.

The familiar total station is still the backbone of engineering survey work. Although GPS is not a replacement for optical surveying, interoperability between optical equipment and GPS is growing, and GPS is gaining ground slowly among engineers. GPS is an ideal addition to the surveying toolbox for a variety of applications, such as locating the starting point for a new stakeout. The conventional total station survey will require traversing a large distance for this work. GPS can do this work much faster by navigating a person right to the point where the stake should be placed (ASCE, 2001).

Fleet Management

An efficient fleet management system is essential to improving customer service. A wireless system that uses GIS and GPS technologies can substantially and economically improve the efficiency of fleet management. An integrated GIS/GPS procedure can be used to track multiple moving vehicles from a command center. It can show the location, speed, and movement of each vehicle on the tracking display. Off-the-shelf mobile devices such as Web-enabled cell phones and personal digital assistants (PDAs) can be used in conjunction with a GIS/GPS to provide the information needed for fleet management, such as dispatching and tracking the maintenance vehicles, generating driving directions, and trip routing. For example, Gearworks customized MapInfo's MapMarker J Server and Routing J Server to calculate driving directions and travel statistics. A client-side custom mapping application was created, which communicates with a MapXtreme server to create the data requested by the dispatcher and to deliver the information to the drivers mobile device. This application allows the dispatcher to view a map of the entire fleet to better assign work orders and deliveries, perform real time tracking, and deliver accurate status updates through the Web interface (GEOWorld, 2001).

GIS/GPS applications have consistently lowered the cost of fleet management by 10 to 15%.

GPS APPLICATIONS IN GIS

GPS technology represents a space-age revolution in GIS data collection. It is providing an efficient and increasingly popular way for collecting both the location (coordinates) and the attributes data in the field. The new line of GPS receivers brings technology to water- and wastewater system operators and managers, who can populate their existing maps with precise location of features such as manhole covers, catch basins, overflow points, hydrants, valves, pumps, flow meters, and rain gauges.

The most basic application of GPS is collection of (x,y) coordinates for the GIS features. Initially, these coordinates were manually entered into the GIS database. Most GPS receivers record their data in an American Standard Code for International Interchange (ASCII) format that can be imported into a GIS without having to type the coordinates' values. Recent GPS equipment can provide data in a GIS-compatible format, such as the ESRI Shapefile format.

Attributes are usually collected in two phases:

1. Initial attribution phase (Phase I) or first pass: Limited attributes can be collected by the GPS survey crew when a structure is visited or surveyed for the first time. These attributes — such as manhole cover type, catch basin condition (clean, debris, etc.), or outfall condition (submerged, flowing, dry, etc.) — should be visible from outside and should not require confined space entry or opening of structures. This phase may also include ID-marking of structures for the second phase or future visits. In this phase, typically 250 to 300 points can be surveyed per day.
2. Final attribution phase (Phase II) or second pass: Additional attributes can be collected when a structure is revisited by the attribute crew. These attributes may require field measurements, detailed inspections, confined space entry, or opening of structures. Examples include manhole depth, pipe size, and structural or hydraulic deficiencies. The final attributes depend on the mapping application. For example, if the maps will be used to develop an H&H model, weir height in an overflow structure might be measured. If the GIS will be used for NPDES permit reporting, the condition of the sanitary outlet (clean, clogged, surcharged, etc.) in the overflow structure might be a critical attribute. In this phase, final attributes for 50 to 75 structures can be surveyed per day. New real time differential GPS (DGPS) receivers can navigate the attribute crew back to the structures visited in Phase 1.

GPS SURVEY STEPS

The data conversion process for municipal utilities often begins with GPS data. Features such as manholes and hydrants are collected using a GPS unit. These point features are combined into a GIS layer and then attributed. At the conclusion of each day, GPS corrections are applied if necessary. The data set is massaged to ensure its stability including calculating the IDs and building the data set. The data set is then exported and saved as a backup file. After the process is completed for an entire area, check plots are created. These check plots show the lines and points along with their attributes. The flow direction of the lines is also displayed to ensure that proper connectivity has been established for the system. The plots are then delivered to the municipal engineer for review.

Generally, the following steps are required to collect GIS data, using GPS (Zimmer, 2001b):

1. Prepare a project plan to define the scope, resource inventory (paper maps, databases, etc.), and data dictionary. The data dictionary is a template for field data

collection. It specifies the features to be surveyed and their attributes. A simple data dictionary for Phase-1 GPS survey of a sewer system is given below:

- Sanitary features
  - San_mh
    - Lid_size
    - Lid_type
    - Condition
    - ID
- Storm features
  - Catchbasin
    - Size
    - Type
    - Condition
    - ID
- Endwall
  - Condition
    - ID

2. Conduct field work.
3. Download data from GPS receiver to computer.
4. Perform quality assurance/quality control (QA/QC) for completeness and attribute information.
5. Apply differential corrections if not using RTK equipment.
6. Calculate the spatial error.
7. Export GPS data and attributes to GIS.
8. Create a metadata file to document GPS survey information, such as survey date, equipment used, horizontal and vertical accuracy, etc.

GPS EQUIPMENT

GPS equipment selection depends on the purpose or application of GPS data. For example, for utility mapping projects, mapping-grade GPS equipment equipped with code-based receivers is suitable. For engineering design and projects that require highly accurate survey data, survey grade GPS equipment equipped with carrier-phase receivers is needed. Table 5.1 provides a summary of GPS receiver types. Accuracy specifications are based on the ideal operating conditions described earlier.

Table 5.1 GPS Receivers

Receiver Type	Horizontal Accuracy (m)	Price (US$)
Recreation	5–10	100–300
Basic	1–5	3,000–5,000
Advance	<1	8,000–10,000
Survey and geodetic	<0.2	25,000 and upward

Recreational GPS Equipment

These types of receivers are intended to support recreational activities such as hiking and camping. Commonly referred to as the "*$200 receivers*," they provide low accuracy and are, therefore, not suitable for most engineering, mapping, or surveying work. They can be used as basic navigation and reconnaissance tools.

Basic GPS Equipment

These lightweight and handheld systems are about the size of a calculator. They are less expensive ($4000) and less accurate (1 to 5 m; least accurate in woods). Differential corrections are applied in a postprocessing step after the collected field data have been downloaded to a computer running a GPS-processing software package. Because of their low accuracy, they are not suitable for utility surveys. They are more suitable for educational use. Trimble's Geoexplorer III handheld GIS data collection system shown in Figure 5.1 is an example of this category. Geoexplorer

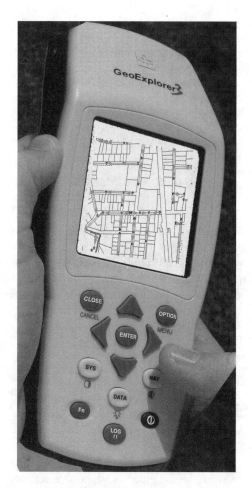

Figure 5.1   Basic GPS equipment example (Trimble Geoexplorer III).

provides support for popular GIS and CAD software packages such as ArcInfo, Intergraph, MGE, MapInfo, ERDAS, and AutoCAD. Geoexplorer provides for GIS export of GPS data, using Trimble's Pathfinder Office software. The steps for using Geoexplorer III GPS data in GIS are given below:

- Install Pathfinder Office software on your PC
- Create a data dictionary, i.e., define attributes for each feature type
- Transfer data dictionary from PC to receiver
- Enter attributes during GPS survey
- After the GPS survey, transfer the location and attribute data from receiver back to Pathfinder Office on your PC
- From Pathfinder Office, export data to your GIS in ArcView Shapefile or MapInfo format

Advanced GPS Equipment

Advanced-grade GPS receiver types include the so-called professional, mapping-grade, and resource-grade products. These systems are more expensive ($10,000) but provide the submeter horizontal accuracy generally required in utility mapping. Differential corrections, which are especially applicable for accurate navigation, can be applied in real time if the GPS receiver is linked to a radio receiver designed to receive broadcast corrections; Trimble's GPS Pathfinder Pro XR/XRS GIS data collection system is an example. In Pathfinder, attributes can be entered using Trimble's handheld Asset Surveyor, or a pen-based notebook or Tablet PC running Trimble's ASPEN software. As with Geoexplorer, GIS export is accomplished using Pathfinder Office software. The XRS models have an integrated differential correction receiver and can provide real time submeter accuracy. Reportedly, Pathfinder also features patented Everest multipath rejection technology to reduce or eliminate multipath signals in reflective environments. Figure 5.2 shows Trimble's Pathfinder Pro XR system, which is also referred to as the "Backpack" because it stores the system components in a convenient backpack. Figure 5.3 shows the components stored inside the Backpack.

Leica Geosystems (Heerbrugg, Switzerland) is a surveying and geomatics company that provides systems for accurate 3D data capturing, visualization, and modeling of space-related data to customers in surveying, engineering, construction, GIS, and mapping. In the GIS community, Leica Geosystems is known for its GPS and field data collection related equipment. ERDAS, Inc. (Atlanta, GA), is the world's leading provider of geographic-imaging solutions. The company's products and related services help organizations visualize, manipulate, analyze, measure, and integrate geographic imagery and geospatial information into 2D and 3D environments. In July 2001, ERDAS's geographic-imaging software company was acquired by Leica Geosystems to form a new GIS and Mapping Division for growth in the GIS and remote sensing markets. Congestion builds up as too many users crowd the same radio frequencies. Leica's new GPS equipment, Leica System 500 (GS50), provides uninterrupted RTK data link through cell phones. Also referred to as the *GIS Receiver*, GS50 is Shapefile-compatible and boasts a new "maximum tracking technology" that allows users to work under trees and in deep urban or natural canyons.

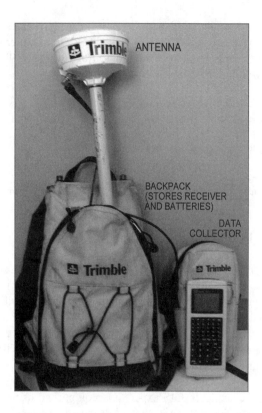

Figure 5.2   Advanced GPS equipment example (Trimble Pathfinder Pro XR Backpack).

Mapping-grade GPS equipment vertical accuracy of ± m is not adequate if accurate elevation data (e.g., manhole rim elevations) are required to support specific applications, such as H&H modeling and engineering design that require precise elevation data.

SURVEY GRADE GPS EQUIPMENT

Survey grade equipment features RTK technology, which can deliver ±1 cm horizontal and vertical accuracy in real time, using dual frequency receivers. Due to their high accuracy, they are expensive ($25,000). Trimble's GPS Total Station 4800 integrated real time kinematic survey system is an example of this category. Due to their more-than-needed accuracy and high cost, they are not essential for wastewater system mapping applications. However, survey grade GPS can save substantial amount of time compared with the conventional ground survey method when high accuracy is required. For example, the RTK survey for a sewer construction project in Fitchburg (Massachusetts) was completed in a quarter of the time that a ground survey would have taken (Mische, 2003). Survey grade GPS equipment's centimeter-level vertical accuracy is adequate for collecting accurate elevation data.

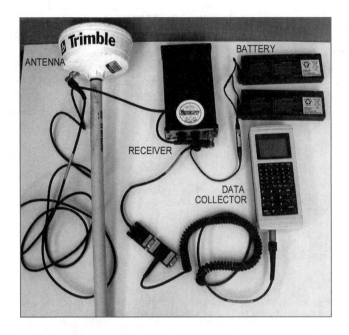

Figure 5.3   Components of Trimble Pathfinder Pro XR Backpack.

USEFUL WEB SITES

Professional Surveyor Magazine	www.profsurv.com
Trimble Navigation	www.trimble.com
Leica Geosystems Software	www.leica-geosystems.com

CHAPTER SUMMARY

Today's GPS equipment is like a GIS mapping device. GPS receivers are being referred to as the GIS receivers because they can collect both the geographic data (coordinates) and the attribute data for the water industry infrastructure. The water system and sewer system feature attributes can be captured from a GPS survey by using field computers, handheld computers, and field data collectors. The GPS equipment should be selected carefully to meet both the horizontal and vertical accuracy requirements of the utility mapping projects. The so-called $200 receivers designed for recreation activities that cannot provide accuracy of 5 m or better are not suitable for GPS mapping of water and sewer systems. A differential GPS that can provide real time submeter horizontal accuracy is suitable if accurate elevations are not necessary. Survey sgrade RTK GPS equipment that provides centimeter-level horizontal and vertical accuracy is suitable if accurate elevations are important.

CHAPTER QUESTIONS

1. What is GPS accuracy, and what factors can affect it?
2. List the typical applications of GPS technology in the water industry.
3. How are GIS and GPS related? How are they helping each other?
4. Describe the typical steps for conducting a GPS survey.
5. List the various types of GPS equipment and their suitability for different applications.

CHAPTER 6

Internet Applications

Internet-based mapping applications are changing the way people use and manage geographic information for their water, wastewater, and stormwater systems.

Johnson County (Kansas) Web-based wastewater mapping application.

LEARNING OBJECTIVE

The learning objective of this chapter is to familiarize ourselves with the applications of Internet technology in GIS for the water industry.

MAJOR TOPICS

- Internet GIS applications
- Internet GIS software
- Internet GIS case studies

LIST OF CHAPTER ACRONYMS

AM/FM/GIS Automated Mapping/Facilities Management/Geographic Information System
DXF Drawing Exchange Format
PDA Personal Digital Assistant
RDBMS Relational Database Management System
WWW World Wide Web

DUBLIN'S WEB MAP

Application	Web-based mapping
GIS software	AutoCAD, AutoCAD Map, and MapGuide
GIS data	1-m, natural color, orthorectified, precision pan-sharpened imagery of the entire city from Space Imaging (Thornton, Colorado) and vector layers for school districts, corporate limits, county boundaries, subdivisions, streets, parcels, and water (mains and hydrants), stormwater (sewers, structures, and streams), and sanitary sewer (sewers and manholes) systems
Study area	Dublin, Ohio

One of the fastest growing communities in the Midwest, Dublin's 22 mi^2 area is home to about 30,000 people and 2,500 businesses. The city received the year 2000 Best Practices Award from the Ohio Geographically Referenced Information Program, a state agency responsible for educating local government about GIS technology. The city is constantly enhancing its GIS capabilities and serves as a model medium-sized municipal GIS for the state. This application was developed after City personnel started to realize that retrieving data from hard-copy records was inefficient because it reduced staff time to perform engineering review and analysis. The city needed an automated system that could recall, evaluate, and reproduce data more quickly. The City implemented a Web-based GIS data server using Autodesk's (San Rafael, California) Web-based MapGuide technology

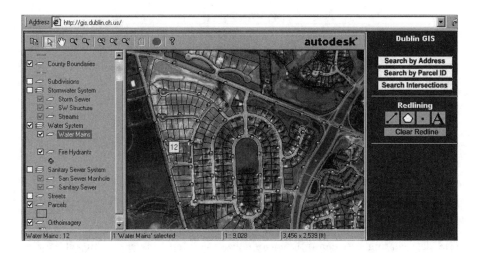

Figure 6.1   Screenshot of Dublin (Ohio) Internet GIS Web Site.

(Cerasini and Herskowitz, 2001). Figure 6.1 shows a screenshot of Dublin's Internet GIS Web site.

In addition to serving the development department, Dublin's GIS serves all other city departments and divisions, with data distributed to more than 200 users. Dublin's GIS Web site (gis.dublin.oh.us) provides quick access to utility data for maintenance, planning, and emergency response activities. The users range from planning and engineering professionals to grounds maintenance staff. A major challenge for any GIS is deciding who will maintain the data and how. Dublin's management structure encourages departments and divisions to exchange information freely, allowing departments to maintain their own data within the enterprise framework while feeding the GIS. Such an approach benefits everyone.

In 2001, the City started to implement GIS based asset management systems for the water, wastewater, and stormwater systems. Dublin also moved much of the GIS mapping and city assets to a mobile environment with wireless technology and GPS to allow data products to be distributed to a mobile workforce with personal digital assistants (PDAs) and wireless notebooks. The future is expected to bring even more integration between traditional municipal applications and GIS because the City also plans to integrate other applications, including work orders, asset management, crime analysis, public safety, and emergency dispatch.

Determining how much money Internet GIS technology is saving the City of Dublin is difficult to assess. However, the City feels that in most cases the time and effort required to respond to utility data requests have decreased from several hours to minutes, which translates to increased staff productivity and enhanced responsiveness to customers.

This chapter provides information on the GIS applications for serving water, wastewater, and stormwater system maps on the Internet.

INTERNET GIS

The Internet is the place where an increasing number of people are going for their mapping needs. In fact, the word *map* is one of the most commonly searched keywords on the Web (ESRI, 2002).

Imagine going to a GIS Web site, zooming to a piece of property, defining the building footprint, and displaying various cut and fill configurations. You could then simulate a 10-year, 24-hour design storm to model where the runoff will go and how you would control it to meet the local stormwater management ordinance.

The Internet is the fastest growing and most efficient technology for distributing GIS data to the public. The Internet and the World Wide Web (WWW or Web) are simplifying how the maps are created and maintained. The Internet is facilitating GIS data sharing between different organizations. Intranets are making it easy to share GIS data between coworkers of the same organization. Until the mid-1990s spatial information was generally distributed via the Internet as static map images. In the late 1990s dynamic GIS Web pages started allowing the public to interactively display and query GIS databases, using what is now referred to as Internet GIS technology (Wild and Holm, 1998).

Web browsers now provide both experts and novices with a common, powerful, inexpensive, and intuitive interface for accessing a GIS. Because casual public users can now access GISs through Web browsers, the water industry GIS staff is free to spend more time on improving the database for providing better customer service (Irrinki, 2000). Many large utilities are now using the Internet and GIS to improve the operation and management of their water and sewer systems. They are using Internet and GIS-based work order management and customer service software to accomplish this.

Figure 6.2 shows a screenshot of an Internet GIS Web site for the municipality of Penn Hills, Pennsylvania, created using ESRI's ArcIMS software. The tools shown at the top of the screen can be used to navigate (pan, zoom, etc.) and query the maps. Users can also apply the measure tool to determine distance and area from the map. The right side of the screen helps the users select the active layer for analysis (e.g., query) and turn it on (visible) or off (invisible). The bottom of the screen shows the query results.

Cincinnati Water Works reported a $2-million cost savings in the first year of using Internet GIS (Anderson, 2000).

GIS and enterprise database management systems can be combined to create geospatial data warehouses. For example, using the power of Oracle's spatial extensions, a data mining software (e.g., KnowledgeMiner from SpatialAge Solutions, Atlanta, Georgia, www.byers.com) can mine a network inventory model from an automated mapping/facilities management/geographic information system (AM/FM/GIS) and make it readily available to the enterprise using the Web. The object

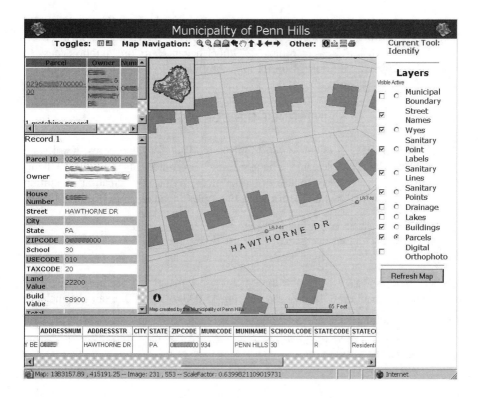

Figure 6.2   Screenshot of Penn Hills, Pennsylvania ArcIMS Web site.

relational database management system (RDBMS) spatial extensions, such as Oracle Spatial, facilitate linking data warehouses with GIS software packages. RDBMS is especially well suited for Internet and intranet data sharing. A data mining application can translate Internet requests into RDBMS queries and send RDBMS responses back to the Internet as Web pages. Utilities that rely on large and complex spatial data sets could greatly benefit from the recent fusion of object RDBMS and Internet technologies (Lowe, 2000a).

Internet Security

Security issues and concerns should be evaluated when posting GIS maps on the Internet. Precautions need to be taken to assure application security and data integrity (Miller et al., 2004). Appropriate security measures such as firewalls, data backups, user authentication, and password access can be used to reduce the chances of security breaches by hackers to harm the GIS data or corrupt the application software. Sensitive information that could help terrorists should not be posted or should be restricted to authorized users only. According to a Rand Corporation (www.rand.org) study released in April 2004, less than 1% of publicly available federal Web sites and databases in the U.S. contain geospatial information that could help terrorists and other hostile forces mount attacks in the U.S. (Geospatial Solutions, 2004). In the wake of the September 11, 2001, terrorist attacks in the U.S., many Internet GIS sites were removed. Others began restricting public access to some of their sensitive information through the Internet.

INTERNET GIS SOFTWARE

Internet GIS uses a client/server architecture. Therefore two types of software are required to Web-enable a GIS: client-side and server-side (Shamsi, 2003). Client-side software is very easy to use but does not provide data creation or editing capability. It is usually available for free. Examples include:

- ArcExplorer, ESRI
- GeoMedia Viewer, Intergraph
- MapGuide Viewer, Autodesk

Server-side software, accessible through browsers, makes GIS usable over the Web. Based on the capabilities, the cost ranges from $5,000 to $25,000. Examples include:

- ArcIMS, ESRI
- GeoMedia Web Map, Intergraph
- MapGuide, Autodesk
- MapXtreme, MapInfo

INTERNET GIS APPLICATIONS

Representative applications of Internet GIS technology are described in the following subsections.

Data Integration

Imagine an organization with watershed data in ESRI Shapefiles, parcel data in an Oracle Universal Data Server with Spatial Cartridge, and water main data in an AutoCAD file. Imagine further that different data are stored on different network drives in different projections. Software is now available to integrate different data formats and convert them into a common projection system on the fly. Intergraph's GeoMedia Web Map is an example of such a software (Lowe, 2000).

Project Management

A project Web site promotes collaboration and efficient completion of utility construction or rehabilitation projects by providing all involved parties instant access to current information and allowing them to disseminate information automatically and immediately. A Web site also eliminates software and hardware compatibility problems among project team organizations. Redline (edit) capability can be added to a project Web site to expedite the design review process (Irrinki, 2000).

These possibilities are the result of a new generation of software from the leading GIS and CAD software companies. For example, in addition to the traditional CAD work, Autodesk's AutoCAD 2000i enables users to collaborate with design professionals anywhere in the world and move the designs and processes to their companies' intranets and the Web. Similarly, Intergraph's GeoMedia, in conjunction with Oracle Spatial, enables all the users to see the latest engineering project information against the base-map images. The various projects are represented as polygons overlayed on the base-map image. A user can simply click on one of the project polygons to obtain the most recent information about that project. After a project has been completed, the engineering information (including redlines) is used to update the spatial data stored in Oracle Spatial. This approach was successfully implemented by the Public Works Department of the City of Minneapolis. The City used GeoMedia and an Oracle Spatial data warehouse to provide future users the most current information about the city's infrastructure (Murphy, 2000). Figure 6.3 shows a project portal designed using this approach for a wastewater improvement project for the Borough of California (Pennsylvania), created by Chester Engineers, Pittsburgh, Pennsylvania.

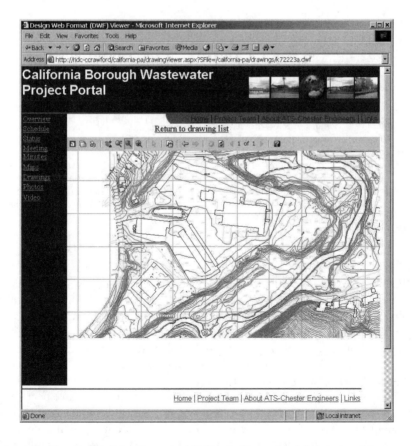

Figure 6.3   Screenshot of the Borough of California (Pennsylvania) project portal.

3D Visualization Applications

Internet GIS applications can also provide 3D visualization capability that converts 2D aerial photographs and satellite images into 3D terrain. The 3D terrain can be used as a base map for draping other information such as realistically textured buildings. For example, Skyline Software Systems' (www.skylinesoft.com) TerraExplorer provides realistic, interactive, photo-based, 3D maps of many locations and cities of the world on the Internet. Skyline's Internet technology enables creation, delivery, and interaction with content-rich, 3D geographic environments. Skyline enhances the Web experience for a variety of location-specific applications such as travel and tourism, real estate, entertainment, mapping, tracking, planning, and engineering. Standard CAD or GIS data layers such as DXF or Shapefiles can be easily incorporated to create dramatic 3D movies. Such Internet-based mapping applications are designed for a layperson and do not require GIS expertise.

CASE STUDIES

Two case studies illustrating the application of Internet GIS technology are described in the following subsections.

Tacoma's Intranet and Mobile GIS

Application	Intranet and Wireless GIS
GIS software	AutoCAD, AutoCAD Map, MapGuide, and GE Smallworld
Other software	Microsoft Access and SQL Server databases
Hardware	Laptop computers from Dell Computer Corp. and Gateway, Inc., 14.4K CDPD wireless PC card modems from Sierra Wireless
GIS data	Black and white orthophotos and planimetric base map layers from Bentley ADR, 3-ft and 6-ft color digital orthophotos from Triathlon, Inc.
Study area	Tacoma, Washington
Organization	Tacoma Public Utilities and Tacoma Water

Tacoma Water, a division of Tacoma Public Utilities (TPU) located in the state of Washington, serves 146,000 customers in the city of Tacoma's 52 mi^2 area and surrounding parts of Pierce and King counties. Tired of flipping through outdated map books and clumsy blueprints to locate water and sewer lines, Tacoma Water implemented intranet and wireless GIS applications.

Because of their familiarity with CAD, which was used in their engineering projects, Tacoma Water began a $750,000 multiyear project in 1990 to create CAD-based digital base maps and facility layers. The base map consisted of digital orthophotos and planimetric layers, such as building footprints, pavement edges, rights-of-way, lots and blocks, and easements, in a CAD format. Data conversion was completed in 1997 producing 6.5 GB of data with a horizontal accuracy of 6 in. Initially these data were distributed on CD-ROMs. In 1998, Tacoma Water implemented an intranet GIS using Autodesk's intranet and Internet software suite consisting of AutoCAD, AutoCAD Map, and MapGuide for an initial investment of $12,000. Using these packages, they integrated their CAD data in a GIS and hosted it on a central server from where it was accessible to any computer using a standard Internet browser. The intranet GIS enabled the users to query the map features by simply placing the mouse cursor over that feature. By putting the cursor over a valve, for instance, the user will see a pop-up window showing the size, type, and number of turns required to open or close the valve.

According to Tacoma Public Utilities, "Having multiple layers of information available on mobile computing devices is like a holy grail to the water field crews."

In order to provide the available intranet GIS data to crews in the field, Tacoma Water implemented a wireless GIS in 1999. Consisting of off-the-shelf laptop computers and wireless modems, the truck-mounted wireless GIS was initially used by 18 field crews. The cost of unlimited wireless airtime was $50 per month. The mobility allowed by such a system puts field crews light years ahead of the time when finding a water main meant rummaging through a map book or reading microfiche (Teo, 2001).

Montana's Watershed Data Information Management System

Application	GIS-based watershed data management system
GIS software	ArcView
Organization	Montana Department of Environmental Quality

Montana Department of Environmental Quality (DEQ) is responsible for developing integrated water, air, waste management, and energy plans to protect Montana's environmental resources. To comply with the Montana Water Quality Act and the Federal Clean Water Act, an ArcView-based watershed data management system called MontanaView was developed in 2000. The system was designed to meet the specific information management needs of the state of Montana, focusing on water resource management issues. The project goals were: (1) integrating extensive and diverse data into one system, covering demographic, land-use and land-cover, infrastructure environmental, and natural resource information and (2) orienting programs and organizations around watersheds as the primary unit for assessing and managing resources.

These goals were accomplished by developing MontanaView, a PC-based application that runs under Windows operating systems, and linked to distributed databases over the Internet. The system provided online access to U.S. EPA mainframe databases and United States Geological Survey (USGS) real time and historical stream-flow data. For example, it provided access to EPA's national water quality database, STORET, to identify water quality monitoring stations that met user-defined requirements for selected parameters. It could then go to the Montana USGS Internet site (montana.usgs.gov) and import stream-flow data for the user-specified station. The maps generated by MontanaView proved to be excellent visual aids during public meetings (Samuels et al., 2000).

USEFUL WEB SITES

Dublin, Ohio, Internet GIS Web site	gis.dublin.oh.us
Johnson County, Kansas, Internet GIS Web site	aims.jocogov.org
Town of Blacksburg, Virginia, Internet GIS Web site	arcims2.webgis.net/blacksburg/default.asp

CHAPTER SUMMARY

This chapter provided information on the GIS applications for serving water, wastewater, and stormwater system maps on the Internet. The Internet is the fastest growing and most efficient technology for distributing GIS data. Internet-based mapping applications are changing the way people use and manage geographic information in the water industry. The Internet can be used to both obtain and distribute GIS data. Two types of software are required for creating Internet GIS Web sites: client-side software and server-side software. Many cities and counties in the U.S. have developed Internet GIS sites for sharing their GIS data and maps.

CHAPTER QUESTIONS

1. What is Internet GIS?
2. What are the applications of Internet GIS technology?
3. What kind of software is required for developing Internet GIS applications?
4. Search the Internet to describe three Internet GIS Web sites that provide water, wastewater, and stormwater system maps on the Internet.

CHAPTER 7

Mobile GIS

Mobile GIS is helping the field workers by providing them with GIS maps on their hand-held computing devices. Field workers equipped with mobile GIS can update their enterprise GIS database from the field.

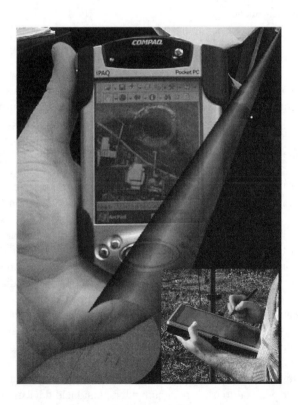

Mobile GIS brings GIS maps to field workers using mobile computers and PDAs.

LEARNING OBJECTIVE

The learning objective of this chapter is to determine the applications of mobile GIS technology in water industry.

MAJOR TOPICS

- Mobile GIS basics
- Mobile GIS applications
- Wireless Internet technology
- GPS integration

LIST OF CHAPTER ACRONYMS

LBS Location Based Services
PDA Personal Digital Assistant
RTK Real Time Kinematic

MOBILE GIS BASICS

Integration of GIS with consumer electronics such as cell phones and automobile navigation systems has spawned a whole new industry called location-based services (LBS). LBS services are used for managing dispatch and routing, fleets, work orders, and field crews — all of which improve customer service. Telegeoinformatics is also a new discipline resulting from the integration of mobile computing with wired and wireless communications, GIS, GPS, and remote sensing technologies (Karimi and Hammad, 2004). GIS lies at the heart of most mobile information resource systems. For example, it is now possible to wirelessly transmit a GPS location from the field, which can then be viewed on a GIS map via the Internet.

Using GIS directly in the field for data collection into a GIS database is referred to as mobile or field GIS. According to a recent study, about 40% of all mobile data applications in utilities are using a GIS (Wilson, 2001). With mobile GIS, feature location and attribute data can be verified, augmented, and immediately updated. New features can be added, existing features moved to their correct location, and nonexisting features deleted in the field. The ultimate goal of a mobile GIS is to link mobile workers with data to make their jobs easier and more efficient.

Mobile computers are the field devices used by mobile workers and field personnel to run GIS software (or its field version) and to collect location and attribute data in the field. Examples include notebooks, laptops, and pocket computers; tablet PCs and slate tablets; pen and touch-screen computers; handheld devices and PDAs; and smart phones (a combination of a PDA and a cell phone). Due to their small screen; limited memory, storage, processing speed, and battery life; and lack of a keyboard for field data entry, PDAs may not be suitable for large inspection projects. Currently, in the U.S., most PDAs use three operating systems: Linux, Palm OS, and Windows

Figure 7.1   Tablet PC being used for GPS and field data collection.

CE (Lowe, 2002). Basically a hybrid between a laptop and a PDA, the tablet PC combines the mobility and handwriting recognition of PDAs with the larger screen size and full-featured operating systems provided by laptops. Tablet PC screens can rotate 180° to create digital writing pads that lie flat against the keyboard. This feature allows field workers to sketch notes and diagrams or redline existing maps using a digital pen. Figure 7.1 shows a tablet PC connected to a mapping-grade GPS being used for field inspections. Figure 7.2 is a tablet PC screenshot showing notes and sketches taken in the field.

This field editing feature requires taking appropriate GIS layers and base maps in the field on a mobile computer. Custom forms can be created to mimic the paper forms that field crews are accustomed to using. The main advantage of a mobile GIS is that it eliminates postprocessing of paper data in the office.

Mobile GIS is the result of advances in four key areas: (1) field GIS software (e.g., ESRI's ArcPad); (2) lightweight, reliable, and affordable field computing devices; (3) more accurate and affordable GPS receivers; and (4) wireless communications (ESRI, 2004).

MOBILE GIS APPLICATIONS

Mobile GIS is helping cities in their field data collection and asset inventory activities. For example, many local governments are using ESRI's ArcPad field GIS software to improve speed and accuracy and lower the costs associated with the field inventory of their infrastructure assets (ESRI, 2004a). ArcPad, now a mobile client

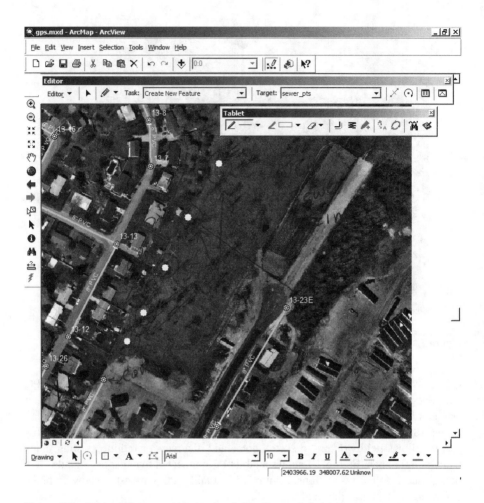

Figure 7.2   Tablet PC screenshot showing field notes and sketches.

for ArcGIS, allows data collectors to collect asset locations and attributes in the field as ESRI-compatible GIS files. ArcPad Application Builder allows creation of custom field inspection forms using Visual Basic Scripting (VBScript). OnSite from Autodesk, InService from Intergraph, and MapXtreme from MapInfo are other examples of mobile GIS software.

Mobile GIS applications can benefit from wireless and Internet technologies for live import or export to office or enterprise data. Wireless connections allow real time access to office GIS data in the field and real time updates to office data. When a persistent network connection is not feasible, mobile GIS can transport the field data back to the parent software in an office for automatic validation and delivery to the target system at the end of each inspection day. Intergraph's IntelliWhere OnDemand vector-based mapping product for PDAs also runs on some smart phones. Among other features, this product allows workers to red-line maps on their mobile devices and then wirelessly send the edits back to GIS servers in the office for immediate inclusion into the system. It also has an "occasionally connected" mode that allows field workers to extract data from the office GIS, view it, and disconnect until it is time to synchronize updates back to the office GIS (Lowe, 2002a).

WIRELESS INTERNET TECHNOLOGY

Wireless Internet solutions consist of four key technology components: (1) applications software running on the Internet and mobile devices; (2) mobile middleware software that resides on a server and a mobile device; (3) wireless data network to deliver data to and from mobile devices; and (4) mobile devices. In 2000, one of Southern California's largest wholesale water providers, the Calleguas Municipal Water District began a program to improve its infrastructure and operations management. It used Fieldport™ mobile operations management solution from Spacient Technologies, Inc. (Los Angeles, California). The system provides both Web-based and wireless asset management as either a traditional on-site software installation or as an Internet-hosted service. Fieldport allows GIS integration using its GIS/Mapping Module with an industry-standard database design powered by ESRI's ArcIMS software (Stern and Kendall, 2001).

GPS INTEGRATION

Mobile GIS applications typically use a GPS receiver; however, using a GPS is optional. Tiny GPS receivers are now available from various vendors that clip into most mobile devices and enable real time positioning on the map display. Without GPS integration, mobile GIS collects only attribute data; with a GPS receiver, data such as coordinates, elevation, and attributes can be collected. Inexpensive recreation-grade GPS receivers that provide typical 10-m horizontal accuracies can be used for navigation purpose. This arrangement would eliminate the need to pan and zoom around the map file to find the inspection location; the GPS connection can automatically center the map on the inspection location. Note that recreation-grade GPS receivers cannot and should not be used for collecting accurate horizontal and vertical positions. Figure 7.3 shows a sample GPS integration with a mobile device. The top part of the figure shows Hewlett Packard's iPAQ Pocket PC and Navman's recreation-grade GPS receiver for iPAQ. The GPS unit comes in the form of a sleeve that the iPAQ can slide into, as shown in the bottom part of the figure. According to the manufacturer, this GPS receiver typically provides better than 5 m (16 ft) horizontal accuracy for 95% of the time under optimal conditions.

Another technology that is helping mobile GIS is the Bluetooth wireless technology. Bluetooth is an international specification for a compact, low-cost radio solution that links mobile devices without using cables. Bluetooth devices can find each other when they are within their working range, which is typically 10 m (33 ft). For example, an RTK GPS rover on a pole can be wirelessly connected to the GPS receiver mounted inside a vehicle, or RTK GPS corrections can be received via a Bluetooth-enabled cell phone.

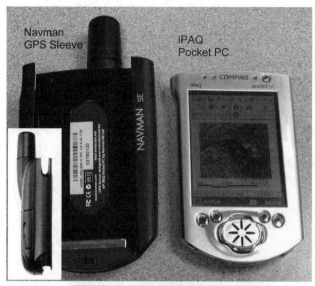

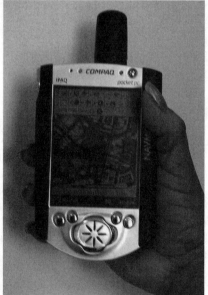

Figure 7.3   Sample mobile GIS device. Above: Navman GPS sleeve and iPAQ Pocket PC shown separately; below: iPAQ inserted into the sleeve.

USEFUL WEB SITES

Hewlett Packard's iPAQ Pocket PC   www.hp.com
Navman's recreation grade GPS for iPAQ   www.navman.com

CHAPTER SUMMARY

Mobile GIS plays an important role in simplifying the jobs of field workers and reducing the cost of field inspections and maintenance. For example, GIS allows users to identify valves to be closed for repairing or replacing broken water mains by simply clicking on a mobile field device such as a tablet PC.

CHAPTER QUESTIONS

1. What is mobile GIS?
2. What is the main application of mobile GIS in the water industry?
3. How is mobile GIS related to Internet, GPS, and wireless technologies?

CHAPTER 8

Mapping

GIS provides powerful and cost-effective tools for creating intelligent maps for water, wastewater, and stormwater systems.

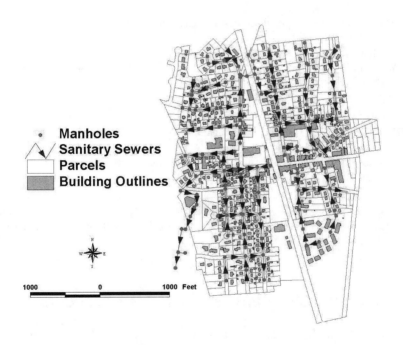

A sewer system map created by GIS (Borough of Ramsey, New Jersey).

LEARNING OBJECTIVE

The learning objective of this chapter is to understand how to create GIS maps for water, wastewater, and stormwater systems.

MAJOR TOPICS

- Mapping basics
- Map types
- Advantages of GIS maps
- GIS mapping steps
- Mapping case studies

LIST OF CHAPTER ACRONYMS

AM/FM Automated Mapping/Facilities Management

AML Arc Macro Language

DRG Digital Raster Graphics (USGS topographic maps)

NAD-27 North American Datum of 1927

NAD-83 North American Datum of 1983

QA/QC Quality Assurance/Quality Control

SPC State Plane Coordinate (Map Projection System)

TIGER Topologically Integrated Geographic Encoding and Referencing System (U.S. Census Bureau Mapping System)

UTM Universal Transverse Mercator (Map Projection System)

VBA Visual Basic for Applications

This book focuses on the four main applications of GIS, which are mapping, monitoring, maintenance, and modeling and are referred to as the "4M applications." In this chapter we will learn how to implement the first *m* (mapping).

LOS ANGELES COUNTY'S SEWER MAPPING PROGRAM

In the 1980s, the Sanitation Districts of Los Angeles County, California, envisioned a computerized maintenance management system that would provide decision makers with essential information about the condition of the collection system. A sewer and manhole database was subsequently developed, but investments in GIS technology were deferred until the early 1990s when desktop PCs became powerful enough to run sophisticated GIS applications. In 1993, a GIS needs analysis study was performed, which recommended implementation of a large-scale enterprise-wide GIS. An in-house effort was started to implement the recommendations of the study. Several sections in the Districts formed a project committee to pilot test GIS technology that could be duplicated in all 25 districts. The pilot project developed a mapping application for a small district that acted as a front-end to the large database of sewerage facilities developed in the 1980s and 1990s. At this point, much of the information was nonspatial, including multiple databases in a variety of formats and paper maps. Converting this information into GIS proved to be the most time consuming and costly operation. Creation of the layers for sewers and manholes was the most laborious of all the layers that had to be created. Manhole data were represented by approximately 24,000 points or nodes digitized from the paper maps, using a base map. The GIS layers were created in CAD software and linked to the legacy databases. Once linked, detailed data such as sewer pipe and manhole construction material, size, condition, flow, capacity, and inspection data were available for query and analysis through an intuitive map-driven interface. By 2003, the pilot project had grown to become the first enterprise-wide solution deployed by the Districts. Called "Sewerage Facilities GIS," the system allowed users to access and view data from the legacy databases by selecting sewers and manholes on a map or using standard queries. The mapping application also provided sewer tracing functionality that proved helpful in delineating study area boundaries for design projects, annotating sewers with flow direction, and tracking potential discharge violations (Christian and Yoshida, 2003).

MAPPING BASICS

The basic concepts essential for understanding GIS mapping are summarized in the following subsections.

Map Types

There are two major types of GIS maps: vector and raster. In *vector* format, objects are represented as points, lines, and polygons. Examples of the vector format are maps of water mains, hydrants, and valves. Scanned maps, images, or aerial photographs are examples of *raster* format. Raster data are also referred to as grid, cell, or grid–cell data. In raster format, objects are represented as an image consisting of a regular grid of uniform size cells called *pixels*, each with an associated data value. Many complex spatial analyses, such as automatic land-use change detection, require raster maps. Raster maps are also commonly used as base maps (described later in this chapter). Existing paper maps that are used to create GIS maps are called *source maps*.

Topology

Topology is defined as a mathematical procedure for explicitly defining spatial relationships between features. Spatial relationships between connecting or adjacent features, such as a sewer tributary to an outfall or the pipes connected to a valve, which are so obvious to the human eye, must be explicitly defined to make the maps "intelligent." A topological GIS can determine conditions of adjacency (what is next to what), containment (what is enclosed by what), and proximity (how near something is to something else). Topological relationships allow spatial analysis functions, such as network tracing, that can be used to facilitate development of hydraulic models for water and sewer systems.

Map Projections and Coordinate Systems

Because the Earth is round and maps are flat, transferring locations from a curved surface to a flat surface requires some coordinate conversion. A *map projection* is a mathematical model that transforms (or projects) locations from the curved surface of the Earth onto a flat sheet or 2D surface in accordance with certain rules. Mercator, Robinson, and Azimuthal are some commonly used projection systems. Small-scale (1:24,000 to 1:250,000) GIS data intended for use at the state or national level are projected using a projection system appropriate for large areas, such as the Universal Transverse Mercator (UTM) projection. The UTM system divides the globe into 60 zones, each spanning 6° of longitude. The origin of each zone is the equator and its central meridian. X and Y coordinates are stored in meters. Large-scale local GIS data are usually projected using a State Plane Coordinate (SPC) projection in the United States.

A *datum* is a set of parameters defining a coordinate system and a set of control points with geometric properties known either through measurement or calculation. Every datum is based on a spheroid that approximates the shape of Earth. The North American Datum of 1927 (NAD27) uses the Clarke spheroid of 1866 to represent the shape of the Earth. Many technological advances, such as the global positioning system (GPS), revealed problems in NAD27, and the North American Datum of 1983 (NAD83) was created to correct those deficiencies. NAD83 is based on the GRS80 spheroid, whose origin is located at the Earth's center of mass. The NAD27 and NAD83 datum control points can be up to 500 ft apart.

Coordinates are used to represent locations on the Earth's surface relative to other locations. A *coordinate system* is a reference system used to measure horizontal and vertical distances on a map. A coordinate system is usually defined by a map projection. The GIS and mapping industries use either latitude/longitude- or geodetic-based coordinate grid projections. Because much of the information in a GIS comes from existing maps, a GIS must transform the information gathered from sources with different projections to produce a common projection.

Map Scale

Map design addresses two fundamental map characteristics: accuracy and depicted feature types. Both characteristics vary with map scale. Generally, larger scale maps are more accurate and depict more detailed feature types. Smaller scale maps, such as U.S. Geographical Survey (USGS) quadrangle maps, generally show only selected or generalized features. Table 8.1 summarizes the relationships among map scale, accuracy, and feature detail.

Data Quality

The famous computer industry proverb "garbage in, garbage out" conveys very well the importance of GIS data quality. A GIS map is only as good as the data used to create it. Data quality roughly means how good the data are for a given application. Data quality is important because it determines the maximum potential reliability of the GIS application results. Use of inappropriate data in a GIS map may lead to

Table 8.1 Relationships among Map Scale, Accuracy, and Feature Detail

Map Scale	Minimum Horizontal Accuracy, per National Map Accuracy Standards	Examples of Smallest Features Depicted
1 in. = 50 ft	±1.25 ft	Manholes, catch basins
1 in. = 100 ft	±2.50 ft	Utility poles, fence lines
1 in. = 200 ft	±5.00 ft	Buildings, edge of pavement
1 in. = 2000 ft	±40.00 ft	Transportation, developed areas, watersheds

misleading results and erroneous decisions, which may erode public confidence or create liability.

Data Errors

There are two types of data errors: inherent errors embedded in the source of data and operational errors introduced by users during data input, storage, analysis, and output. Inherent errors can be avoided by using the right kind of data. Operational errors can be prevented by quality control and training.

A data conversion team should be aware of sources and magnitudes of data error. For example, spatial information in USGS 1:24,000-scale (7.5-min) topographic maps is certified to have 90% of its features within 50 ft (15 m) of their correct location. 50 ft is large enough to underestimate the runoff from a new development and undersize a detention pond for adequate stormwater management.

Map Accuracy

A primary factor in the cost of data conversion is the level of positional accuracy. Required map accuracy and resolution depend on the application in which the maps will be used. A 2000 survey conducted by the Geospatial Information and Technology Association (GITA) indicated that the water utilities were seeking more landbase accuracy of 5-ft compared with other utilities, such as the 50-ft accuracy sought by the gas companies (Engelhardt, 2001; GITA, 2001). The same survey for 2002 indicated that the water industry required the highest accuracy in their GIS projects. Among the water organizations, 27% were using a 6-in. landbase accuracy compared to electric (12%), gas (17%), pipeline (17%), and telecom (0%) organizations (GITA, 2003). These data reveal that a trend toward increasing accuracy may be emerging in the water industry.

Engineering applications usually require ±1 to 2 ft accuracy. For planning and regional analysis applications, ±5 to 10 ft accuracy is generally appropriate (Cannistra, 1999). Sometimes relative accuracy (e.g., ±1 ft from the right-of-way line) is more important than an absolute level of accuracy (e.g., ±1 ft from the correct location). For the applications where positional accuracy is less important, supposedly low-resolution data, such as USGS digital orthophoto quadrangles (DOQs), may be acceptable. In other applications where features must be positioned within a foot of their actual position, even the presumably high-resolution data, such as 1-m IKONOS imagery, may not be accurate enough. As a rule of thumb, a database built from a map will have positional inaccuracies of about 0.5 mm at the scale of the map because this is the typical line width of the drawing instrument. This can cause inaccuracies of up to 12 m in a database built from 1:24,000 mapping, such as USGS DRGs (Goodchild, 1998).

Precision and accuracy are two entirely different measures of data quality and should not be confused. A GIS can determine the location of a point feature precisely as coordinates with several significant decimal places. However, many decimal places in coordinates do not necessarily mean that the feature location is accurate to a 100th or 1000th of a distance unit. Once map data are converted into a GIS environment, the data are no longer scaled, as the data can be scaled as desired to create any output map scale. However, the spatial data can never be any more accurate than the original source from which the data were acquired. GIS data are typically less accurate than the source, depending on the method of data conversion. Therefore, if data were captured from a source map scale of 1 in. = 2000 ft, and a map was created at 1 in. = 100 ft, the map accuracy of features shown would still be 1 in. = 2000 ft (PaMAGIC, 2001).

MAP TYPES

Various map types used in GIS are discussed in the following subsections.

Base Map

The map layers are registered to a coordinate system geodetic control framework and a set of base information, often referred to as a base map. The foundation for a successful GIS mapping project is an appropriately designed base map. The base map is the underlying common geographic reference for all other map layers. The common reference provides registration between various layers and allows them to be overlayed, analyzed, and plotted together. Because the base map serves as the reference layer for other layers, its accuracy can affect the accuracy of other layers. This is especially true if the base map is used to create other layers by on-screen (heads-up) digitization.

Selection of an appropriate base-map scale is largely determined by the earlier choice of GIS applications. Each application inherently requires a certain minimum base-map accuracy and certain map features. For engineering and public-works applications, the required map accuracy is in the range of ±1 ft, as dictated by the need to accurately locate specific physical features, such as manholes and catch basins. Planning applications, which most often deal with areawide themes, do not generally require precise positioning. Accuracies of ±5 ft, or perhaps as much as ±40 ft, are often acceptable. Less detailed maps, showing nothing smaller than roads and buildings, for example, may be adequate for many planning applications.

Whatever the range of mapping requirements, the base map must be accurate and detailed enough to support applications with the most demanding map accuracies of better than ± 2 ft. Utility asset location also requires mapping that depicts specific small features such as manholes and catch basins. As shown in Table 8.1, these requirements are met by a map scale of 1 in. = 50 ft.

There are three common types of base maps: digital orthophotos, planimetric maps, and small-scale maps.

Digital Orthophotos

For laypersons, digital orthophotos (or orthophotographs) are scanned aerial photos. For GIS professionals, they are orthorectified raster images of diapositive transparencies of aerial photographs. Creation of a digital orthophoto requires more than a photo and a scanner, and includes surveyed ground control points, stereo plotters, and a digital elevation model. In fact, the digital orthophoto creation process involves many steps, which are listed below:

- Establish ground control
- Conduct aerial photography
- Perform analytical aerotriangulation
- Set stereo models in stereo plotters
- Capture digital elevation models
- Scan aerial photographs
- Digitally rectify the scanned photographs to an orthographic projection
- Produce digital orthophotos

Digital orthophotos are popularly used as base maps that lie beneath other GIS layers and provide real-life perspectives of terrain and surroundings that are not available in the vector GIS layers. Typical vector data do not show vegetation. The vector layers can show the manhole location but may not include the vegetation hiding the manhole. High-resolution orthophotos with submeter accuracy can guide the public-works crews directly to a manhole hidden behind bushes. Knowing the land-cover characteristics before leaving for an emergency repair of a broken water main will allow the crews to bring the appropriate tools and equipment. Knowing whether the job will be on a busy intersection or in somebody's backyard will determine the kind of equipment, material, and personnel required for the job. Figure 8.1 shows a water system map overlayed on a digital orthophoto base map with an accuracy of ±1.25 ft. Typical digital orthophotos cost $800 to $1600 per mi^2.

Planimetric Maps

Like digital orthophotos, planimetric base maps are also created from aerial photographs. However, instead of scanning the aerial photos, the features are digitized from them. Thus, whereas digital orthophotographs are raster files, planimetric maps are vector files. Planimetric maps generally show building footprints, pavement edges, railroads, and hydrography. Parcels digitized from existing maps are often added to the mix. Figure 8.2 shows a sample planimetric map for the Borough of Munhall, Pennsylvania, extracted from the Allegheny County land base. The borough's sewer lines and manholes are overlayed on the planimetric base map. The cost of planimetric maps depends on the level of detail and, therefore, varies significantly from project to project. The typical cost range is $2,500 to $10,000 per mi^2.

Figure 8.1   A water distribution system overlayed on a digital orthophoto base map.

Small-Scale Maps

Small or rural systems often use small-scale street maps or topographic maps as base maps. Street maps can be created by digitizing the existing maps, obtained from a government agency (e.g., U.S. Census Bureau, USGS, or state department of transportation) or purchased from commercial data vendors. In the U.S., 1:24,000-scale raster topographic map layers called digital raster graphics (DRG) are provided by USGS. Shamsi (2002) provides detailed information about the sources of small-scale maps. Users should be aware of the scale, resolution, accuracy, quality, and intended use of small-scale base maps before using them. Most maps at scales of 1:100,000 and smaller are not detailed enough to be used as site maps or engineering drawings, but they can be used for preliminary studies and planning projects. Figure 8.3 shows interceptor sewers and pumping stations for the Kiski Valley Water Pollution Control Authority in Pennsylvania, on a base map

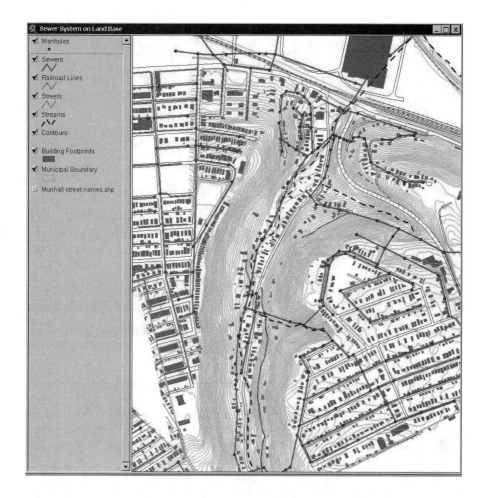

Figure 8.2 A sewer system overlayed on a planimetric base map.

of streets. The 1:100,000-scale solid roads are from the U.S. Census Bureau's 1990 Topologically Integrated Geographic Encoding and Referencing System (TIGER) data. The 1:24,000-scale dashed roads are from the Pennsylvania Department of Transportation. Unlike double-line pavement edges shown on the planimetric maps, these road layers show the single-line street center lines. The difference in the position of the roads in the two layers can be attributed to the resolution, scale, and accuracy of the two layers.

ADVANTAGES OF GIS MAPS

The most challenging part of a GIS application project is to obtain the right kind of maps in the right format at the right time. Therefore, maps are the most important component of a GIS. Without maps, you simply have a computer program, not a GIS.

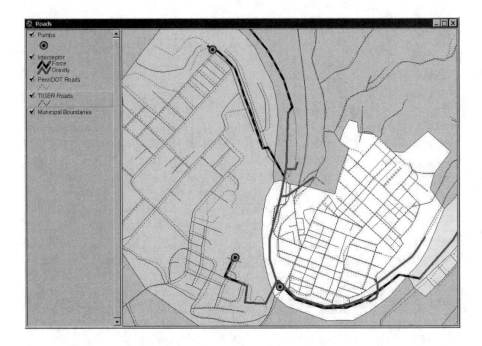

Figure 8.3   A sewer system overlayed on a streets base map.

In many water and wastewater systems, there is a backlog of revisions that are not shown on the maps and the critical information is recorded only in the memories of employees. However, there is no longer any excuse to procrastinate because GIS-based mapping is easy and affordable.

In the past, users have selected computer-aided drafting (CAD) and automated mapping and facilities management (AM/FM) systems to map their water and sewer systems, CAD being the most common method. Although a map printed from CAD or AM/FM might look like a GIS map on paper, it does not have the intelligence of a GIS map. GIS maps are intelligent because they have attributes and topology. Most conventional CAD maps do not have attributes; they simply print data as labels or annotations. For example, the mapmaker must manually write the pipe diameter next to a pipe or must manually change the pipe color or line type to create a legend for pipe size. This is a very cumbersome process. On the other hand, GIS stores the attributes in a database and links them to each feature on the map. This capability allows automatic creation of labels and legends at the click of a mouse button. In GIS, map labels and legends are changed automatically if an attribute changes. In CAD, the mapmaker must manually delete the old label and retype the new label. Only a GIS map knows the spatial relationships among its features. Called *topology*, this capability makes the GIS maps intelligent. For example, a GIS map is intelligent enough to know which watershed is adjacent to which. Although both the CAD and AM/FM offer map layers to store different types of objects, only a GIS map has the capability to relate data across layers. The spatial relations among layers allow spatial analysis operations such as identifying the gate valves that must be closed to isolate a broken water main for repair.

A commercial map atlas company may use a CAD system because its applications are primarily for cartographic products. A telephone company will use an AM/FM system to support its telephone system operations and maintenance, because it must be able to quickly trace a cable network and retrieve its attributes (Korte, 1994). For a water or sewer system, a GIS map is most suitable because the system must conduct many types of spatial analyses, asking questions such as how many customers by type (residential, commercial, industrial) are located within 1000 ft of a proposed water or sewer line.

In addition to water and sewer system management, GIS-based maps can also support other needs of the municipality. For example, a planning department can generate 200-ft notification lists as part of its plan review process. A public works department can conduct maintenance tracking and scheduling. A public-safety department can perform crime location analysis. GIS-based mapping is therefore the most appropriate mapping technology to meet all the mapping needs of a municipality.

GIS MAPPING STEPS

GIS mapping consists of five typical steps:

1. Needs analysis
2. Data collection
3. Data conversion
4. Data processing
5. Map production

These steps are intended for GIS technicians who work in a GIS lab (or production shop) equipped with map-making equipment such as digitizers, scanners, and plotters. Some professionals, such as civil and environmental engineers who are users rather than creators of GIS maps, generally do not perform all of these steps.

Needs Analysis

Mapping work should begin with needs analysis as described in Chapter 2. The needs analysis study describes the features that should be captured during the data conversion step, mapping specifications (accuracy, resolution, scale, etc.), and source documents. For example, the needs analysis determines whether or not the customer meters should be mapped. If the answer is yes, then how and to what accuracy should they be mapped. Are precise coordinates needed or can they be drawn at the end of service line? For large systems (populations greater than 50,000), a pilot project should be conducted as described in Chapter 2 (Needs Analysis). The pilot project allows potential fine-tuning of mapping specifications in a two- to four-sheet area before starting the map production phase.

Data Collection

When existing maps do not exist or are inadequate, mapping data should be collected using a field survey with or without GPS. Required data are usually scattered among a multitude of different organizations and agencies. For instance, it is claimed that 800 worker-years would be needed to convert England's sewer system to digital format, including field verification work (Bernhardsen, 1999). Field data collection using mobile GIS and GPS technology that employs handheld devices and tablet PCs is becoming common for collecting attributes (Chapter 7, Mobile GIS). The question of quantity should be evaluated carefully before the data conversion is started. Too little data limits the mapping applications. Too much data typical of a "data-driven" approach might be wasteful.

Data Conversion

Also referred to as the production work, this step converts hard-copy maps into digital files using digitization with or without scanning. The data conversion process is typically the most expensive and time-consuming part of a GIS mapping project. Depending on the level of detail, data conversion can cost $500 to $5000 per map sheet. Approximately 75% of typical GIS costs are related to data conversion and creation. The data conversion work, therefore, should be structured to ensure consistency, integrity, and accuracy.

Captured data are stored in *layers* also referred to as coverages or themes. For example, manholes are captured as a point layer, sewer pipes are captured as a line layer, and sewersheds are captured as a polygon layer. The layers that support development of other layers should be given a higher development priority. For example, geodetic control and base-map layers should be developed first because most other layers depend on them.

Data conversion includes capturing both the graphic (geometry and coordinates of features) and nongraphic (attributes) data. Graphic and nongraphic data may be entered simultaneously or separately.

Capturing Attributes

The annotations (labels) shown on the source map are the most common source of attributes. The source document ID (e.g., drawing number) is one of the most important attributes that should always be captured during data conversion. Features such as valves and hydrants have unique IDs, which can be easily captured from labels (annotations) on source maps. The features without IDs should be assigned new IDs during data conversion. Some data conversion application software allow automatic ID creation during data conversion.

Most source maps show pipe sizes as labels or legends, which can also be easily captured during data conversion. Geometric properties like length, area, and perimeter of features can be internally calculated by GIS and entered as an attribute. For example, if pipe length is labeled, it can be entered as a Length_Map (source map) attribute. If pipe length is not shown, it can be automatically calculated by GIS and entered as Length_GIS (GIS-calculated) attribute.

Additional attributes are collected from other sources such as legacy databases, tables, and field and GPS surveys. Some existing databases and tables can be linked to GIS databases to avoid manual entry of attributes. Once the linkage has been established, attributes from the linked documents are available in the GIS database for query, analysis, and thematic mapping. For example, unique hydrant IDs captured during data conversion can be used to link to a hydrant-testing database. Applications that require locating the facilities by customer address such as work orders and customer complaints should also capture the address attribute in at least one layer (buildings, parcels, or meters). GIS attributes can also be found in some unexpected places, such as customer billing records. With a technique called address geocoding, GIS can convert the billing records or any postal address to a point on a map. Alternatively, facilities can be located by customer address using a geocoded streets layer.

Capturing Graphics

The data conversion methodology for capturing graphics depends on database design, source materials, and project budget. Digitization and scanning are the two most common data conversion methods.

Digitization

Digitization is a process of converting a paper map into a vector file by a computer using a digitizer or digitizing table (or tablet). The source map is placed on the usually backlit surface of the digitizer, and map features are traced using the digitizer puck that looks like a computer mouse with a crosshair. The source map should be registered (or calibrated) to control points. Figure 8.4 shows map conversion work using a digitizer. Digitizers cost approximately $10,000. Digitized files, when properly prepared, are ready for immediate use in a GIS.

Conventional table digitization is a laborious process. New "heads-up" or "on-screen" digitization is a less cumbersome alternative in which the visible features from scanned maps, digital aerial photos, or satellite imagery can be traced with a standard computer mouse on the computer monitor. If a digitizer is not available but a scanner is, data conversion can be done by scanning the source maps followed by heads-up digitization. The use of this process for certain specialized mapping projects, such as land-use/land-cover mapping, requires familiarity with photo interpretation techniques. For easy identification, certain utility assets that are difficult to see on the aerial photographs (e.g., valves, hydrants, manholes) can be premarked. Premarking is done by placing (or painting) targets with special symbols and colors over or adjacent to the asset to be captured. Premarking cost is usually $2 to $10 per target.

As-built or construction drawings usually have dimensional or offset information on utility assets (e.g., a sewer pipe 10 ft from the street right-of-way). This information can be used to position the utility assets using the automated drafting capabilities of modern GIS packages. This method is more accurate than digitization (Cannistra, 1999).

Figure 8.4   Data conversion using digitization.

Scanning

Scanning is a process of converting a paper map into a raster file (or image) by a computer using a map-size scanner. Black-and-white scanners are most common because they cost about the same as a digitizer. Color scanners cost twice as much. Figure 8.5 shows map-conversion work using a scanner. Scanning is generally most efficient when converting maps for archival purposes. Sometimes, as-built drawings are scanned and linked to vector features. Because many applications (e.g., hydraulic modeling and field inspections) require the utility features in vector format, scanned images must be converted to a vector format. This can be done by heads-up digitization of the scanned maps or by using a process called *vectorization* that automatically converts the raster data to vector format. Vectorization is a cost-effective data conversion method but it is based on new technology that is still improving. Vectorization may not always produce reliable results, especially in the unsupervised (unmanned) mode. Linework and symbols vectorize better than annotations. Complex, faded, or unclear maps may require extensive postprocessing and manual touch-up.

Data Conversion Software

Data conversion requires GIS development software that draws objects as points, lines, or polygons or represents them as pixels. ESRI's ArcGIS® and Intergraph's GeoMedia Pro® are examples of GIS development software (Shamsi, 2002). For large projects, data conversion application programs are very useful. These programs

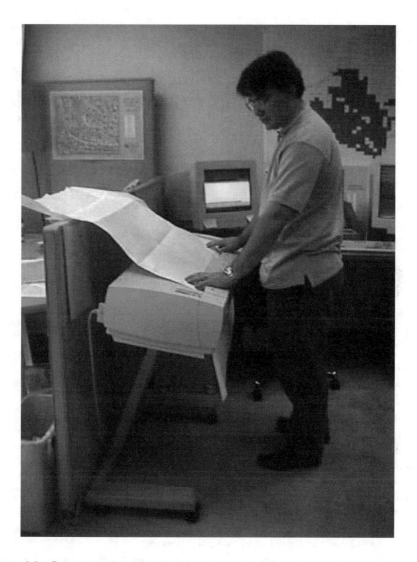

Figure 8.5   Data conversion using scanning.

are usually developed as extensions to standard GIS packages to supplement specific mapping needs. For example, ESRI's ArcFM water data model supports water and sewer system mapping in a geodatabase. Chester Engineers' (Pittsburgh, Pennsylvania) water editing tool is an application that allows user-friendly mapping of water and sewer systems for ESRI software users. It eliminates attribute entry and editing via command prompts or database tables, which require data conversion skills. Users without prior data conversion experience can enter attributes from sources documents (e.g., as-built drawings) in a menu environment simply by clicking on points (e.g., manholes) and lines (e.g., sewers). The water editing tool was developed in several

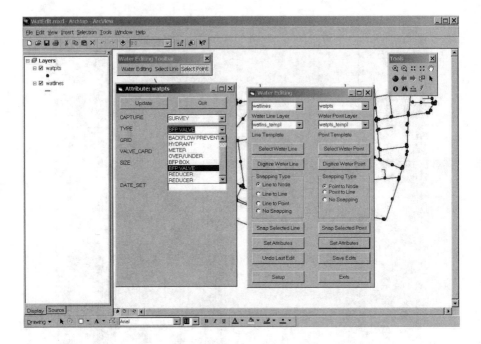

Figure 8.6   Sample water system data conversion application for ArcMap.

versions for different users. The ArcView version was developed using Avenue and ESRI's Dialog Designer Extension to support water system and sewer system mapping needs of the ArcView 3.x users. The ArcInfo 7.x version was written in Arc Macro Language (AML) to support water system and sewer system mapping in an Arcedit session. The latest ArcMap version that adds a new water editing toolbar in ArcMap was developed using Visual Basic for Applications (VBA) for ArcGIS users. Figure 8.6 shows a screenshot of the ArcMap version of the water editing tool. The right window shows the form for adding and editing point and line features. The left window shows the form for editing point attributes.

Data conversion is started by drawing and attributing the points. After the points are attributed, the connecting lines are added from point to point. The application prompts the user to select the upstream and the downstream points. After both are selected, the line is added to the data set. By selecting upstream and downstream manholes to add the lines, the flow direction of a gravity sewer is incorporated into the data set, ensuring the proper connectivity of the system. Additionally, the lines are snapped to the point features, ensuring that the points of connection correspond to manholes.

Data conversion applications like the water editing tool increase the speed, efficiency, and accuracy of data conversion for water, wastewater, and stormwater systems. They reduce data-entry errors and can even be used to validate the input data. If desired, the applications can be modified to meet the project-specific mapping requirements.

Data Processing

Data processing includes some or all of the following tasks:

1. Data preparation
2. Topological structuring
3. Data management
4. Quality control

Data Preparation

This task makes the raw data from source maps or field/GPS surveys available for GIS use or GIS-ready. Typical activities include postprocessing of capture data, changing data formats (e.g., from DXF to Shapefile), applying map projections, georeferencing the image data, and/or clipping or mosaicking the aerial photos.

Postprocessing of captured data is warranted if the source maps and the base map have different accuracy, and captured data do not align with the base map. In this case, captured data is edited (moved, stretched, or resized) to fit the base map.

GIS data are stored in various file formats. The number of data formats has increased exponentially with the growth in the GIS industry (Goodchild, 2002). According to some estimates, there might be more than 80 proprietary geographic data formats (Lowe, 2002b). Why are there so many geographic data formats? One reason is that a single format is not appropriate for all applications. For example, a single format cannot support both fast rendering in a command and control system and sophisticated hydraulic analysis in a water distribution system. Different data formats have evolved in response to diverse user requirements.

GIS software cannot read all the data formats simply because there are so many formats. Disparate data formats should be converted to one of the formats compatible with a particular GIS software. Although many GIS packages provide data conversion for the most common data formats, no GIS can support all possible conversions. Many government agencies and data and software companies provide data translators for this purpose (Shamsi, 2002).

Data translation from one format to another can potentially lead to the loss or alteration of data. For instance some platforms can support better numeric precision than others. Depending on the mapping scale and the coordinate system, this could seriously affect data quality. In one system, curves representing a water main might be defined mathematically, but in another described as a series of straight line segments.

Topological Structuring

How much topology should be captured in a GIS map depends on the intended applications of the map. For example, in a sewer system modeling application, correct topology (e.g., flow direction) is generally more important than the correct pipe location on the street (e.g., left, right or center of the street). As much topology as possible should be captured during data conversion. For example, flow direction

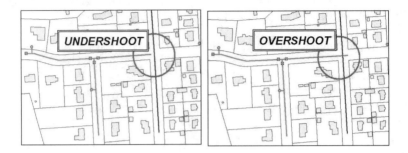

Figure 8.7   Common digitizing errors: undershoot (left) and overshoot (right).

should be captured if flow direction arrows are shown on source maps. Flow direction can be automatically captured if sewer pipes are digitized from upstream to downstream direction. In ArcInfo 7.x "coverage" model, the topology of an arc includes its upstream and downstream nodes (fnode, tnode). ArcInfo 7.x performs computationally intensive topological relationship calculations up front and stores that information as attributes (fnode, tnode). These attributes are used to determine flow direction of arcs. If source maps do not have topological relationships or if the data conversion technicians are not adequately skilled to understand them, topology can be added after the data conversion by knowledgeable staff. For example, in a water distribution system mapping project, valves and hydrants are logically linked to their appropriate pipe segments by the people who are knowledgeable about system characteristics (Cannistra, 1999).

Conventional GIS packages, such as ArcInfo 7.x and ArcCAD, provide the legacy *BUILD* and *CLEAN* commands to build topology and eliminate topological digitizing errors such as undershoots and overshoots. As shown in Figure 8.7, an undershoot occurs when a line stops short of an intended intersection point and an overshoot occurs when a line extends past an intended intersection point. Traditionally, these "cleaning" operations were done after the data conversion step. Today's modern GIS packages allow enforcing the topological rules (spatial constraints) during data conversion and editing. Thus, when a manhole is moved, the topological rules ensure that the sewers connected to that manhole move simultaneously. For example, ArcInfo 8.3 provides new tools to define, validate, and maintain topology in a geodatabase. Users define which layers participate in a topology and which rules are appropriate. Topology is stored in a commercial off-the-shelf DBMS.

Data Management

This step organizes attribute data in a database management system (DBMS) format. A GIS database, though not as visible as a GIS map, is an equally critical GIS component. The most difficult job in creating a GIS is the enormous effort required to enter the large amount of data and to ensure both its accuracy and proper maintenance (Walski and Male, 2000). Conventional file-based GIS databases (e.g., ArcInfo 7.x) store geographic data in vendor-specific proprietary format files. Attribute data are stored in a separate file, which generally has a common nonproprietary format (e.g., a .dbf file). The two separate files are linked using pointers. Modern spatial or object-oriented databases do not require links between features and attributes because they store both the features and attributes in the same file. Incorporating spatial data with attribute data in a relational database provides a platform for efficient management of huge data resources, as well as fast and sophisticated spatial query and retrieval methods (Dalton, 2001).

Quality Control

A comprehensive and well-defined quality assurance/quality control (QA/QC) program should be implemented for large data conversion projects. Data conversion QA/QC can be ensured by checking data completeness, content, symbology, format, structure, and compliance with database specifications during data production. Positional accuracy can be checked using field trips, survey, GPS, photographs, and video tapes (Cannistra, 1999).

QA/AC can employ both manual checks and automatic methods. For example, the water editing application shown in Figure 8.6 provides an automatic QA/QC capability. It displays the points that are completely defined using a white box. Points that have missing attributes are displayed as a box with an X through it. Points that are added from source drawings are shown as a green box with an X through it. After each point is defined, the list of attributes for each point is displayed on the screen, allowing the user to double-check the attributes entered for each point.

Data-entry QA/QC can be accomplished by restricting the user input to valid data ranges and rejecting the data entries outside the valid ranges. This capability generally requires development of custom data-entry forms using computer programming.

Map Production

The previous steps enable display of GIS layers on the computer screen. The map production step allows plotting of GIS data on paper and other print media. It includes creation of thematic maps using appropriate symbols, colors, patterns, and legends. For example, water main sizes are shown by different line widths, pressure zones by different colors, and valve types by different symbols. Map production also involves printing the scale bar, north arrow, and organization logo. Sometimes tables, charts, and photographs are also inserted. Every map should have a title. Other information such as filename, map number, project name and number, plot date, mapmaker, and map source should also be printed on each important map.

This last step is important because it shows the final result of all the previous steps. Plotting of GIS maps in the correct style, orientation, scale, size, and format is a time-consuming process. Plotter speed and network speed are often the main bottlenecks. Templates for symbology, legends, and layouts should be developed in-house or obtained from other organizations to expedite map production. Some vendors package templates and symbol libraries specific to water, wastewater, and stormwater system with their software products.

Maps are needed for map books, wall mounting, archives, presentations, reports, etc. The purpose of the map and the target audience should be considered to design an appropriate map layout. For example, a map for an engineering application should be plotted differently from a map for a water utility newsletter. This step also includes distribution of the maps to the staff of an organization and optionally to the general public using the Internet, intranet, or wireless technology.

CASE STUDIES

Borough of Ramsey, New Jersey

The Ramsey Board of Public Works is responsible for the operation and maintenance of water distribution and sanitary sewer collection systems throughout the Borough of Ramsey, New Jersey. In an effort to more efficiently manage its geographically referenced data, the board started to explore the benefits and applications of GIS technology and computer mapping in early 1993. As a first step, with the assistance of Chester Engineers, Inc. (Pittsburgh, Pennsylvania), the board completed a needs analysis and a GIS pilot project described in Chapter 2 (Shamsi et al., 1995; Shamsi et al., 1996). The Borough did not have any complete large-scale mapping resources suitable for utility asset location. System design recommended the following mapping specifications:

- A new digital orthophoto base map should be developed accurate to the tolerances normally associated with mapping at a scale of 1 in. = 50 ft.
- Orthophoto base mapping should be established on the New Jersey State Plane Coordinate System. Survey monuments should be established to enable subsequent mapping efforts to be tied to the basemap coordinate system.
- Lot and block boundaries should be included as a unique layer in the base map.
- Additional map layers for the water distribution system, the sanitary sewer collection system, the stormwater collection system, and census geography should be developed to support other needs of the Borough.
- Use GPS to verify the locations of valves, hydrants, and manholes.

Table 8.2 summarizes the required map layers and the features depicted in each layer. Table 8.3 presents the data dictionary for Ramsey's pilot project.

Table 8.2   Map Layers and Depicted Features

Layer	Features
New Jersey State Plane Grid System	Grid tics, survey monuments
Photogrammetric base map	Building outlines, street outlines and centerlines, vegetation, water courses, topography
Parcels	Lot and block boundaries
Water distribution system	Mains, valves, hydrants
Sanitary sewer collection system	Sanitary sewers, manholes, cleanouts
Stormwater collection system	Storm sewers, manholes, catch basins
Census geography	Blocks, block groups, tracts

Table 8.3   Data Dictionary

Feature Type	Required Attributes
Water mains	Segment ID
	Pipe material
	Pipe size
	Repair date(s)
	Repair type(s)
Water valves	Valve ID
	Installation date
	Type
	Date last cycled
	Normal position
Fire hydrants	Hydrant ID
	Associated valves
	Type
	Installation date
	Flow data
	Preferred status
Sewers	Sewer ID
	Pipe material
	Pipe size
	From manhole ID
	To manhole ID
Manholes	Manhole ID
	Manhole type
	Lid type
	Depth
	Inflow problem status
Catch basins	Catch basin ID

All of the borough's sewer system data were stored in a nondigital form. This information was manually entered into the GIS database. Map features and attribute information were gleaned from a variety of sources. Table 8.4 lists recommended sources for specific map features and attributes. After reviewing the Borough's available maps and considering the mapping requirements of the priority applications, the use of digitization was recommended for data conversion.

ESRI's ArcInfo (Version 6) and ArcView (Version 2) software products were the two main software programs used in this project. Because the Borough did not have the technical staff to maintain an ArcInfo GIS, data conversion work was done by the consultant using ArcInfo. Due to its low cost, ease of use, and compatibility

Table 8.4   Source Documents for Map Features and Attributes

Types of Features	Source Documents			
	Base Map	Utility Markouts	Water System as Built, 1970, 1:50-Scale	Sewer System as Built, 1970, 1:100-Scale
Water mains		☑	☑	
Fire hydrants	☑	☑	☑	
Water valves		☑	☑	
Sewer pipes		☑		☑
Manholes	☑	☑		☑
Catch basins	☑	☑		☑

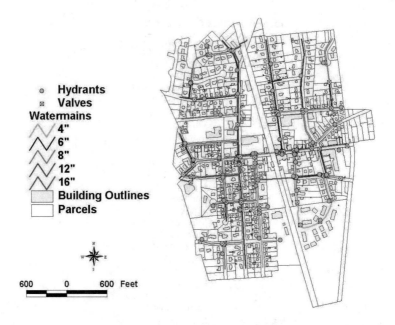

Figure 8.8   Water distribution system map showing pipes classified by diameter.

with the ArcInfo file format, ArcView was installed in the Borough for routine display and plotting of maps and querying the GIS database.

Figure 8.1 shows the Ramsey digital orthophoto base map with an accuracy of ± 1.25 ft. The figure on the first page of this chapter shows a map of sanitary sewers and manholes overlaid on the parcel and buildings layers. The sewer features were taken from as-built drawings and adjusted to locations of surface features that were field verified. Figure 8.8 shows a water system map with water mains, hydrants, and valves overlaid on the parcel and buildings layers. Water mains have been classified by the diameter attribute of the water main layer.

City of Lake Elsinore, California

Elsinore Valley Municipal Water District located in the city of Lake Elsinore, CA, provides water, wastewater, and agricultural services to approximately 25,000 customers. It maintains an infrastructure consisting of 305 mi of water mains, 290 mi of sewer pipes, and 246 mi of irrigation pipelines. An ESRI software user since 1992, the District migrated to ESRI's new ArcGIS geodatabase platform in 2000. The migration data conversion work, which included 5000 as-built drawings and 300 atlas pages, was completed in 8 months. The source documents were scanned and linked to GIS features. The system has 35 users and employs SQL Server 7 relational database management system software, Gateway 7400 and XI NetRAIDer dual 800 MHz servers, and Pentium III workstations. The system maintains approximately 75,000 maps in an enterprise GIS environment. GIS allows operations personnel to create atlas maps and perform asset maintenance, a customer service department to analyze meter-reading routes and identify illegal connections, and an engineering department to generate as-built construction drawings (ESRI, 2002a).

Allegheny County, Pennsylvania

Each year, sewage overflows affect Pittsburgh's rivers up to 70 days during the boating season, making water unacceptable for recreational contact. In addition, the overflows affect Allegheny County's primary source of drinking water. Municipalities in the region could face a potential $2-billion investment in order to correct the problem. In 2003, the U.S. Environmental Protection Agency (USEPA) issued administrative consent orders to 83 municipalities of Allegheny County for controlling the sewage overflows. The orders required communities to address the sewage overflow issue through mapping, cleaning, and televising the system, and flow monitoring. This information will help communities develop a long-term wet-weather control plan. The orders required the following GIS mapping requirements:

- Create an ESRI-compatible base map that includes streets, street names, municipal boundaries, and streams.
- Create an updated ESRI-compatible comprehensive sewer map of the sanitary sewers with the following content:
  - Location of the sewer lines
  - Direction of flow
  - Size of the sewer lines
  - Sewer line material
  - Locations of interconnections with other municipal sewer systems
  - Field-verified location of manholes and catch basins (identified by a comprehensive numbering or lettering system)
  - Location of pump stations, force mains, and siphons
  - Location of streams or drainage ways tributary to the sewers
  - Location of overflow (combined sewer overflow [CSO] and sanitary sewer overflow [SSO]) structures
- Locate all significant structures to a minimum horizontal accuracy of 3 ft.
- Use as-built drawings, GPS, or traditional land surveying methods to determine structure locations.
- Use State Plane Pennsylvania South NAD83 projection system for coordinates and NAVD88 datum for elevations.
- Manhole inverts and rim elevations to a minimum vertical accuracy of 0.10 ft for overflow structures, sewers with greater than 10-in. diameters, and other critical sewers.
- Incorporate field inspection data into comprehensive sewer maps.

USEFUL WEB SITES

Geospatial Information and Technology Association (GITA)	www.gita.org
USGS GIS Maps for Water Resources	water.usgs.gov/maps.html
USEPA GIS Information	www.epa.gov/epahome/gis.htm

CHAPTER SUMMARY

GIS provides powerful capabilities for creating cost-effective and intelligent maps for water, wastewater, and stormwater systems. The quality of maps depends on data quality, errors, accuracy, scale, and resolution. GIS maps can be developed in many different ways. A mapping approach that best meets the application needs of an organization should be selected. GIS mapping requires five steps: needs analysis, data collection, data conversion, data processing, and map production. Needs analysis and pilot project testing is highly recommended, as explained in Chapter 2 (Needs Analysis). Data conversion is the most time-consuming and costly part of mapping projects and should be performed carefully. Data conversion application software can be used to reduce data conversion cost. Digitization and scanning are the two popular methods of data conversion but users can benefit from innovative variations of these methods. GIS maps should always be constructed in reference to a base map that has the required accuracy and resolution to support the intended applications of GIS maps.

CHAPTER QUESTIONS

1. What are the different types of maps?
2. What are the different types of base maps and their characteristics?
3. How is a GIS map different from a CAD drawing?
4. What is the relationship between map accuracy, scale, and resolution?
5. What are the various steps for making GIS maps? Which step is most critical and why?
6. What is data conversion and how is it done?
7. What is topology, and how does it affect GIS maps?
8. Identify a GIS application in your organization and recommend the appropriate mapping layers and source documents to implement that application.

CHAPTER 9

Mapping Applications

A map is worth a thousand numbers but you must go beyond mapping to sustain your GIS program. This requires putting your GIS maps to beneficial use.

3D display of sewer system for the Municipality of Bethel Park, Pennsylvania.

LEARNING OBJECTIVE

The learning objective of this chapter is to understand the applications of GIS maps in the water industry.

MAJOR TOPICS

- Common mapping functions
- Mapping application examples for water systems
- Mapping application examples for wastewater systems
- Mapping application examples for stormwater systems

LIST OF CHAPTER ACRONYMS

2-DTwo Two-Dimensional
3-DThree Three-Dimensional
COGO Coordinate Geometry
CSO Combined Sewer Overflow
DEM Digital Elevation Model
DTM Digital Terrain Model
I/I Inflow/Infiltration
MS4 Municipal Separate Storm Sewer System
NPDES National Pollution Discharge Elimination System
SQL Structured Query Language
TIN Triangular Irregular Network/Triangulated Irregular Network

---

This book focuses on the four main applications of GIS, which are mapping, monitoring, modeling, and maintenance and are referred to as the "4M applications." In this chapter we will learn the applications of the first *M* (mapping).

CUSTOMER SERVICE APPLICATION IN GURNEE

The Village of Gurnee, located near Chicago (Illinois), has seen explosive growth since the 1970s. Population more than doubled from 13,701 in 1990 to 28,834 in 2000. This growth bogged down Gurnee personnel with public meetings, rezoning hearings, code violations, road construction, and water line and sewer line repairs. Several departments were legally required to notify property owners of meetings, pending changes, or future disruptions. The Village used GeoMedia GIS software from Intergraph to automate the resident notification process (Venden and Horbinski, 2003). The automated system — known as Notifast™ — allows staff members to enter a street address or select a parcel, specify a distance radius, and identify all property owners within that distance. Notifast then generates a list of mailing labels

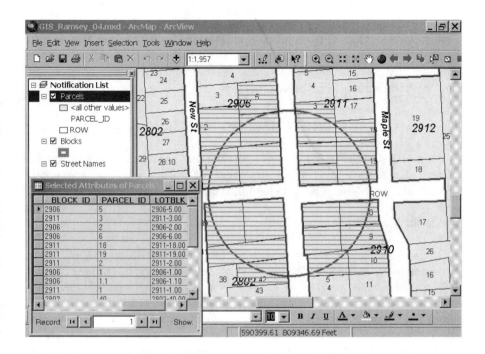

Figure 9.1   Selecting notification parcels in ArcGIS.

and mail-merged form letters. Each letter includes an embedded map that clearly shows the area in question. This application of existing GIS maps has improved efficiency and customer service. What used to take hours is now completed in minutes.

Figure 9.1 shows a screenshot of how parcels can be selected in ESRI's ArcGIS software by drawing a circle. The 200-ft radius circle is drawn with its center at the project site. All the parcels that are intersected by the circle are selected as shown in the hatched pattern. The attribute data for the selected parcels are shown in the table to the left of the circle. These data can be copied in a word processor's mail-merge feature to generate notification letters.

As explained in Chapter 8 (Mapping), creation of GIS maps is a huge undertaking. Unfortunately, in today's competitive world, despite the enormous effort required to create them, colorful maps alone can no longer justify the existence of a GIS program. This is because, after spending large sums of money on a mapping project, many water and sewerage agencies are wondering what else to do with their GIS maps. To garner continued financial support for a GIS program, GIS programs must go beyond conventional map printing activities and demonstrate the usefulness (applications) of GIS maps. This chapter shows how to do that.

It does not occur to most people that a good map raises more questions than it answers — that the question of why things are located where they are raises important intellectual issues, with immediate serious implications. (Lewis, 1985).

A paper map provides the power of place but a map in GIS can automate "the why of where." Most GIS applications originate from GIS maps. This chapter describes simple mapping applications. Advanced applications like hydrologic and hydraulic modeling, facilities management, inspection and maintenance, and planning are described in other chapters.

COMMON MAPPING FUNCTIONS

Most GIS packages provide simple data analysis capabilities that can be used to create basic mapping applications without custom programming. Typical examples of these GIS functions are given in the following subsections.

Thematic Mapping

Thematic mapping is one of the easiest ways to develop mapping applications. Map components can be logically organized into sets of layers called *themes*. A theme is a group of similar geographic features, such as hydrants (points), water mains (lines), or pressure zones (polygons). These themes can be manipulated using classification and legend (symbology) to create different thematic maps. Classification places attribute values into groups. For example, water mains for a city can be classified by diameter to create groups such as pipes smaller than 6 in., pipes 8 to 12 in., and pipes greater than 12 in. These groups can then be displayed using different legends reflecting different colors, line thicknesses, and symbols (e.g., solid, dashed, or dotted lines). Each combination of a classification method and legend will produce a different thematic map. Figure 9.2 shows the difference between an unclassified and nonsymbolized sewer map (top) and the same map classified and symbolized by pipe diameters (bottom). Obviously, the bottom map will be more useful in applications that require pipe diameters, such as flow monitoring and hydraulic modeling.

Spatial Analysis

Many GIS packages also allow users to select the features of one or more themes using the features of another theme. Referred to as spatial or overlay analyses, these *theme-on-theme* selection functions identify whether one feature lies within another, whether it completely contains another, or whether it is within a specified distance of another, and so on. For example, these mapping operations can identify the manholes in a given drainage basin, or census blocks within a watershed, or the properties within a certain distance of a water main.

Buffers

A buffer can be created around one or many features by using the buffer function typically found in most GIS packages. Creating a buffer provides a visual representation on a map of the area within a certain distance of one or more features. Buffer can also be used to select features in other layers that fall within the buffered area.

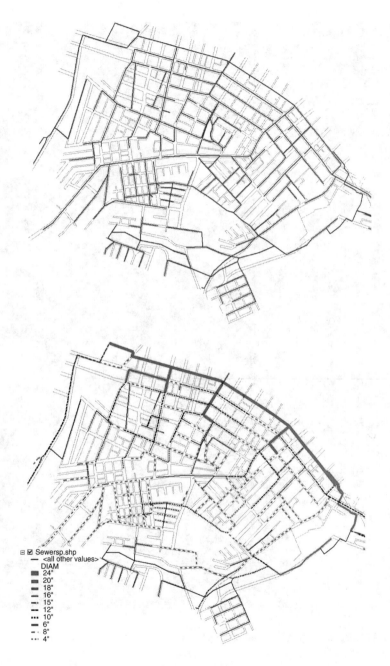

Figure 9.2   A sewer system map. Above: without attribute classification and symbology; below: with attribute classification and symbology for pipe diameter.

Figure 9.3   Buffer for a sewer separation project.

For instance, a buffer can be drawn to show the area around a well that is contaminated or to represent a floodplain around a river. Figure 9.3 shows a buffer created around existing combined sewers in the Borough of Kittanning, Pennsylvania. It identifies the area targeted for sewer separation in the Borough's combined sewer overflow (CSO) control plan.

Hyperlinks

Some GIS packages such as ArcView and ArcGIS have Hotlink or Hyperlink functions respectively to provide additional information about the features. For example, when a feature is clicked with the Hyperlink tool, a document or file is launched using the application associated with that file type. This function can be used as a document management system from within a GIS. For example, users can click on a valve to display the valve card, on a sewer pipe to launch a televised video file, or on a water main to display its as-built drawing. Figure 9.4 shows a display of a scanned image of a water distribution system valve parts list using ArcView's Hotlink function. Additional information on hyperlinks is provided in Chapter 15 (Maintenance Applications).

WATER SYSTEM MAPPING APPLICATIONS

Representative mapping applications for water distribution systems are given in the following subsections.

MWRA Water System Mapping Project

In the late 1990s the Massachusetts Water Resources Authority (MWRA) conducted a major in-house effort to develop a linked system schematic set. These overview maps did not have a scale or base, but showed how the elements of the system connected hydraulically. In addition to developing the digital version of these maps, the GIS staff created links to a number of real time databases and the location plans for each valve. Thus, the schematic showed the status of all mainline valves for emergency response planning. If a valve was inoperable or had been opened or closed, the GIS updated the color of the pipe line on the schematic, alerting the users to its condition.

MWRA then consolidated their water distribution facility data and source document references into a manageable set of data tables and digital maps that were easily accessible to MWRA staff. They also linked the records of every pipe, valve, meter, and fitting in the distribution system to existing databases of record plans, land takings, hydraulic conditions, leaks, maintenance activities, and schematic representations in the GIS. This effort allowed MWRA users to click on a pipe drawing, and instead of being able to access just the record drawing, choose from a menu of options regarding the land records and model scenarios, as well as the physical condition and maintenance history of the pipe. This provided access to a variety of maintenance and data development activities throughout the water division (Estes-Smargiassi, 1998).

Service Shutoff Application

To strengthen enforcement activities on delinquent accounts, the Philadelphia Water Department (PWD) developed an ArcView GIS tool to support its shutoff program. The shutoff tool identifies priority sites and automates crew assignments

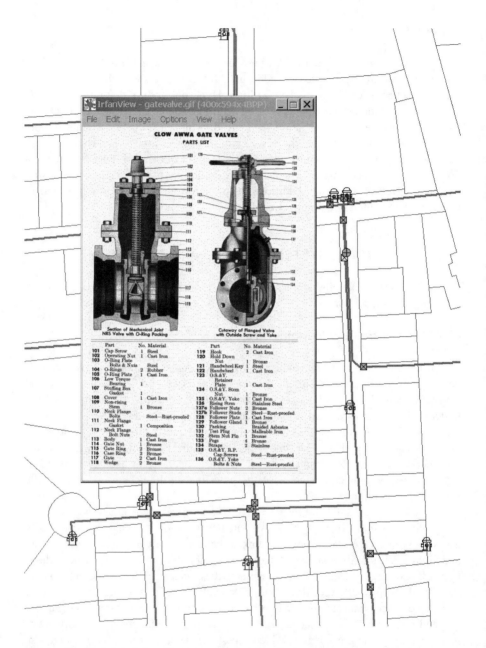

Figure 9.4   Hotlink function displaying valve information for a water distribution system.

to increase crew productivity. A prioritizing algorithm identifies delinquent customers based on age and amount in arrears. These priority accounts identify clusters of shutoff sites. Clusters are formed using ArcView Network Analyst extension to locate nearby sites reachable by the street network. The application generates maps and tables for each crew's daily schedule. With the help of this application and other measures, PWD improved enforcement efficiency by 60% (Lior et al., 1998).

Generating Meter-Reading Routes

Seattle Public Utilities (SPU) has more than 174,000 customers. In an effort to save meter-reading cost and improve customer service, GIS was used to create more efficient meter-reading routes. In 1995, SPU's Metering and GIS departments joined forces to explore the feasibility of rerouting meter-reading routes. In 1997, a pilot project was completed using software from ESRI and RouteSmart Technologies, which convinced the SPU management to implement a citywide rerouting project. The project implementation resulted in cost savings for the ratepayer, especially because of SPU's prior investments in GIS data conversion (Lee, 1998).

Map Maintenance Application

In Ohio, Cincinnati Water Works (CWW) developed a GIS application for water distribution system maintenance, hydraulic modeling, and customer service. A major component of this application was the development of a Map Maintenance Application (MMA), which is an update developed using ArcInfo and Oracle. It allows simultaneous maintenance of water system layers and related RDBMS tables. The MMA features COGO data entry, scroll lists for table inserts, and feature abandonment. It also utilizes the integrity constraint checking capability of the RDBMS (Guetti, 1998).

WASTEWATER SYSTEM MAPPING APPLICATIONS

Representative mapping applications for wastewater systems are given in the following subsections.

Public Participation with 3D GIS

At the present time many cities in the U.S. are planning to spend billions of dollars to meet the regulatory requirements for correcting their wastewater problems, such as wet-weather overflows from the sewer systems. Given the potential for significant expenditures of public funds, a key to the future success of capital projects is public participation. The regulatory agencies are mandating public participation as a required element of these projects. Therefore, establishing early communication with both the public and the regulatory agencies should be an important first step in a long-term planning approach, and crucial to the success of a wastewater capital improvement program. By informing the public early on in the planning process about the scope and goals of the program, and continuing this public involvement during development, evaluation, and selection of alternatives, issues and potential conflicts can be identified and addressed more expeditiously, minimizing the potential for prolonged delay or additional cost.

Three-dimensional (3D) maps are an effective means of public education, notification, and communication. They provide impressive "real-life" visions of proposed projects. GIS, CAD, and multimedia technologies can be combined to produce 3D models and 3D animations.

3D maps can be created by adding elevation data to existing 2D vector layers. Elevation data can be obtained from contour maps, digital elevation models (DEM), digital terrain models (DTM), aerial surveys, engineering drawings, or GPS surveys. The quality of 3D analysis largely depends on the accuracy and resolution of elevation data. For engineering design, highly accurate data are required. For presentation mapping, less accurate data can be used. For example, in-lieu of accurate heights, buildings can be elevated in proportion to the number of floors.

GIS fly-through scenes allow the public to experience a 3D walk-through of a proposed project. Animation can animate objects within a model, e.g., cars moving down the road or people walking on the curbs. Realistic materials and textures can be added to depict the actual field conditions, e.g., wet streets. Climatic conditions, e.g., snow cover or sunlight can be added to portray the specific design conditions.

3D capability is usually added to standard GIS packages by installing additional 3D extensions or modules. For example, 3D Analyst Extension enables ESRI's ArcGIS users to visualize and analyze surface data. Using 3D Analyst, one can view a surface from multiple viewpoints, query a surface, determine what is visible from a selected location on a surface, and create a realistic perspective image draping raster and vector data over a surface.

The figure on the first page of this chapter shows a 3D map of a sewer system for the Municipality of Bethel Park, Pennsylvania. Created using 3D Analyst Extension, the map shows buildings, contours, and sewer lines overlaid on an aerial photograph. Because wastewater must flow down the hill, design of gravity sewers depends on topography. GIS can help create 3D maps to facilitate efficient engineering of a new sewer system, quick assessment of an existing sewer system, or presentation of the project layout to the general public.

Figure 9.5 shows a 3D fly-through animation of an Allegheny County Sanitary Authority (ALCOSAN) sewer construction project in the City of Pittsburgh, Pennsylvania. Completed in 1999, this $24-million project involved construction of a 5.2-mi-long parallel relief interceptor to relieve sewer overflows and basement flooding in the Saw Mill Run Watershed of the city. The pipe sizes for the relief interceptor ranged from 36 to 54 in. in diameter. The parallel interceptor was designed to follow the alignment of the Saw Mill Run Creek because the right-of-way corridor along the main street (Route 51) was not conducive to a major interceptor construction project. Animation was used to present the project design in a general public meeting. Figure 9.5 shows the sewers and manholes exposed above the ground surface to help the general public envision the proposed interceptor layout. The animation clearly indicated that the new sewers will mostly avoid backyards, parking lots, and roads — an issue that was very important to the residents of the community. This approach helped ALCOSAN to win the public support for the project.

Mapping the Service Laterals

Recent inflow and infiltration (I/I) studies have shown that service laterals are one of the most significant I/I contributors, generating as much as 60% of the total

Figure 9.5   3D fly-through animation of a sewer construction project.

I/I quantity. Unfortunately, most collection systems do not have maps of their service laterals. TV inspection logs provide accurate lateral location information, which can be used to prepare service lateral GIS maps. For example, Chester Engineers (Pittsburgh, Pennsylvania) developed a Wyemaker ArcView extension to add service wyes (Ys) to sewer lines. Developed using ArcView Dialog Designer extension, this utility is a good example of GIS customization.

Figure 9.6 shows a typical two-page TV inspection log. The first page provides general information about a sewer pipe. The second page details the specific pipe defects and their location. This page also includes the location of service laterals measured in feet from one manhole to the next. The Wyemaker extension will first create a database containing the wye information for each sewer main. The wye information is entered in a wye input form shown in Figure 9.7 that basically replicates the data fields on the log sheets. The upstream and downstream manhole identification numbers are entered. In this case, the sewer segment ID is created by concatenating the upstream and downstream manholes. The sewer length from the inspection log is entered as well as the beginning distance (usually zero). The user then indicates the direction of the measurements, the wye measurement and the side of the line in which the wye is located. The direction of the measurements is the direction in which the line was inspected. The direction is either "against the flow" (from the downstream manhole to the upstream manhole) or "with the flow" (from

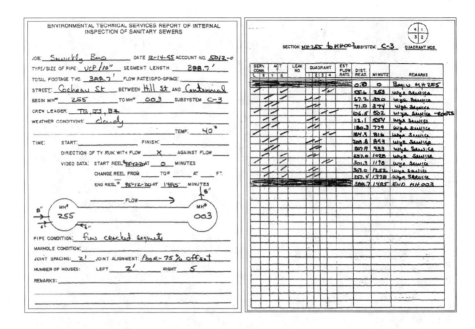

Figure 9.6   Sample TV inspection log.

Figure 9.7   Service laterals view in ArcView.

the upstream manhole to the downstream manhole). Entering the direction of measurement and the side of the line eliminates the need for the user to recalculate distances if the line was measured against the flow. The calculations are automatically done by the extension and the position of the wye (right or left) is automatically switched if necessary. Once the form is completed, the user simply clicks on "Save Record" and the wye record is automatically saved to the wye table.

When all records are entered, the database table is used to create a Shapefile of the wyes. All the wyes in the Shapefile are exactly the same length and at the exact same angle from the sewer line, creating an aesthetically appealing data set. The wyes are kept in a separate Shapefile rather than being added to the existing sewer theme in order to maintain the connectivity of the main lines. A sample lateral map created using this method is shown in Figure 9.7.

Additional examples of wastewater system mapping applications are included in Chapter 15 (Maintenance Applications).

STORMWATER SYSTEM MAPPING APPLICATIONS

A representative mapping application for stormwater systems is presented below.

Stormwater Permits

In the U.S., many cities with storm sewers and outfalls must comply with a National Pollution Discharge Elimination System (NPDES) Stormwater Permit under the U.S. Environmental Protection Agency's Municipal Separate Storm Sewer System (MS4) permit program. These permits require U.S. municipalities to develop stormwater pollution prevention programs, track the compliance schedules, and monitor stormwater system maintenance and water quality sampling.

Some cities finance the operation of a stormwater utility that pays for permit compliance by assessing a stormwater fee (drainage tax) on the properties. Initially, some U.S. municipalities chose to implement a flat tax for all property owners, based on the tax value of their property. Because only the impervious area of a property contributes to stormwater runoff, many cities used a more equitable approach that assessed the tax in proportion to the impervious area of each property. Accurate calculation and tracking of impervious areas from conventional methods (e.g., from survey or paper maps) is cumbersome and costly. GIS allows accurate and inexpensive capture of impervious surfaces from aerial photographs, orthophotos, scanned site plans, and parcel maps.

GIS can be used to develop stormwater user fees using the following steps:

1. Delineate property land covers using digital orthophotos.
2. Assign a percent imperviousness factor to each land cover. For example:
   - Roofs, 100%
   - Concrete driveway, 100%
   - Gravel driveway, 75%
   - Parking lot, 90%

3. Estimate land-cover area-weighted mean percent imperviousness of each property.
4. Determine stormwater rates based on mean percent imperviousness of a property rather than the total impervious area.

For example, the Texas legislature passed a law allowing municipalities to create drainage utilities that assessed utility fees associated with each land parcel based on the amount of stormwater runoff it contributed. The City of Irving, Texas, started using GIS and color-infrared aerial photography in 1993 to examine and map the sources of the city's stormwater runoff. The City assumed that a simple distinction between pervious surfaces such as vegetative cover and impervious surfaces such as buildings and parking lots would be sufficient to determine the stormwater runoff of each parcel with acceptable accuracy. To determine the area of each surface type with individual parcels, the tax-parcel layer was overlaid with the pervious/impervious layer. This operation created a list of each parcel's tax identification number and the associated pervious/impervious area. This table was then linked to the City's water utility billing system to calculate the amount due from each property owner and include that amount in the monthly utility bill (*American City&County*, 1994).

The City and County of Denver, Colorado, has a storm drainage utility that assesses a fee on all properties in Denver based on the amount of impervious surface present on each property, as well as on the percentage of the total area of each property that the impervious surface represents. Initially, Denver used field measurements but did not find them cost-effective. In 2002, they hired the Denver firm of Elroi Consulting, Inc., to implement an impervious surface GIS for the County. Initially, the system was implemented using ArcGIS 8.1 and a personal geodatabase that required only ArcView licenses. The system was later upgraded to work with an ArcSDE geodatabase, complete with versioning and multiuser editing. The system builds on the ArcMap component of ArcGIS, both removing unnecessary toolbars from the user interface and adding several custom toolbars. The user-friendly interface enables even the non-GIS users (e.g., field inspectors) to use the system. Impervious surfaces are digitized directly into the GIS, thus making impervious surface fee assessments defensible. The impervious surface GIS was subsequently linked to Denver's stormwater billing system for rapid database update. With this system, Denver now collects over $12 million annually in impervious surface fees (Elroi, 2002).

Oakland County, Michigan, which has more than 440,000 parcels, uses high-density spot elevation and break-line data in GIS to create triangular irregular network (TIN) surface models. These models help the county drain commissioner's office to visualize the terrain and delineate catchment areas. This approach identifies precise drainage patterns and the land that contributes runoff to a storm drain. The catchments define the drainage district boundaries that are used to assess the appropriate drainage tax in an equitable manner (Gay, 2003).

The conventional paper-based permit compliance approach can be very hectic for large cities. GIS is being used in many U.S. cities to meet the stormwater permit requirements system in a user-friendly environment. For example, SQL, Avenue, and Dialog Designer have been utilized to develop an interface between ArcView GIS and Microsoft Access for developing an NPDES compliance application (Huey, 1998). This application is helping both the management of stormwater features such

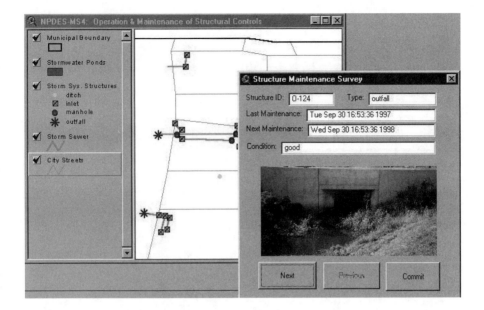

Figure 9.8   Utilization of dialogs to monitor and update structure-maintenance activities.

as sewer lines and outfalls, and preparation of complex NPDES Annual Reports. The report requires summarizing all stormwater management program goals and efforts toward meeting those goals. An example of the use of dialogs and custom-programmed icon buttons for query is shown in Figure 9.8.

CHAPTER SUMMARY

Paper maps are useful but their applications are limited. In most cases, mapmaking alone cannot justify the continued financial support for a GIS department. GIS maps and associated databases can be manipulated in a variety of ways to develop applications that automate and simplify the routine tasks. GIS maps are the starting point for developing GIS applications: no maps, no applications. In this chapter we learned about some basic applications of GIS maps. Advanced applications will be presented in subsequent chapters.

CHAPTER QUESTIONS

1. What is the meaning of "beyond mapping?" Why is it important to go beyond mapping?
2. Compile a list of ten mapping applications.
3. Using literature review or the Internet, write a 200-word example of a mapping application for a water distribution system not described in this chapter.
4. Repeat Question 3 for a wastewater collection system.
5. Repeat Question 3 for a storm drain system.

CHAPTER 10

Monitoring Applications

GIS is ideally suited to install, maintain, and query monitoring equipment such as, rain gauges, flow meters, and water quality samplers for system physical and hydraulic characterization. GIS allows display and analysis of monitoring data simply by clicking on a map of monitoring sites.

Gauging and inspection stations in the city of Los Angeles, California.

LEARNING OBJECTIVE

The learning objective of this chapter is to familiarize ourselves with GIS applications in monitoring data for effective operation and management of water, wastewater, and stormwater systems.

MAJOR TOPICS

- Rainfall monitoring
- Satellite and radar rainfall data
- Flow-monitoring applications
- Permit reporting applications
- Internet monitoring applications
- Infrastructure monitoring applications

LIST OF CHAPTER ACRONYMS

CCTV Closed Circuit Television
GML Geography Markup Language
HRAP Hydrologic Rainfall Analysis Project (a map projection system)
NCDC National Climatic Data Center (U.S.)
NEXRAD Next Generation Weather Radar
NOAA National Oceanic and Atmospheric Administration (U.S.)
NPDES National Pollution Discharge Elimination System (U.S.)
NWS National Weather Service
OGC Open GIS Consortium
SCADA Supervisory Control and Data Acquisition Systems
WSR Weather Surveillance Radar

This book focuses on the four main applications of GIS, which are mapping, monitoring, modeling, and maintenance and are referred to as the "4M applications." In this chapter we will learn about the applications of the second *M* (monitoring).

MONITORING REAL TIME RAINFALL AND STREAM-FLOW DATA IN AURORA

The city of Aurora, Colorado, is a community of approximately 247,000 people located in the southeast Denver metro area. The City's Utilities Department used ESRI's ArcInfo and ArcView software to analyze and display real time rain-gauge and stream-flow data from the City's ALERT Flood Warning System. Data were obtained by querying a remote real time data-collection database. The Utilities Department was able to access and analyze both historical and current real time rainfall and stream-flow data from an easy-to-use graphical interface. ArcView's Spatial Analyst Extension was used to develop a continuous areal rainfall surface from the point rainfall data. This allowed a better and clearer understanding of a particular storm and allowed a more complex analysis of the true impact of the storm event. Applications of this program included studies in emergency flood response, location of flood reports, routine maintenance for storm sewers, and National Pollution Discharge Elimination System (NPDES) compliance for water quality (Rindahl, 1996).

MONITORING BASICS

Monitoring of various types of data is essential to the effective management of water, wastewater, and stormwater systems. Generally, two types of data are required: (1) physical characterization data and (2) hydraulic characterization data. Physical characterization data describe the physical condition of the infrastructure, such as pipe and manhole conditions. Examples of physical characterization data sources include closed-circuit television (CCTV) inspection of pipes, manhole inspections, and smoke-testing of buildings. GIS applications for these types of data are described in Chapter 15 (Maintenance Applications).

Hydraulic characterization data describe the quantity and quality of flow through pipes and open channels as well as meteorological factors impacting the flow, such as precipitation.

Wastewater and stormwater systems typically require data on flow quantity (depth, velocity, flow rate, and volume), quality (e.g., suspended solids and bacteria), and rainfall. Figure 10.1 shows a flowmeter and weir installation in a combined sewer system overflow manhole. The flowmeter (Flo-Dar from Marsh-McBirney, Inc.) shown on the left records incoming combined sewage depth, velocity, and flow data. The weir shown on the right collects outgoing overflows. Some wastewater and stormwater hydrologic and hydraulic (H&H) models require data on additional meteorological parameters such as ambient temperature, evaporation rate, and wind speed. Water distribution systems typically require water pressure and water quality data. In many projects, monitoring tasks make up a significant portion of the scope of work and could cost 20 to 30% of the total budget. Careful installation of monitors and effective management of monitoring data is, therefore, highly desirable for the on-time and on-budget completion of monitoring projects.

GIS is ideally suited for selecting the best sites for installing various hydraulic characterization monitors. Once the monitors have been installed, GIS can be used to query the monitored data simply by clicking on a map of monitoring sites. GIS can also be used to study the spatial trends in the monitored data. GIS is especially useful in processing and integrating radar rainfall data with H&H models of sewage collection systems and watersheds. This chapter will present the methods and examples of how to use GIS for installing and maintaining the monitors, and for querying and analyzing the monitoring data.

REMOTELY SENSED RAINFALL DATA

Many watersheds, especially those smaller than 1000 km^2, do not have recording rain gauges capable of recording at hourly or subhourly intervals. Sometimes rain gauges exist, but data are found missing due to equipment malfunction. Quite often,

Figure 10.1   Flowmeter and weir installation in a manhole for monitoring incoming and outgoing flows.

rain-gauge density is not adequate to accurately capture the spatial distribution of
storm events. Such data gaps can be filled by the rainfall data provided by weather
satellites and radars, as described in the following subsections.

Satellite Rainfall Data

Direct measurement of rainfall from satellites is not feasible because satellites
cannot penetrate the cloud cover. However, improved analysis of rainfall can be
achieved by combining satellite and conventional rain-gauge data. Meteorological
satellites such as the NOAA-N series, those of the Defense Meteorological Satellite
Program, and U.S. geostationary satellites can observe the characteristics of clouds
with precipitation-producing potential and the rates of changes in cloud area and
shape. Rainfall data can now be estimated by relating these cloud characteristics to
instantaneous rainfall rates and cumulative rainfall over time. Cloud area and temperature can be used to develop a temperature-weighted cloud cover index. This
index can be transformed to estimate mean monthly runoff values. Satellite rainfall
estimating methods are especially valuable when few or no rain gauges are available
(ASCE, 1999).

Radar Rainfall Data

Weather radars provide quantitative estimates of precipitation, which can be used as input to H&H models. Radar rainfall estimates augmented with data from sparse rain-gauge networks are useful in H&H modeling. Weather radars provide real time, spatially distributed rainfall data that can be extremely valuable for flood forecasting and flood warning.

NEXRAD Rainfall Data

The U.S. National Weather Service (NWS) has a group of weather radars called the Next Generation Weather Radar (NEXRAD) system. NEXRAD comprises approximately 160 Weather Surveillance Radar–1988 Doppler (WSR-88D) sites throughout the U.S. and selected overseas locations. This system is a joint effort of the U.S. Departments of Commerce (DOC), Defense (DOD), and Transportation (DOT). The controlling agencies are the NWS, Air Weather Service (AWS), and Federal Aviation Administration (FAA), respectively. Level II data provide three meteorological base data quantities: reflectivity, mean radial velocity, and spectrum width. These quantities are processed to generate numerous meteorological analysis products known as Level III data. Level II data are recorded at all NWS and most AWS and FAA WSR-88D sites. Level III products are recorded at the 120 NWS sites. The data are sent to the National Climatic Data Center (NCDC) for archiving and dissemination.

NEXRAD Level III Data

There are a total of 24 Level III products routinely available from NCDC, including 7 graphic products in clear-air mode, 11 in precipitation mode, 5 graphic overlays, and 1 alphanumeric product. Each product includes state, county, and city background maps. Level III graphic products are available as color hard copy, grayscale hard copy, or acetate overlay copies. A brief description and possible uses of these products are given below:

- Base Reflectivity (R): A display of echo intensity measured in dBZ (decibels of Z, where Z represents the energy reflected back to the radar). This product is used to detect precipitation, evaluate storm structure, locate boundaries, and determine hail potential.
- Base Spectrum Width (SW): A measure of velocity dispersion within the radar sample volume. The primary use of this product is to estimate the turbulence associated with mesocyclones and boundaries.
- Base Velocity (V): A measure of the radial component of the wind either toward the radar (negative values) or away from the radar (positive values). Negative values are represented by cool colors (green), whereas positive values are represented by warm colors (red). This product is used to estimate wind speed and direction, locate boundaries, locate severe weather signatures, and identify suspected areas of turbulence.

- Composite Reflectivity (CR): A display of maximum reflectivity for the total volume within the range of the radar. This product is used to reveal the highest reflectivities in all echoes, examine storm structure features, and determine the intensity of storms.
- Echo Tops (ET): An image of the echo top heights color-coded in user-defined increments. This product is used for a quick estimation of the most intense convection and higher echo tops, as an aid in identification of storm structure features, and for pilot briefing purposes.
- Hail Index Overlay (HI): A product designed to locate storms that have the potential to produce hail. Hail potential is labeled as either probable (hollow green triangle) or positive (filled green triangle). Probable means the storm is probably producing hail and positive means the storm is producing hail.
- Mesocyclone Overlay (M): This product is designed to display information regarding the existence and nature of rotations associated with thunderstorms. Numerical output includes azimuth, range, and height of the mesocyclone.
- One-Hour Precipitation (OHP): A map of estimated 1-h precipitation accumulation on a 1.1 nmi × 1.1 nmi grid. This product is used to assess rainfall intensities for flash flood warnings, urban flood statements, and special weather statements.
- Severe Weather Probability Overlay (SWP): A measure of a storm's relative severity as compared with those around it. The values are directly related to the horizontal extent of vertically integrated liquid (VIL) values greater than a specified threshold. This product is used for quick identification of the strongest storms.
- Storm Structure (SS) (alphanumeric product): A table displaying information on storm attributes that include maximum reflectivity, maximum velocity at lowest elevation angle, storm overhang, mass-weighted storm volume, storm area base and top, storm position, and storm tilt.
- Storm Total Precipitation (STP): A map of estimated storm total-precipitation accumulation continuously updated since the last 1-h break over the entire scope. This product is used to locate flood potential over urban or rural areas, estimate total basin runoff, and provide rainfall data 24 h a day.
- Storm Tracking Information Overlay (STI): A product that shows a plot of the past hour's movement, current location, and forecast movement for the next hour or less for each identified thunderstorm cell. This product is used to determine reliable storm movement.
- Tornadic Vortex Signature Overlay (TVS): A product that shows an intense gate-to-gate azimuthal shear associated with tornadic-scale rotation. It is depicted by a red triangle with numerical output of location and height.
- VAD Wind Profile (VWP): A graphic display of wind barbs plotted on a height staff in 500-ft or 1000-ft increments. The current (far right) and up to ten previous plots may be displayed simultaneously. This product is an excellent tool for meteorologists in weather forecasting, severe weather, and aviation.
- Vertically Integrated Liquid (VIL): The water content of a 2.2 nmi × 2.2 nmi column of air, which is color-coded and plotted on a 124 nmi map. This product is used as an effective hail indicator to locate most significant storms and to identify areas of heavy rainfall.

NEXRAD data and various visualization and analysis software tools are available from NCDC and commercial vendors.

Estimating Rainfall Using GIS

Because rainfall is a critical component in conducting H&H analyses, the quality of rainfall data is critical for accurate system hydraulic characterization. It is often the case that the spatial distribution of rain-gauges over a collection system is too sparse to accurately estimate the rainfall over a given basin. Radar rainfall technology can be used to obtain high-resolution rainfall data over an approximately 2 km × 2 km area (called a pixel). GIS can be used to display and process the radar rainfall data.

NEXRAD Level III data from Doppler radar measurements provide spatially dense rainfall data. These data are similar to those commonly seen on weather maps. Such data do not require interpolation between point data from widely scattered rain-gauges because they provide continuous rainfall measurements throughout a watershed or sewershed (Slawecki et al., 2001).

WRS-88D radar images have a mean resolution of 4 km × 4 km. They are registered to the Hydrologic Rainfall Analysis Project (HRAP) map projection system. The radar rainfall data can be incorporated in a GIS-based distributed hydrologic model by importing it in a raster grid format. However, this may require reprojecting the rainfall grid from the HRAP coordinate system to the coordinate system being used by the model.

The accuracy of radar rainfall estimates can be improved substantially by calibrating the radars using the point rain-gauge observations. Generating NEXRAD rainfall estimates requires extensive expertise and computational resources (ASCE, 1999). Hourly NEXRAD rainfall data called hourly digital precipitation array (DPA) are available from several NWS-authorized commercial data vendors, such as WSI Corporation and Paramax Systems Corporation. Weather data vendors such as DTN Weather Services (now part of Meteorlogix, Inc.) provide continuously updated, real time, GIS-ready, georeferenced weather data in ESRI GRID (raster) and Shapefile (vector) formats and georegistered TIFF format to weather-enable the GIS applications for water and wastewater utilities. The DTN data include 5-min NEXRAD updates and storm cell type, severity, speed, and direction. Meteorlogix provides weather extensions for ESRI's ArcGIS software package. NEXRAIN Corporation (Orangevale, California) provides georeferenced polygon Shapefiles of radar pixels over a given study area. Each radar pixel is given a unique ID field, which can be used to visualize and analyze rainfall data from a separate text or database file. Shapefiles are projected into the user-specified coordinate system. NEXRAIN-2k product is extracted from a national mosaicked data set of 2-km radar data, updated every 15 min. Figure 10.2 shows a Shapefile of the NEXRAIN-2k radar pixels in Colorado State Plane Central (NAD83 ft) over Colorado Springs, Colorado, for a storm event on August 31, 2001. The top-left window shows rainfall distribution at 2 A.M.; the top-right window shows rainfall distribution at 5 P.M.; the bottom-left window shows rainfall distribution at 8 P.M.; the bottom-right window shows the attribute table; the top-center window shows the results of a pixel query; and the bottom-center insert shows a rainfall map overlaid on a street map.

WSI Corporation, a supplier of weather data in the U.S., provides the WSI PRECIP data over these pixels in 15-min time increments. These data can be obtained

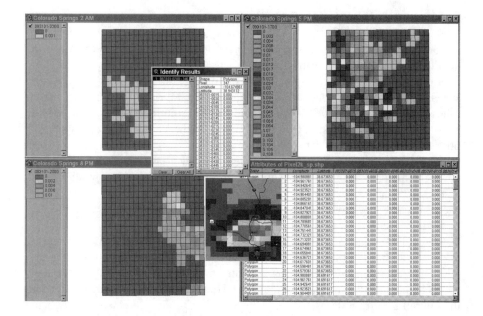

Figure 10.2   Shapefile of the NEXRAIN-2k radar pixels for Colorado Springs, CO, on August 31, 2001. Top left: rainfall at 2 A.M.; top right: rainfall at 5 P.M.; bottom left: rainfall at 8 P.M.; bottom right: attribute table; top center: pixel query window; bottom center: rainfall overlaid on street map.

through commercial vendors and adjusted to reliable ground gauges, if necessary. Hence, the radar rainfall volume adjusted through calibration with ground gauge, combined with high-resolution spatial distribution, provides modelers with a very accurate local estimate of rainfall. These estimates can be used with their flowmetering to characterize the basin of interest (Hamid and Nelson, 2001).

Radar Rainfall Application: Virtual Rain-Gauge Case Study

Precipitation varies with time and across a watershed. GIS can easily create a rainfall contour map from point rainfall values. Unfortunately, point rainfall values are generally scarce due to inadequate rain-gauge density. Many watersheds have few or no rain gauges to accurately measure the spatial rainfall distribution. In this case, a hydrologist may have to use the precipitation data from a nearby watershed. Many watersheds do not have recording-type rain gauges to measure the hourly variation of precipitation. In this case, a hydrologist may need to apply the temporal distribution of a nearby watershed. The recent availability of radar precipitation data via the Internet eliminates these problems by providing subhourly precipitation data anywhere in the watershed.

Three Rivers Wet Weather Demonstration Program (3RWWDP) is a nonprofit organization located in Pittsburgh, Pennsylvania. It was created in 1998 to help Allegheny County (Pennsylvania) municipalities address the region's aging and deteriorating sewer infrastructure to meet the requirements of the federal Clean Water Act.

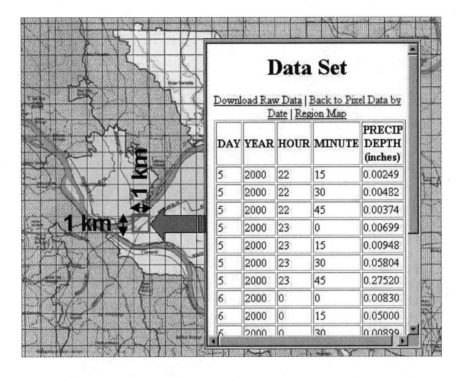

Figure 10.3   Radar-based virtual rain-gauge map and retrieved rainfall data.

Funded by federal grants and local matching funds, 3RWWDP strives to assist communities through education, financial grants, and outreach efforts. In 1999, 3RWWDP developed an Internet-based precipitation atlas for Allegheny County. The atlas consisted of clickable maps that provided 15-min precipitation data every 1 km^2 from the CALAMAR system. CALAMAR is a software and NEXRAD radar service package that integrates high-resolution radar data with precipitation measurements collected from a network of 21 rain-gauges to produce accurate, dependable precipitation maps. Each 1 km × 1 km pixel of the rainfall atlas is equivalent to a virtual rain-gauge. The users can click on any of the 2276 pixels to retrieve the 15-min rainfall data in HTML or spreadsheet (Excel) format. This is equivalent to having 2276 virtual rain gauges in the County. This powerful tool revolutionizes collection system planning, modeling, and management by providing the critical missing link in urban hydrology — accurate local precipitation data. The rainfall atlas for a watershed near the city of Pittsburgh is shown in Figure 10.3. The figure also shows the clickable pixel cells and retrieved 15-min-interval rainfall data for November 2000 for one pixel.

Vieux & Associates (Norman, Oklahoma) provides basin-averaged radar hyetographs at user-defined time increments (5, 15, 30, 60 min, or daily). Spatial resolutions smaller than 1 km with data precision at 256 levels can be achieved. A real time archive provides access to the radar rainfall estimates. Data are provided in

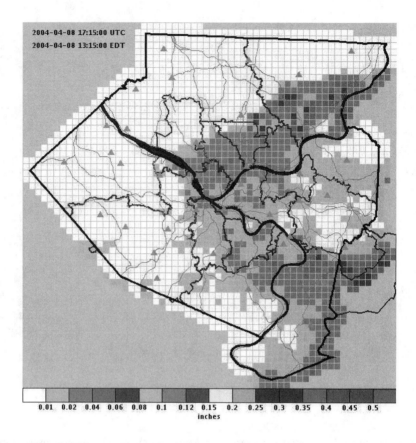

Figure 10.4   RainVieux radar rainfall data for Pittsburgh.

GIS-ready (Shapefile) and hydrologic-model-ready formats. Figure 10.4 shows a sample image from the RainVieux Web site tailored for 3RWDDP. It provides historic and real time radar rainfall measurements that can be queried from a Web site. Radar and rain-gauge measurements are used for cost-effective design of sewer rehabilitation projects in the Pittsburgh Metro area.

Mandated by an EPA consent decree, Miami-Dade Water and Sewer Department (MDWASD), which operates a 317-MGD wastewater treatment plant, implemented a hydraulic computer model to reduce sanitary sewer overflows (SSOs). In order to provide rainfall input to the computer model, MDWASD implemented a virtual rain-gauge system similar to 3RWWDP. A grid-system GIS coverage was created, which yielded 256 virtual rain gauges of 2 km × 2 km size each. The rainfall data are captured every 15 min and downloaded to a database. The grid is joined with the rainfall database, which allows quantification of impacts of individual rainfall events on specific regions within the system. This approach improves the accuracy of modeled flow estimates that are used to design appropriate SSO control measures (Day, 1998).

FLOW-MONITORING APPLICATIONS

GIS is ideally suited to selecting the best sites for installing flowmeters, water quality samplers, and rain gauges. The following case study illustrates a GIS application for selecting appropriate flow-monitoring sites.

The city of Huntington, WV, has a combined sewer system with 23 combined sewer overflow (CSO) discharge points permitted under the NPDES program. The sewer system is operated by the Huntington Sanitary Board (HSB). One of the CSO requirements specified in the NPDES permit was to monitor each CSO event for cause, frequency, duration, quantity, and quality of flow. As a first alternative to complying with this requirement, the feasibility of monitoring all the 23 CSO locations was studied in 1993. This option required purchase, installation, monitoring, and maintenance of flow monitors, water quality samplers, and rain gauges for each CSO site. This option was ruled out because of its excessive cost, estimated at over a million dollars, and other problems such as adverse environmental conditions in the CSO structures, difficult access issues, and new federal requirements for confined space entry. The second option consisted of a combination of monitoring and modeling. In this option, a representative subset of CSOs can be monitored temporarily to collect sufficient calibration data and develop a calibrated model for each monitored CSO area. Calibrated model parameters could subsequently be applied to the models of unmonitored CSO areas. CSO models can eventually be used to predict the quantity and quality of CSO discharges from observed rainfall data. This option, costing approximately one third of the first option, was preferred by both the HSB and the EPA (Region V), and was selected for implementation.

In 1994, ArcView was used to select a representative subset of the 23 CSO sites to be monitored. Selecting the representative sites is a complex process. The objective is to maximize the extent and value of the data for use in model calibration and provide a uniform distribution of flowmeters, samplers, and rain gauges throughout the system so that all parts of the service area are equally covered. Other factors such as providing coverage to all land uses, known problem CSO areas, and all types of diversion chambers, as well as access for installation and maintenance of monitoring equipment should be thoroughly studied. Using these selection criteria, a total of six sites were finally selected for sampling and monitoring. Figure 10.5 presents an ArcView plot showing the monitored vs. unmonitored CSOs and CSO areas. The map indicates that by monitoring only 25% of the total number of CSOs, approximately 70% of the study area was covered.

SCADA Integration

Real time monitoring systems called supervisory control and data acquisition systems (SCADA) monitor field operations or remote sites for water and wastewater utilities. Also referred to as telemetry, SCADA is a sophisticated means of monitoring, reporting, and controlling the operation of modern water and wastewater systems and treatment plants.

GIS stores the system design and operation data. A link between SCADA and the GIS database can be established to use the GIS for real time decision making. For

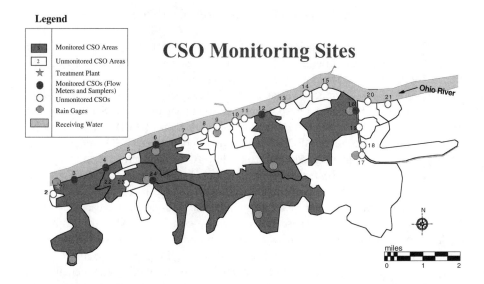

Figure 10.5 Flow-monitoring sites selected by using GIS.

example, real time or near-real time rainfall data can be used in a raster GIS to estimate the subbasin mean aerial rainfall. Real time subbasin rainfall data can be used to estimate real time surface runoff, for flood forecasting, or to estimate real time sewer flows to optimize storage and pumping operations for wet-weather overflow control.

Pidpa is a European company that produces and distributes drinking water to more than 1.1 million people throughout the Antwerp province in Belgium (Flanders). Pidpa used ESRI's ArcIMS technology to create a Web site named GeoLink as the central application to retrieve GIS data linked to corporate data. In the GIS viewer of GeoLink, water storage tanks have a hyperlink that opens the SCADA Web interface. The interface shows the real time water level or 24-hour history of the water level for the selected tank (Horemans and Reynaert, 2003).

SCADA Connect from Haestad Methods (acquired by Bentley Systems in August 2004) is a commercially available software product for dynamically connecting a wide range of SCADA systems with Haestad Methods' WaterCAD, WaterGEMS, and Sewer-CAD software programs. Using SCADA Connect, information can be automatically queried from the SCADA system and passed to the WaterGEMS database, eliminating manual processes. For example, tank levels and pump and valve settings can be updated in WaterGEMS based on the SCADA database, so that the next time WaterGEMS is run, it will reflect the current field conditions. SCADA Connect has been deployed for the City of Bethlehem, Pennsylvania's Bristol Babcock SCADA system and City of Casselberry, Florida's Citect SCADA system (Haestad Methods, 2004).

NPDES-PERMIT REPORTING APPLICATIONS

The well-known GIS magazine *Geospatial Solutions* holds an "Applications Contest" each year (Geospatial Solutions, 2001a). The entries are judged based on interest, technology, and importance. The first place winner for the 2001 contest was a stormwater pollutant load modeling application for Los Angeles, California.

As a condition of its NPDES Municipal Stormwater permit, the Los Angeles County Department of Public Works is required to calculate and report pollutant load for more than 256 contaminants. The County maintains a network of stormwater monitoring devices to determine and report pollutant load for more than 256 contaminants. But, in addition to providing data about monitoring system locations and readings, the County also has to estimate the pollutant load to the Pacific Ocean from unmonitored drainage areas. The manual calculation of this information took engineers 250 working hours for an area one quarter of the required study area. The County, therefore, turned to GIS to create an automated model for calculating pollutant load. Relying on an existing public-works GIS, a collaborative team added several data layers to the system to integrate information about rainfall depth, drainage area, land use, imperviousness, and mean countywide concentrations for all the 256 pollutants based on monitoring-station data.

With that information, the team developed a modeling algorithm and refined its operations (the weighting and correlations of different variables) by running models for areas with known pollutant loads and comparing the results with monitoring-station readings. After several iterations, the calibrated model estimated pollutant loads with 95% accuracy. The team then developed an integrated menu-driven interface that walks the user through the process of creating and setting parameters and generating reports of pollutant load.

After specifying that he or she wants to perform a new load run, the user identifies a study area by interactively defining a polygon on the screen. The program then processes information in the background, extracting land-use information for the study area stored in a land-use/imperviousness table. Next, the user selects a rainfall year and event, which specifies the tabular database to use for the modeling, to identify the inches of rainfall at each monitoring station. The application then generates a rainfall grid using the tabular data for the specified rainfall year and event to calculate the pollutant load. The final parameter the user specifies is the pollutants. A dialog box enables interactive selection of the 256 pollutants (the program automatically selects a default set of 64 pollutants). Once all the parameters are defined, the user can calculate the pollutant load and generate maps and reports for print using built-in templates. The program also automatically saves the selected parameters and model results in a database table so that others can retrieve the load run without having to reset all the parameters.

The project team used ESRI's ArcView and Spatial Analyst for GIS analysis and an automated report generation process from Crystal Decisions' Crystal Reports.

MONITORING VIA INTERNET

Integration of GIS and Internet technologies is allowing users to click on a GIS map and see the real time data from any type of Web-connected sensor, from stream gauges to space and airborne Earth-imaging devices. Referred to as the Sensor Web, the sharing of sensor data via the Internet represents a revolution in the discovery and use of live monitoring data. For example, data from the SAIC (a research and engineering company in San Diego, California) Weather Monitoring System is supplied from an Open GIS Consortium (OGC) Web Collection Service to a mapping client, overlaying geographic information coming from an OGC Web Map Service. The weather data are encoded in OGC Observation and Measurements Language, which is an extension of Geography Markup Language (GML) (Reichardt, 2004).

MONITORING THE INFRASTRUCTURE

From 1998 to 2000, the City of Toronto (Ontario, Canada) collected water pipe and soil samples to investigate the condition of its water distribution network. The City hoped that this effort would explain the causes of water main deterioration and guide the City in its water main rehabilitation efforts. The GIS data used in this project consisted of three layers: (1) Geocoded street center lines, (2) sample locations, and (3) surface soils. Eighty-four pipe and associated soil samples (resistivity, pH, and sulfide content) were collected, tested, and plotted on the street and soil maps using the GIS' address-matching capability. Using the GIS, a spatial analysis of the data collected was performed to identify any trends between the data samples and their locations. For example, water mains located in the clay-silt deposit near lakeshores were suspected to be at greater risk for corrosion, and there was no evident relationship between pH and soil formation (Doyle and Grabinsky, 2003).

Address geocoding is a GIS process that locates street addresses as points on a map, using a reference layer (usually a geocoded street layer). *Address matching* is the comparison of a street address to the address ranges in the reference layer. When an address matches the address range of a street segment, an interpolation is performed to locate and assign real-world coordinates to the address.

A Massachusetts water utility, puzzled by abnormally high lead and copper levels in its distribution system, found that interpretations of water quality problems based on aggregate monitoring data can be very misleading unless analysis is performed at the appropriate scale. Incorporation of system and monitoring-site physical characteristics data as well as the monitoring results into a GIS could have saved them considerable investigatory effort (Schock and Clement, 1995).

USEFUL WEB SITES

Crystal Decisions	www.crystaldecisions.net
Geospatial Solutions	www.geospatial-online.com
Meteorlogix	www.meteorlogix.com/GIS
NCDC Radar Products	www.ncdc.noaa.gov/ol/radar/radarproducts.html
NEXRAIN Corporation	www.nexrain.com
Vieux & Associates	www.vieuxinc.com
WSI Corporation	www.wsi.com

CHAPTER SUMMARY

This chapter presented the methods and examples of GIS applications in installing and maintaining various types of hydraulic characterization monitors (e.g., flowmeters and rain gauges) and querying and analyzing the monitoring data. The examples and case studies presented in this paper indicate that GIS is ideally suited for selecting the best sites for installing the monitors. Once the monitors have been installed, GIS allows user-friendly query of the monitoring data simply by clicking on a map of monitoring sites. GIS facilitates studying the spatial trends in the monitoring data. GIS is especially useful in processing and integrating radar rainfall data with H&H models of sewage collection systems and watersheds.

CHAPTER QUESTIONS

1. What are radar rainfall data? How are they used in H&H modeling? How does GIS help in processing them for H&H models?
2. How does GIS help in selecting the feasible flow-monitoring sites?
3. How does GIS allow a trend analysis of spatial patterns in monitoring data?

CHAPTER 11

Modeling Applications

GIS and H&H model integration allows the users to be more productive. Users can devote more time to solving the problems and less time on the mechanical tasks of inputting data and interpreting reams of model output. More than just text outputs, models become automated system-evaluation tools. GIS integration saves time and money.

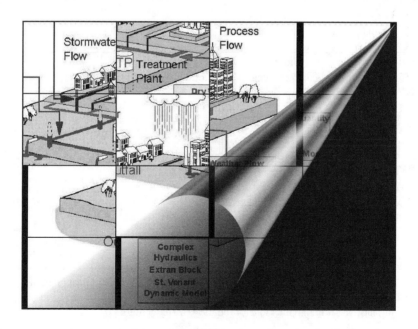

GIS integration is ideally suited to solve the computer modeling puzzle.

LEARNING OBJECTIVES

The learning objectives of this chapter are to classify the methods of linking water, wastewater, and stormwater system computer models with GIS, and to understand the differences between various linkage methods.

MAJOR TOPICS

- GIS applications in H&H modeling
- Modeling application methods
  - Interchange method
  - Interface method
  - Integration method
- Examples and case studies of the preceding methods

LIST OF CHAPTER ACRONYMS

AML Arc Macro Language
BASINS Better Assessment Science Integrating Point and Nonpoint Sources
COE Corps of Engineers (U.S. Army)
DEM Digital Elevation Model
GUI Graphical User Interface
H&H Hydrologic and Hydraulic
HEC Hydrologic Engineering Center (U.S. Army Corps of Engineers)
HSG Hydrologic Soil Group
HSPF Hydrologic Simulation Program — FORTRAN
NPS Nonpoint Source
NRCS Natural Resources Conservation Service (U.S.)
SCS Soil Conservation Service (U.S.) (Now NRCS)
SWAT Soil and Water Assessment Tool
SWMM Storm Water Management Model
TMDL Total Maximum Daily Load
TR-20 Technical Release 20
VBA Visual Basic for Applications
WMS Watershed Modeling System

This book focuses on the four main applications of GIS, which are mapping, monitoring, modeling, and maintenance and are referred to as the "4M applications." In this chapter we will learn about the applications of the third *M* (modeling).

TEMPORAL-SPATIAL MODELING IN WESTCHESTER COUNTY

GIS software   ArcInfo
Modeling software   Rational method and kinematic wave model

GIS data	Land use (from low-altitude infrared color photography), USGS DEM (1:24,000 scale), NRCS soils (1:12,000 and 1:24,000 scales), streams, and watersheds
Study area	0.14 mi^2 (0.36 km^2) Malcolm Brook watershed, Westchester County, New York
Organization	New York City Department of Environmental Protection

Lateral flow is an explicit component of the kinematic wave routing equation, yet some kinematic wave models such as HEC-HMS do not include lateral flow. The lateral flow component can be accounted for by using GIS. Until recently, it was considered impossible to conduct time-varying computations within a GIS. In this case study, lateral flows were derived from GIS output for each segment of the stream and at each time interval of the rain storm and were routed using the kinematic wave routing method. Rather than a raster GIS that uses a constant cell size, a vector GIS was used to define hydrologic response units that divide the stream channel into segments that vary in size according to the combined characteristics for land use, slope, and soil type. This approach permitted vector-based spatially distributed modeling of stream flow during storm events. GIS was used to map and visualize contributing areas around a stream channel. During each calculation of the discharge, a graphical image of the watershed and contributing areas was captured as a Graphics Interchange Format (GIF) image. A series of these images were displayed in sequence to produce a continuous animation (Gorokhovich et al., 2000).

H&H MODELING

No one said mathematical modeling would be easy, but the preparation of input data and interpretation of output results required by the ever-changing complexities of sophisticated hydrologic and hydraulic (H&H) models have been mind boggling. The good news is that model building and interpretation of results are now easier than ever before thanks to recent advances in computer hardware and software. The assembly of model input data by traditional, manual map measurements is just too time consuming and difficult to justify, now that personal computers are affordable and powerful. It is much easier to see a color-coded map of surcharged sewers to pinpoint the areas of flooding and surcharging than to digest reams of computer output, especially for nonmodelers. Nonmodelers are those people who are not expert modelers but who need to know the modeling results, such as clients of consulting firms, project managers, elected officials, and regulatory agencies. After all, as advocated in Chapter 1 (GIS Applications), GIS technology is an effective means of bridging the gap between information and its recipients.

Rapidly developing computer technology has continued to improve modeling methods for water, wastewater, and stormwater systems. GIS applications provide an accurate and manageable way of estimating model input parameters such as node demands, sewage flows, and runoff curve numbers. GIS-based modeling, as a side benefit, also provides an updated database that can be used for nonmodeling activities such as planning and facilities management.

There are two fundamental requirements in most H&H modeling projects: a suitable model and input data for the model. It is often difficult to select a modeling approach because of trade-offs between models and data. For instance, a detailed model requires a large amount of input data that is often too difficult to obtain or is too expensive. On the other hand, a simple model that requires little data may not provide a detailed insight into the problem at hand. Modelers must, therefore, use an optimal combination of model complexity (or simplicity) and available data. The recent growth in computational hydraulics has made it increasingly difficult for practitioners to choose the most effective computational tool from among a variety of very simple to very complex H&H models. Fortunately, thanks to the advances in GIS applications, creation of input data sets is easier than ever before.

This chapter serves as a guide to help professionals select the most appropriate GIS applications for their modeling needs. It presents an overview of the GIS and computer modeling integration approaches and software. The chapter also shows how to estimate the physical input parameters of H&H models using GIS. The chapter largely uses watershed hydrologic modeling examples to explain modeling integration concepts. However, the integration methods presented here are equally applicable to modeling of water and wastewater systems. Water system modeling applications and examples are presented in Chapter 12 (Water Models). Sewer system modeling applications and examples are presented separately in Chapter 13 (Sewer Models).

APPLICATION METHODS

There are two types of hydrologic models: lumped and distributed. Lumped-parameter models lump the input parameters of a study area over polygons and use vector GIS applications. Distributed models distribute the input parameters of a study area over grid cells and use raster GIS applications. Application of GIS technology to H&H modeling requires careful planning and extensive data manipulation work. In general, the following three major steps are required:

1. Development of spatial database
2. Extraction of model layers
3. Linkage to computer models

H&H models, databases, and GIS applications are critical in efficiently and effectively completing large modeling studies. The models and GIS can be linked to other databases for data sharing purposes. For example, data can be imported from other sources like AutoCAD, edited and modified within the application, and exported or linked to other databases. Database files can be saved in .dbf format and linked or imported into Microsoft Access for further data manipulation. For example, ArcView GIS provides Open Database Connectivity (ODBC) features that can be used to link ArcView tables with other tables and queries in Microsoft Access or other database programs, without actually going through any import and export exercises. Such procedures eliminate the data redundancies and user errors typically associated

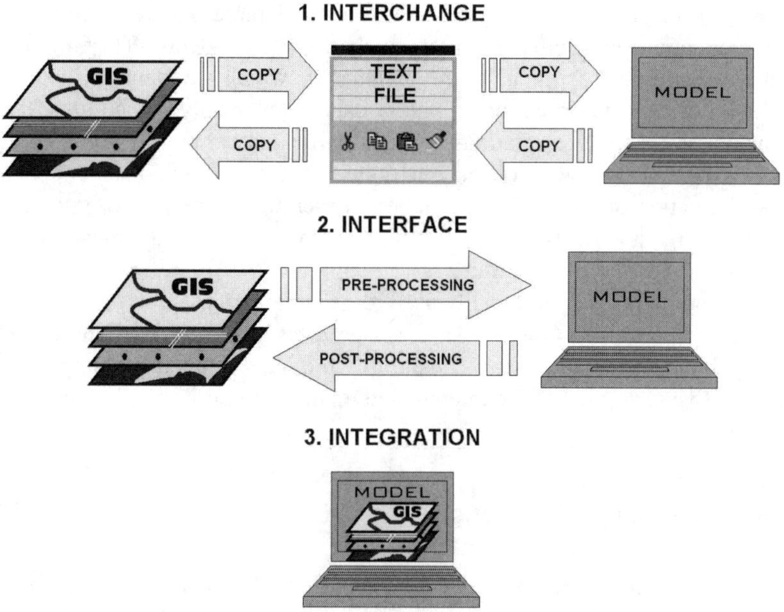

1. INTERCHANGE

2. INTERFACE

3. INTEGRATION

Figure 11.1 Three methods of GIS applications in H&H modeling.

with such cumbersome tasks. This has proven to be beneficial and time-saving during alternative analysis of systems where a large number of scenarios are modeled and reviewed using the same data connectivity and templates for maps (Hamid and Nelson, 2001).

A useful taxonomy to define the different ways a GIS can be linked to computer models was developed by Shamsi (1998, 1999). The three methods of GIS linkage defined by Shamsi are:

1. Interchange method
2. Interface method
3. Integration method

Figure 11.1 shows the differences between the three methods. Each method is discussed in the following text with the help of examples and case studies. As described in Chapter 1 (GIS Applications), advanced applications such as the capability to automatically create a model input file always require writing a computer program using a scripting language like Visual Basic for Applications (VBA).

INTERCHANGE METHOD

Preprocessing is defined as the transfer of data from the GIS to the model. Postprocessing is defined as the transfer of data from the model to the GIS. The interchange method employs a batch-process approach to interchange (transfer) data between a GIS and a computer model. In this method, there is no direct link between the GIS and the model. Both are run separately and independently. The GIS database is preprocessed to extract model input parameters, which are manually copied into a model input file. Similarly, model output data are manually copied in a GIS as a new spatial layer for presentation-mapping purposes. Script programming is not necessary for this method, but it may be done to automate some manual operations such as derivation of runoff curve numbers (described in the following text). This is the easiest method of using GIS in computer models and is the most commonly utilized method at the present time.

In this method GIS is essentially used to generate model input files and display model output data. Any GIS software can be used in the interchange method. A GIS with both vector and raster capabilities provides more interchange options. Representative examples of the interchange method are described in the following subsections.

Subbasin Parameter Estimation

Most H&H models need input data for subbasin parameters such as area, overland flow width, and slope. If subbasins are represented as polygons, GIS can automatically calculate the area as a layer attribute. Overland flow width can be measured interactively in any direction (e.g., along a stream or a sewer) by using the measurement tools available in most GIS packages. For example, ArcView 3.x provides a Measure Tool

that can be used for on-screen measurement of distance. It also provides a Drawing Tool

that can be used for on-screen measurement of polygon area. Subbasin slope can be estimated from digital elevation models (DEMs) as described in Chapter 4 (DEM Applications).

GIS estimation of subbasin runoff curve number, a critical input parameter in many rainfall-runoff models, is perhaps the best example of the interchange method.

Runoff Curve Number Estimation

Runoff curve numbers are a set of standard empirical curves that are used to estimate stormwater runoff. Accurate estimation of curve number values is critical to accurate runoff modeling because the quantity of runoff is very sensitive to runoff curve number values. GIS estimation of subbasin runoff curve number, a critical input parameter in many rainfall-runoff models, is perhaps the best example of the interchange method. The estimation approach is based on the land use, hydrologic soil group (HSG), and runoff curve number relationships developed by U.S. Natural Resources Conservation Service (NRCS), formerly known as the Soil Conservation Service (SCS). These relationships are available in the form of runoff curve number tables (U.S. Department of Agriculture, 1986). These tables provide runoff curve numbers for a large number of land uses and four hydrologic soil groups: A, B, C, and D. They also list average percent imperviousness values for various land-use classes. Table 11.1 shows selected data from the NRCS tables for percent imperviousness and runoff curve number for typical land-use classes. If existing land-use classifications are not consistent with SCS taxonomy, they may be replaced with an equivalent SCS land-use class shown in Table 11.2.

Table 11.1 SCS Runoff Curve Numbers

Land-Use Class	Average Percent Imperviousness	Runoff Curve Number for Hydrologic Soil Group			
		A	B	C	D
Fully Developed Urban Areas (Vegetation Established)					
Open space (lawns, parks, golf courses, etc.)	—	49	69	79	84
Paved parking lots, roofs, driveways, etc.	—	98	98	98	98
Paved streets and roads	—	98	98	98	98
Dirt roads	—	72	82	87	89
Commercial and business	85	89	92	94	95
Industrial	72	81	88	91	93
Residential: lots 1/8 acres or less (town houses)	65	77	85	90	92
Residential: lots 1/4 acres	38	61	75	83	87
Residential: lots 1/2 acres	25	54	70	80	85
Residential: lots 1 acres	20	51	68	79	84
Residential: lots 2 acres	12	46	65	77	82
Developing Urban Areas					
Newly graded areas		77	86	91	94
Idle lands		77	86	91	94
Nonurban Areas					
Row crops (straight row, poor condition)	—	72	81	88	91
Row crops (straight row, good condition)	—	67	78	85	89
Meadow (continuous grass, no grazing)	—	30	58	71	78
Brush (fair condition)	—	35	56	70	77
Woods (fair condition)	—	36	60	73	79
Farmsteads (buildings, driveways, and lots)	—	59	74	82	86

Table 11.2 Equivalent SCS Land-Use Classes

Land-Use Class	SCS Equivalent Land-Use Class	Average Percent Imperviousness	Runoff Curve Number for Hydrologic Soil Group			
			A	B	C	D
Parks, cemeteries, ball fields, etc.	Open space	0	39	61	74	80
High-density residential	Average lot 506–1012 m² (0.125–0.25 acres)	51	69	80	87	90
Medium-density residential	Average lot 1349–2024 m² (1/3–1/2 acres)	28	56	71	81	86
Low-density residential	Average lot 4047–8094 m² (1–2 acres)	16	49	66	78	83
Schools	Average lot 1349–2024 m² (1/3–1/2 acres)	28	56	71	81	86
Agricultural	Row crops average	0	64	75	83	87
Disturbed	Newly graded areas	35	77	86	91	94

A vector layer for subbasin runoff curve numbers is created by overlaying the layers for subbasins, soils, and land use to delineate the runoff curve number polygons. Each resulting polygon should have at least three attributes: subbasin ID, land use, and HSG. The NRCS landuse-HSG-curve number matrix (Table 11.1 or Table 11.2) can now be used to assign runoff curve numbers to each polygon according to its land use and HSG. Polygon runoff curve number values are then area-weighted to compute the mean runoff curve number for each subbasin. These subbasin-runoff curve numbers can now be entered into the model input file. The runoff curve number estimation technique is shown in Figure 11.2 to Figure 11.5. Figure 11.2 shows the layers for subbasins and land use. Figure 11.3 shows the layers for subbasins and HSGs. Figure 11.4 shows the computed runoff curve number polygons. Figure 11.5 shows the average runoff curve numbers for the subbasins.

Some H&H models also need an input for the subbasin percent imperviousness, which can be estimated in GIS using the NRCS runoff curve number tables. A layer for the subbasin percent imperviousness can be created by overlaying the layers for subbasins and land use to delineate the percent imperviousness polygons. The land-use percent imperviousness matrix can then be used to assign percent impervious values to the polygons. Finally, polygon percent imperviousness values are area-weighted to compute the mean percent imperviousness value for each subbasin.

Water Quality Modeling Data Estimation

Some urban-runoff quality models such as the TRANSPORT Block of SWMM need curb length as a model input parameter, which can be estimated from a GIS layer of road or street centerlines. Subbasin curb length can be estimated by performing

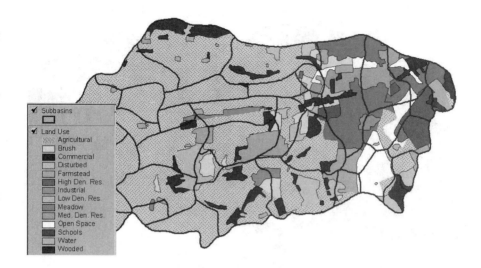

Figure 11.2   Layers for subbasins and land use for runoff curve number estimation.

an overlay analysis of street and subbasin layers. For example, in ArcView 3.x, curb length can be estimated by performing a *theme-on-theme* selection using roads as the target theme and subbasins as the selector theme. Alternatively, for large areas such tasks can be handled more efficiently in ArcView's Network Analyst extension, which has been designed for transportation network applications. Network Analyst is an optional extension that must be purchased separately.

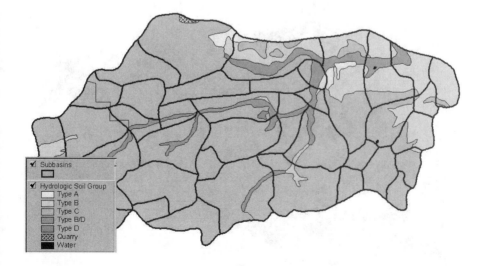

Figure 11.3   Layers for subbasins and HSGs for runoff curve number estimation.

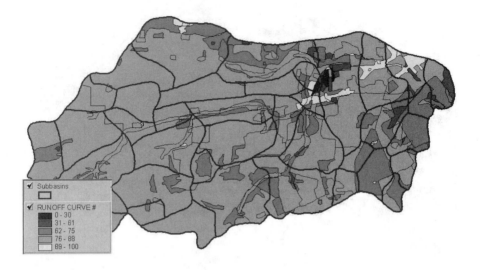

Figure 11.4   Computed runoff curve number polygons.

Demographic Data Estimation

Demographic data can be used for the following modeling tasks:

1. Estimation of quantity and quality sanitary sewage flow
2. Estimation of present and potential future development
3. Creation of auxiliary layers for visualization and presentation purposes

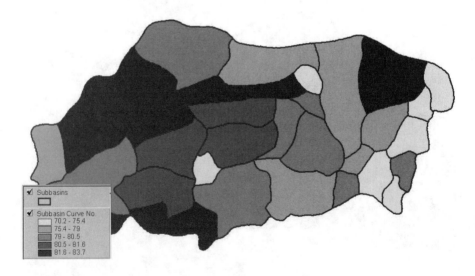

Figure 11.5   Subbasin runoff curve number map.

The first task is required when modeling both combined and sanitary sewer areas in urban sewersheds. For example, SWMM's TRANSPORT Block (Version 4.x) needs inputs for some or all of the following demographic parameters:

1. Dwelling units
2. Persons per dwelling unit
3. Market value of average dwelling unit
4. Average family income

A GIS can compute such demographic parameters from the U.S. Census Bureau's TIGER data (Shamsi, 2002). To estimate subbasin demographic parameters, census-block attribute data files from TIGER files are linked to the polygon topology of the census blocks. Only those attributes pertinent to the model are retained within the working layers. The population (or other demographic parameters) within each subbasin is estimated by determining the census blocks or their portions making up each subbasin. Block data are area-weighted to estimate the mean or total subbasin values of demographic parameters. For example, this can be done in ArcView 3.x from *polygon-on-polygon* selection using census blocks as the target theme and subbasins as the selector theme. Figure 11.6 shows an ArcView map of subbasins, census blocks, and census block groups for a SWMM-based sewer system model.

Figure 11.6 Sample census block and block group layers for estimating subbasin population.

Land-Use Data Estimation

Dry- and wet-weather flows from sewersheds depend on land use. Most rainfall-runoff models, therefore, need land-use information. In some models such as SWMM's TRANSPORT Block, land-use type is input directly. In others, such as Penn State Runoff Model (PSRM) and SWMM's RUNOFF Block, land-use type is needed to derive certain model parameters such as percent imperviousness. TRANSPORT needs input data for the the land-use designation of modeled subbasins. The subbasins should be classified as having one of the five land-use classes:

1. Single-family residential
2. Multifamily residential
3. Commercial
4. Industrial
5. Open space

Most land-use maps and layers have more than these five land-use classes. However, this is not a problem in GIS because the existing classes can be easily consolidated into the five land-use classes mentioned above. Figure 11.7 shows an ArcView 3.x screenshot for subbasin and land-use layers for a SWMM-based sewer system model. The land-use layer was prepared from classification of 30-m-resolution Landsat Thematic Mapper (TM) satellite imagery of the study area. Initially, the layer contained 19 land-use classes. Subsequently, the initial land-use classes were reclassified into the five land-use types required for TRANSPORT input. In order to refine the land-use classification, the layer was combined with the census-block layer

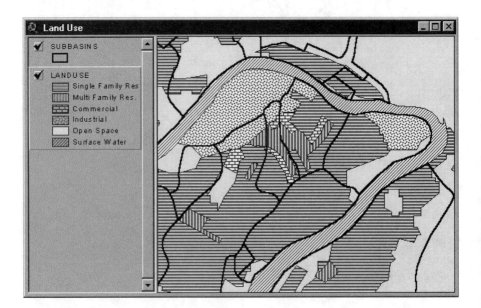

Figure 11.7   Estimating land use using interchange method.

to determine where population densities indicated residential districts. This was done in two ways. First, all nonresidential classed areas were examined to determine if population densities in the local census blocks were below or above a threshold of ten persons per acre. This threshold was chosen as a conservative breakpoint that would allow a reasonable portion of the classification to be correctly reclassified. The areas corresponding to any blocks with densities above the threshold were changed to the residential class. Second, the census blocks of all residential areas were examined to determine in which blocks single-family or multi-family residential units were the majority. On this basis polygons of the residential class were divided into single-family residential or multi-family residential classes.

INTERFACE METHOD

The interface method provides a direct link to transfer information between the GIS and a model. The interface method consists of at least the following two components: (1) a preprocessor that analyzes and exports the GIS data to create model input files and (2) a postprocessor that imports the model output and displays it as a GIS theme. Processing of the model input and output files requires computer programming using the GIS software's scripting language (such as Avenue or VBA). User routines and scripts are incorporated in the GIS by calling them through new pull-down menus and icons.

The interface method basically automates the data interchange method. The automation is accomplished by adding model-specific menus or buttons to the GIS software. The model is executed independently of the GIS; however, the input file is created in the GIS. The main difference between the interchange and interface methods is the automatic creation of the model input file. In the data interchange method, the user finds a portion of a file and copies it. In the interface method, an interface automates this process, so that the pre- and postprocessor can find the appropriate portion of the file automatically.

Learning with examples is an effective method, so let us begin with an interface example. Let us assume that we want to create an input file for a hydrologic model by exporting GIS data to an ASCII text file. Our watershed GIS has the following data:

- Subbasins with attributes: ID, area, overland flow width, slope, and percent imperviousness.
- Reaches (streams) with attributes: ID, upstream subbasin, downstream subbasin, type (natural stream, concrete channel, or concrete pipe), condition (best, good, fair, or bad), depth, width, length, upstream elevation, and downstream elevation.

Our hydrologic model requires a text file with the following input parameters:

- Subbasin data: ID, area, overland flow width, slope, and percent imperviousness.
- Conduit data: ID, upstream subbasin, downstream subbasin, type, depth, width, length, slope, and roughness coefficient.

Table 11.3 Look-up Table for Reach Roughness Coefficient

Reach Type	Reach Condition			
	Best	Good	Fair	Bad
Natural stream	0.025	0.028	0.030	0.033
Concrete channel	0.012	0.014	0.016	0.018
Concrete pipe	0.012	0.013	0.015	0.016

Note that our GIS does not have reach attributes for slope and roughnesss coefficient, but they can be estimated in the GIS from the given attributes. For example, slope can be estimated from the upstream and downstream elevations and length using the following formula:

$$slope = \frac{upstream\ elevation - downstream\ elevation}{length}$$

The roughness coefficient can be estimated using a lookup table like Table 11.3. For example, if the reach type is natural and condition is good, then GIS can estimate the roughness coefficient as 0.028. Now that all the required subbasin and conduit parameters are available in the GIS database, the required fields of subbasin and reach theme tables can be saved as text files.

Let us further assume that we want to create model output maps in a GIS by importing model output files into it. This task can be easily accomplished by linking the model output file to a GIS layer using a common attribute (subbasin and reach IDs). For example, a thematic map of stream flow can be created by linking the modeled stream flow to the streams layer. First of all, a database table of the relevant results (stream flow) should be created from the ASCII output file. The output database file can then be joined to the streams attribute table. The linked output results can now be queried or classified with a legend to make a thematic map.

Typical modeling steps involved in using the interface method are listed below:

1. Start your GIS.
2. Export model input data from GIS to a text or database file (preprocessing).
3. Exit the GIS.
4. Start your model.
5. Import GIS data (saved in Step 2) in the model.
6. Run the model.
7. Exit the model.
8. Reenter the GIS and import the model output (postprocessing).

It can be seen from these steps that the interface method does not run the model inside the GIS. The model must be run outside the GIS by the user. Some interface examples are given in the following text. Additional examples are presented in Chapter 12 and Chapter 13.

HEC-GEO Interface

The Hydrologic Engineering Center (HEC) is an office of the U.S. Army Corps of Engineers (COE) established to support the nation's hydrologic engineering and water resources planning and management needs. To accomplish this goal, HEC develops state-of-the-art, comprehensive computer programs that are also available to the public. HEC-1 and HEC-2 are COE's legacy DOS programs for hydrologic and hydraulic modeling, respectively. Recently HEC-1 and HEC-2 have been replaced with Windows programs called the Hydrologic Modeling System (HEC-HMS) and River Analysis System (HEC-RAS), respectively. HEC Geo-HMS and HEC Geo-RAS have been developed as geospatial hydrology toolkits for HEC-HMS and HEC-RAS users, respectively. They allow users to expediently create hydrologic input data for HEC-HMS and HEC-RAS models. Free downloads of these programs are available from the HEC software Web site.

HEC-GeoHMS

HEC-GeoHMS is an ArcView 3.x GIS extension specifically designed to process geospatial data for use with HEC-HMS. It allows users to visualize spatial information, delineate watersheds and streams, extract physical watershed and stream characteristics, perform spatial analyses, and create HEC-HMS model input files. HEC-GeoHMS uses ArcView's Spatial Analyst Extension to develop a number of hydrologic modeling inputs. Analyzing digital terrain information, HEC-GeoHMS transforms drainage paths and watershed boundaries into a hydrologic data structure that represents watershed's response to precipitation. In addition to the hydrologic data structure, capabilities include the development of grid-based data for linear quasi-distributed runoff transformation (ModClark), the HEC-HMS basin model, and the background map file.

HEC-GeoHMS provides an integrated work environment with data management and customized toolkit capabilities, which includes a graphical user interface (GUI) with menus, tools, and buttons. The program features terrain-preprocessing capabilities in both interactive and batch modes. Additional interactive capabilities allow users to construct a hydrologic schematic of the watershed at stream gauges, hydraulic structures, and other control points. The hydrologic results from HEC-GeoHMS are then imported by HEC-HMS, where simulation is performed. HEC-GeoHMS works with Windows 95/98/NT/2000 operating systems.

HEC-GeoRAS

HEC-GeoRAS (formerly named AV/RAS) for ArcView is an ArcView 3.x GIS extension specifically designed to process geospatial data for use with HEC-RAS. The extension allows users to create an HEC-RAS import file containing geometric attribute data from an existing digital terrain model (DTM) and complementary data sets. HEC-GeoRAS automates the extraction of spatial parameters for HEC-RAS input, primarily the 3D stream network and the 3D cross-section definition. Results exported from HEC-RAS may also be processed in HEC-GeoRAS. ArcView 3D

Analyst extension is required to use HEC-GeoRAS. Spatial Analyst is recommended. An ArcInfo 7.x version of HEC-GeoRAS is also available.

Haestad Methods' HEC-GIS system combines COE's HEC-RAS floodplain modeling program with ArcInfo. HEC-GIS is an Arc Macro Language (AML) program that allows importing floodplain cross-section data from a DEM within ArcInfo, in the form of triangular irregular networks (TINs). It also allows importing HEC-RAS water surface profiles back into HEC-GIS to perform additional spatial analyses and mapping.

Watershed Modeling System

Watershed Modeling System (WMS) is a comprehensive hydrologic modeling package developed by the Environmental Modeling Research Laboratory (EMRL) of Brigham Young University in cooperation with the U.S. Army COE Waterways Experiment Station (WES) (WMS, 2000; Nelson et al., 1993). The focus of WMS is to provide a single application that integrates digital terrain models with industry-standard lumped-parameter hydrologic models such as HEC-1 and TR-20. In this way, hydrologic data developed in ArcInfo, ArcView, or WMS can be directly linked to many common hydrologic models. WMS includes an interface to HEC-1, TR-20, and the Rational Method programs used by many hydrologic engineers to model the rainfall-runoff process.

WMS can import and export GIS data. WMS' GIS interface is accomplished through its map module. Map module data is structured after ArcInfo 7.x, which makes it easy to import and use vector data created inside ArcInfo or any other GIS that can export an ArcInfo GENERATE file. WMS includes an ArcView 3.x extension that facilitates the movement of data from GIS to WMS for hydrologic model development. The interface allows for automatic development of a topological model from TINs. Furthermore, any geometric parameters computed by WMS are supplied to the corresponding model input fields. The extension preprocesses the GIS data to export hydrologic modeling data. WMS then reads a single file (describing the relevant themes exported) to obtain data from the GIS. These data can then be used in WMS to create an initial model for any of the hydrologic programs supported by WMS. WMS itself is integrated with the hydrologic models. Data entry for the hydrologic model — including rainfall, job control, or any other parameters not defined as attributes in GIS Shapefiles — can be completed using WMS' hydrologic modeling interface. WMS can be used to postprocess and then import model results in the GIS software. A screenshot of the WMS interface is shown in Figure 11.8. Unlike some HEC programs such as HEC-HMS and HEC-RAS, WMS is not free. A demo version can be downloaded from the BYU Web site.

GISHydro Modules

GISHydro is a collection of programs, demos, exercises, data, reports, and information designed to support the use of GIS in hydrology and water resources studies. These materials are mostly drawn from the work of Professor David R. Maidment and a community of his graduate students, research scientists, and faculty

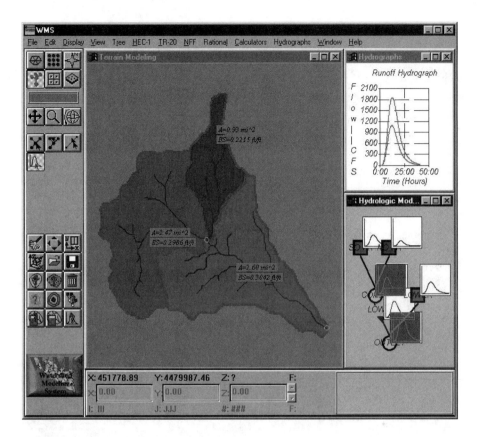

Figure 11.8 WMS interface.

in the Center for Research in Water Resources (CRWR) and the Department of Civil Engineering of the University of Texas at Austin. Initially, CRWR developed a prototype Spatial Hydrology Analysis System (Maidment, 1997) consisting of eight GISHydro modules. These modules, more than just computer programs, are a methodological approach to solving specific types of hydrologic problems using GIS. They contain ArcView and ArcInfo extensions, Avenue and AML scripts, sample data, user instructions, and example exercises. Two GISHydro modules are described in the following subsections:

GISHydro Prepro

This module is a procedure for connecting GIS to external models. Consisting of Avenue and AML scripts, it is designed to transform a subbasins-and-streams network into a schematic model of the hydrologic elements of a watershed. It was initially developed as a preprocessor for COE's new Hydrologic Modeling System (HEC-HMS), which replaced the old HEC-1 model. GH-Prepro takes stream and subbasin GIS data layers and analyzes them to produce a schematic model that defines and spatially connects seven types of hydrologic elements in the landscape: subbasins, reaches, junctions, diversions, reservoirs, sources, and sinks.

GISHydro Runoff

This module provides articles, reports, exercises, and a sample ArcView 3.x extension to compute watershed parameters and runoff at the outlet of a watershed. The ArcView extension uses the Texas Department of Transportation (TxDOT) statewide regional rural regression equations for estimating stormwater runoff. Figure 11.9 shows a screenshot of the TxDOT extension (Olivera and Bao, 1997).

ArcInfo Interface with HEC Programs

Another example of the interface method is the ArcInfo and HEC-1 and HEC-2 interface developed by Woolpert, Inc. (Phipps, 1995). In HEC-1, the interface model uses vector layers for streams and open channels and the ArcInfo networking function to show the locations of and the relationships between different types of flows: sheet flow, shallow channel flow, and concentrated channel flow. The ArcInfo and HEC-1

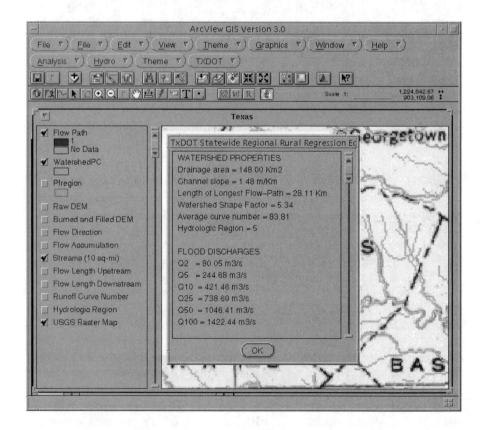

Figure 11.9 GISHydro TxDOT extension.

linkage allows users to see graphically, on a map, how fast the three flow types are moving at different locations within the model (volume/time). The unit hydrographs generated in the HEC-1 hydrologic model are then used to feed the HEC-2 hydraulic model. The ArcInfo and HEC-2 interface allows graphical display of the 2-, 10-, 25-, 50-, 100-, and 500-year frequency flood profiles for the streams and open channels within the watershed. It was found that this interface allowed for easier model development and calibration, which resulted in substantial savings in the cost of modeling tasks.

Intermediate Data Management Programs

Some commercial modeling products use an intermediate data management program to facilitate data transfer between the GIS and a model. This program is written specifically to import data from a variety of common third-party GIS software and export data to a model data set. Under certain conditions this approach could be defined as an interface, but generally it is not. In relation to the Shamsi taxonomy, this method lies between the interchange and interface methods (Heaney et al., 1999). Haestad Methods' GIS connection wizards for the company's WaterCAD, Sewer-CAD, and StormCAD programs (described elsewhere in this book) are typical examples of this method. Figure 11.10 shows WaterCAD's Import Shapefile wizard,

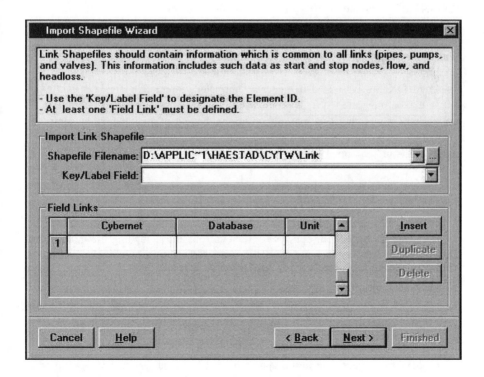

Figure 11.10 WaterCAD's Import Shapefile wizard.

which is used to import WaterCAD model input parameters from ArcView GIS Shapefiles.

Interface Method Case Study

GIS software	ArcInfo
Modeling software	Rhea, an orographic precipitation model developed at the Colorado State University (Rhea, 1977). The Rhea model simulates the interaction of air layers with underlying topography by allowing vertical displacement of the air column while keeping track of the resulting condensate or evaporation.
Other software	Data Explorer (DX), a Scientific Visualization System (SVS) software from IBM
GIS data	DEM grid, city locations, rivers, and rain-gauge locations
GIS format	Raster and vector
Study area	20,530 km² Gunnison River Basin in southwestern Colorado
Organization	U.S. Geological Survey

Very often, the software required to solve complex hydrologic problems is available only in separate pieces rather than as an integrated package. Simulation models and GIS and SVS packages are individually strong in specific aspects of spatial analysis but inadequate in cross-functional analysis of complex hydrologic problems. An integrated system consisting of a hydroclimatic simulation model (Rhea), a GIS (ArcInfo), and an SVS (DX) was developed to facilitate model development, calibration, and verification and visual representations of precipitation across space and time. AML tools translated ArcInfo data into a format that could be read by DX. The converted DX file was used to generate animated, interactive displays using DX software. The integrated system provided improved understanding of hydroclimatic processes in mountainous regions. An additional benefit of the integrated system, the value of which is often underestimated, is the improved ability to communicate model results (Hay and Knapp, 1996).

INTEGRATION METHOD

In the interface method, options for data editing and launching the model from within the GIS software are not available. An interface simply adds new menus or buttons to a GIS to automate the transfer of data between a computer model and a GIS. GIS integration, on the other hand, is a combination of a model and a GIS such that the combined program offers both GIS and modeling functions. This method represents the closest relationship between a GIS and a model. Two integration approaches are possible:

1. GIS-based integration: In this approach, modeling modules are developed in or are called from within a GIS. All the four tasks of creating model input, editing

data, running the model, and displaying output results are available in the GIS. There is no need to exit the GIS to edit the data file or run the model. Because development and customization tools within most GISs provide relatively simple programming capability, this approach provides limited modeling power. It offers full GIS functionality and simplistic modeling capability. It is often more popular because of the availability of computer code for a large number of public-domain legacy computer models, such as SWMM and EPANET, which can be easily customized for GIS integration.

2. Model-based integration: In this method, GIS modules are developed in or are called from a computer model. It provides GIS-like functionality to computer models. Because it is difficult to program all the GIS functions into a computer model, this approach provides limited GIS capability. In this method, GIS is essentially used as a mapping tool. Sophisticated data management and data visualization capabilities are not present. PCSWMM GIS [described in Chapter 13 (Sewer Models)] and Runoff97 (Ritter and Gallie, 1998) are examples of the model-based integration method. In this method, GIS and models are integrated through a common interface that requires extensive computer programming using conventional computer languages (e.g., FORTRAN), and/or GIS macro and scripting languages (e.g., AML, Avenue or VBA).

Representative examples of the integration method are given in the following subsections.

EPA's BASINS Program

EPA's BASINS program is the best example of GIS integration with multiple models.

The best example of GIS integration with multiple models is the Better Assessment Science Integrating Point and Nonpoint Sources (BASINS) program. Originally released in September 1996, BASINS was developed by TetraTech, Inc., for the U.S. Environmental Protection Agency (EPA). BASINS supports the development of total maximum daily loads (TMDLs) using a watershed-based approach that integrates both point and nonpoint sources. BASINS integrates ArcView 3.x GIS with national watershed data and state-of-the-art environmental assessment and modeling tools into one convenient package. The heart of BASINS is its suite of interrelated components essential for performing watershed and water quality analysis. These components are grouped into the following five categories:

1. National databases
2. Assessment tools for evaluating water quality and point source loadings at a variety of scales
3. Utilities, such as land-use and DEM reclassification and watershed delineation
4. Watershed and water quality models including NPSM, TOXIROUTE, and QUAL2E
5. Postprocessing output tools for interpreting model results

---

What Is TMDL?

According to EPA, TMDL is defined as "a calculation of the maximum amount of a pollutant that a water body can receive and still meet water quality standards, and the allocation of that amount to the pollutant's sources." Technically, TMDL can be defined with the help of the following equation:

$$TMDL = WLAP + LAN + RC + FOS$$

where:

WLAP = waste load allocations for point sources
LAN   = load allocations for nonpoint sources
RC   = reserve capacity for future growth
FOS   = factor of safety

BASINS integrates three models within an ArcView GIS environment: NPSM (HSPF), TOXIROUTE, and QUAL2E. Nonpoint Source Model (NPSM) estimates land-use-specific NPS loadings for selected pollutants at a watershed (cataloging unit or user-defined subwatershed) scale. The model uses landscape data such as watershed boundaries and land-use distribution to automatically prepare many of the required input data parameters. NPSM combines a Windows-based interface with EPA's HSPF and is linked to ArcView. HSPF is a model for developing NPS loads and is the backbone of the BASINS modeling system. TOXIROUTE is a screening-level stream-routing model that performs simple dilution and decay calculations under mean- or low-flow conditions for a stream system. QUAL2E is a one-dimensional, steady-state water quality and eutrophication model. It is integrated with ArcView through a Windows-based interface. It allows fate and transport modeling for both point and nonpoint sources. Both TOXIROUTE and QUAL2E can use NPS loadings calculated by HSPF.

Overcoming the lack of integration, limited coordination, and time-intensive execution typical of more traditional assessment tools, BASINS makes watershed and water quality studies easier by bringing key data and analytical components together under one roof. The simulation models run in a Windows environment using data input files generated in ArcView. By using GIS, a user can visualize, explore, and query to bring a watershed to life. This allows the user to assess watershed loadings and receiving-water impacts at various levels of complexity. ArcView geographic data preparation, selection routines, and visual output streamline the use of the models. A postprocessor graphically displays model results. Figure 11.11 shows a screenshot of the BASINS program showing an example of delineated watersheds.

BASINS is distributed on a CD-ROM containing the BASINS program and data for an EPA region of interest. It can also be downloaded from the BASINS Web site. BASINS Version 1.0 was released in September 1996 (Lahlou, et al., 1996) for ArcView Version 2.1. BASINS Version 2.0 was released in December 1998 for ArcView 3.0. It included a national data set on the attributes listed in Table 11.4 (Battin et al., 1998). BASINS Version 3.0 was released in 2001 for ArcView 3.2.

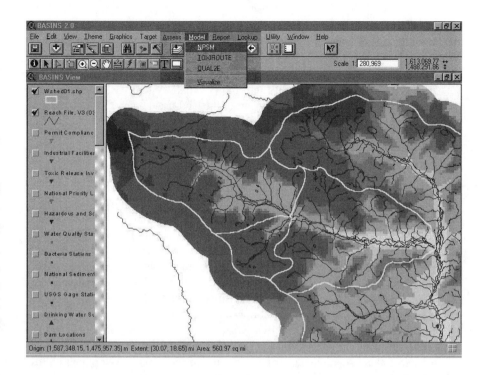

Figure 11.11   BASINS screenshot showing an example of delineated watersheds.

Table 11.4 BASINS Database Attributes

Spatially Distributed Data	
Land use/land cover	USGS hydrologic unit boundaries
Urbanized areas	Drinking water supplies
Populated place location	Dam sites
Reach File, Version 1 (RF1)	EPA region boundaries
Reach File, Version 3 (RF3)	State boundaries
Soils (STATSGO)	County boundaries
Elevation (DEM)	Federal and Indian lands
Major roads	Ecoregions

Environmental Monitoring Data	
Water quality monitoring station summaries	USGS gauging stations
Water quality observation data	Fish and wildlife advisories
Bacteria monitoring station summaries	National Sediment Inventory (NSI)
Weather station sites (477)	Shellfish contamination inventory
Clean water needs survey	—

Point-Source Data	
Permit compliance system	Industrial Facilities Discharge (IFD) sites
Toxic Release Inventory (TRI) sites	Superfund national priority list sites
Mineral availability system/mineral industry location	Resource Conservation and Recovery Act (RCRA) sites

BASINS 2.0 included a watershed delineation tool, a land-use reclassification tool, an HSPF interface, a watershed characterization report on STORET water quality data, DEM elevation data, and USEPA Reach File Version 3 Alpha (RF3 Alpha). Using the import tool, users can add their own data for watershed boundaries, streams, elevation, land use, and observed water quality into the BASINS system. Using the land-use reclassification tool, users can reclassify part of the land-use theme or the entire theme interactively to create more detailed levels of land use. This feature can be used to study the potential impact of land-use changes on water quality. The watershed-delineation tool allows users to interactively subdivide a USGS 8-digit watershed into smaller subwatersheds using mouse point-and-click inputs. Subdelineated watersheds and underlying data are then available for more detailed modeling.

BASINS 3.0 is the latest (year 2001) release at the time of this writing. This release includes additional functional capabilities as well as an updated and expanded set of national data layers. New data and functions include 1° DEM grids, an automatic delineation tool that creates watershed boundaries based on DEM grids, and a new watershed report function for land use, topography, and hydrologic response units. In addition, a new watershed model, the Soil and Water Assessment Tool (SWAT), has been added. SWAT is a watershed-scale model developed to predict the impact of land management practices on water, sediment, and agricultural chemical yields in large complex watersheds with varying soils, land use, and management conditions over long periods of time.

BASINS 3.0 also demonstrates an important change in the development philosophy of BASINS. It is distributed as a core system with several ArcView extensions. All BASINS data management utilities, assessment tools, report tools, and models were converted as extensions. This modular and open architecture offers numerous advantages both to users and developers. For instance, users may customize their BASINS project to include only extensions that are needed. Moreover, it will be easier for developers to maintain and provide updates of individual extensions because it will not involve changing and reissuing a patch or new version for the entire system. This makes it easier for users to upgrade their system. It also encourages the development of extensions for BASINS by users and developers.

BASINS 3.0 installation software contains the new BASINS Version 3.0 system and the tutorial data set (HUC-05010007). In addition to BASINS 3.0, the installation program installs WinHSPF, GenScn, WDMUtil, and PLOAD. WinHSPF is a new interface to HSPF Version 12, which is replacing NPSM from BASINS 2.0. GenScn is a model postprocessing and scenario-analysis tool that is used to analyze output from HSPF and SWAT. WDMUtil is used to manage and create watershed data management files (WDMs) that contain the meteorological data and other time-series data used by HSPF. PLOAD is a GIS and spreadsheet tool to calculate nonpoint sources of pollution in watersheds.

EPA plans to release an update (Version 3.1) in 2004 on the existing ArcView 3.x platform. EPA is currently working on pieces of BASINS for the ArcGIS environment. After the BASINS 3.1 release, EPA plans to port it onto the ArcGIS platform based on the work done by the Center for Research in Water Resources at the University of Texas at Austin, Water Resources Consortium, and ESRI. Future releases of BASINS may include EPA's SWMM program, automatic model calibration, and a reach-indexing tool for georeferencing. A redesign of the BASINS system using RDBMS, Internet, MapObjects, and Java technologies is also anticipated. This move toward a more fully integrated GIS and database system will provide many more BASINS applications. The watershed tool will evolve into a more component-based architecture, so that users will need to install only the components required for their specific analytical needs, including their own legacy models (Whittemore and Beebe, 2000).

BASINS is a valuable tool that integrates environmental data, analytical tools, and modeling programs to support the development of cost-effective approaches to environmental protection. However, BASINS' seamless integration power is not without some disadvantages. BASINS' simplistic modeling approach and user-friendly tools may encourage inexperienced users to become instant modelers, which could be undesirable. The inexperienced users' overreliance on default model segmentation and model coefficients, without an understanding of the underlying assumptions, may result in inaccurate and unrealistic TMDLs (Thuman and Mooney, 2000). BASINS builds upon federal databases of water quality conditions and point-source loadings for numerous parameters where quality assurance is suspect in some cases (Whittemore and Beebe, 2000).

BASINS Examples

Most of the BASINS work and modeling have been on a watershed- or regional-level scale. An example for the BASINS data set for Boulder, Colorado, is shown in Figure 11.12. The size of this file relative to the area it represents reflects a scale

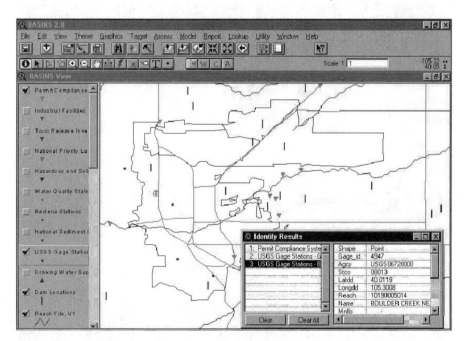

Figure 11.12 BASINS GIS layers for Boulder, Colorado.

of about 1:2000 (Heaney et al., 1999). The Columbus Water Works of Columbus, Georgia, developed a BASINS 2 model for the 2400 mi^2 (3600 km^2) Middle Chattahoochee River Watershed. The main objectives of the model were to formulate TMDL allocation requirements of the Clean Water Act and to demonstrate compliance with EPA's National Combined Sewer Overflow (CSO) Policy. They used monitoring and data management methodologies to calibrate their BASINS model and to perform source-water assessments and protection through provisions of early-warning systems for drinking-water intake operations (Turner et al., 2000).

MIKE BASIN

Danish Hydraulic Institute's (DHI) MIKE BASIN is an ArcView-based decision support system for water resources planning and management. It serves as a generalized river basin network simulation modeling system for regional water resources allocation of complex systems. MIKE BASIN can be used to analyze water supply capabilities in connection with water rights for municipal and industrial water supply, irrigation, and many multipurpose reservoir operations.

MIKE BASIN is fully integrated into ArcView 3.x. It uses a graphical user interface, which links MIKE BASIN to ArcView. It is developed using the object-oriented C++ programming language, which allows for easy implementation of new features. MIKE BASIN permits on-screen creation of models using image themes as backdrops. Figure 11.13 shows the MIKE BASIN interface.

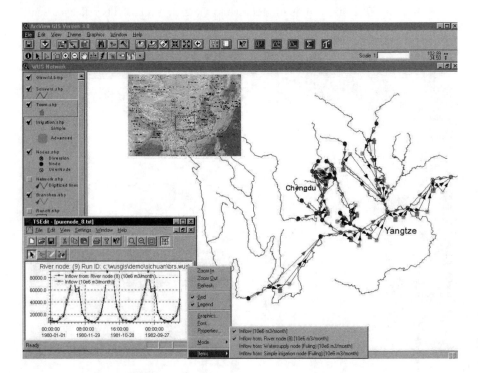

Figure 11.13   MIKE BASIN screenshot.

An important advantage of MIKE BASIN is its vertical compatibility with other DHI modeling systems, which allows the use of several modeling tools of various complexities at different stages of a project cycle. MIKE BASIN can be used in conjunction with other DHI models when detailed studies of specific problems are required. For example, MIKE SHE can be used for modeling runoff quality and quantity, including interaction with surface water systems, MIKE 11 can be used for detailed modeling of river hydrodynamics and water quality, and MIKE 21 can be used for modeling bays and estuaries.

Geo-STORM Integration

An early example of ArcInfo 7.x and modeling integration is the Geo-STORM package developed by Innovative System Developers, Inc., Columbia, Maryland. Geo-STORM computed subbasin runoff, performed flow routing through reach and reservoir networks, and analyzed river hydraulics using SCS Technical Release 20 (TR-20) and Technical Release 55 (TR-55) programs, and COE's HEC-2 computer programs. All these models were contained within a GUI environment within ArcInfo GIS. The software was written in AML, the ArcInfo scripting language. There were external calls to the operating system only to execute the computer programs. The common interface provided easy interaction between the models and switching between the models was hardly noticeable, and output from one model was used as input for another model. The TR-55 model was embedded within the TR-20 model. The output from TR-55 became input for TR-20. The TR-20 model provided discharge values for input to the HEC-2 model, and output from HEC-2 could be directly inserted into TR-20 as rating-table information. All modeling applications were integrated in Geo-GUIDE, a graphical interface that made working with ArcInfo easier. The computer models were embedded within Geo-GUIDE, which provided plotting, reporting, and querying capabilities for the stormwater modeling software suite. The GeoSTORM system was considered to be a database repository for stormwater-related geography and facilities. Using programs like Geo-STORM translated into productivity gains realized by eliminating tedious tasks such as manual map measurements, data entry, and tabular-report interpretation. Geo-STORM automated these time-consuming processes, drawing model inputs from the GIS database. This in time improved the quality and accuracy of model results. Figure 11.14 shows a screenshot of Geo-STORM (Thomas, 1995).

ARC/HEC-2 Integration

ARC/HEC-2 was one of the earliest integrations of ArcInfo and HEC-2, developed at the University of Texas at Austin. ARC/HEC-2 facilitated floodplain analysis in a GIS environment. It allowed direct usage of ArcInfo coverages for development of HEC-2 input, running HEC-2 from within ArcInfo, and for conversion of HEC-2 water-surface elevations computed at each cross section into an ArcInfo coverage representing the floodplain. There were two main data requirements for ARC/HEC-2 to operate: (1) a terrain representation in the form of constructed TIN or contours, and (2) channel centerline (thalweg). ARC/HEC-2 consisted of three separate modules: preprocessor,

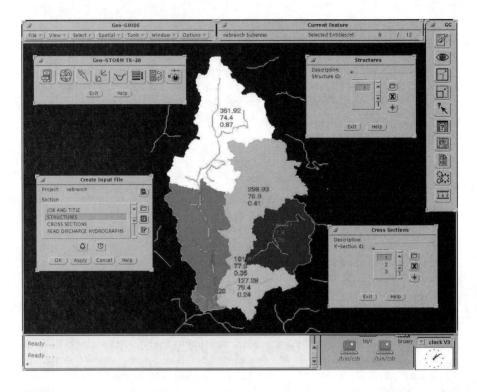

Figure 11.14   Geo-STORM screenshot showing ArcInfo and TR-20 integration.

postprocessor, and viewer. The preprocessor created a complete HEC-2 input file ready to run. For complex HEC-2 models, manual editing of the input file was required, in which case ARC/HEC-2 served as an interface rather than an integration. The postprocessor used the HEC-2 output to determine water surface elevation at each cross section and, based on that information, found the outline of the flooded area in the form of a lattice and a polygon layer. This lattice and coverage could be used in ArcInfo with all other pertinent coverages for the area. Viewer was an AML application that allowed viewing of the main input and output coverages in 2D and 3D formats (Djokic et al., 1993).

Integration Method Case Study

GIS software	ArcView and Spatial Analyst Extension
Modeling Software	Agricultural Nonpoint Source Pollution Model (AGNPS)
GIS data	Soil (STATSGO), USGS DEM (1:250,000 scale), land use, hydrography, watersheds, and crop management
Study area	280 mi² (810 km²) Dowagiac River Watershed located in southwestern Michigan
Organization	Western Michigan University, Kalamazoo, Michigan

Successful implementation of environmental modeling databases involves the integration of GIS, remote sensing, multiple databases, and visualization tools for

extraction of the needed model input parameters and for analysis and visualization of the simulated results. Although it was obvious that GIS technology was ideally suited to accomplish these goals, current GIS packages were found to be difficult to use and offered limited functions to support environmental modeling. It was also noted that in current practice, GIS and hydrologic models were just used together, not really integrated. Therefore, a custom interface called ArcView Nonpoint Source Pollution Modeling (AVNPSM) was developed between ArcView GIS and AGNPS in support of agricultural watershed analysis and nonpoint source pollution management.

AGNPS (Version 5.0) is a single storm event based simulation model for evaluating sediment and nutrient (nitrogen, phosphorus, and chemical oxygen demand or COD) transport from agricultural watersheds. It represents the landscape of a watershed by discretizing the study area into individual grid cells. The PC Windows-based interface developed using Avenue scripts and Dialogue Designer consists of seven modules: the AGNPS utility, parameter generator, input file processor, model executor, output visualizer, statistical analyzer, and land-use simulator. Each module is incorporated into ArcView through pull-down menus and icons. The input file processor menu produces an input file that is compatible with AGNPS input format requirements. The AGNPS model is executed either within Windows using the model executor menu or run separately in simulated DOS mode. The output visualizer menu shows the simulated results of hydrology, sediment, and nutrients in tabular or map format. The calibrated model indicated that development of about 1240 acres of nonurban land caused little increase in the peak flow rate and sediment yield at the mouth of the river and had almost no impact on the runoff volume and nutrient yields. The case study concluded that integration of GIS and simulation models facilitates the development and application of environmental simulation models and enhances the capabilities of GIS packages (He et al., 2001).

WHICH LINKAGE METHOD TO USE?

Each method has its pros and cons. Table 11.5 lists the advantages and disadvantages of the three methods of linking H&H models with GIS. Integrated systems are easy to learn and use and make the excruciating task of H&H modeling a fun

Table 11.5 Comparison of GIS Linkage Methods

Feature	Interchange Method	Interface Method	Integration Method
Automation	Low	Medium	High
User-friendliness	None	Medium	High
Ease of use	Cumbersome	Easy	Easiest
Learning curve	Long	Medium	Quick
Data-entry error potential	High	Medium	Low
Data-error tracking	Easy	Complex	Difficult
Misuse potential by inexperienced users	Low	Medium	High
Development effort	Low	Medium	High
Computer programming	Low	Medium	Extensive

activity. Integrated systems save model development and interpretation time and make modeling cost-effective.

Seamless model integration with GIS is not without some disadvantages. The simplistic modeling approach and user-friendly tools provided by integrated systems like BASINS may encourage inexperienced users to become instant modelers, which could be undesirable (Thuman and Mooney, 2000). GIS can easily convert reams of computer output into colorful thematic maps that can both expose and hide data errors. The user-friendly interface of an integrated package may incorrectly make a complex model appear quite simple, which may lead to misapplication of the model by inexperienced users. Modeling should not be made easy at the expense of good science. Modeling, for all the new tools emerging in the 1990s, is still an art that is done well by only a select few (Whittemore and Beebe, 2000).

The inexperienced users' overreliance on default model parameters may result in inaccurate modeling results. Inexperienced H&H modelers should exercise caution when using integrated systems, make sure to browse the model input and output files for obvious errors, and avoid overreliance on thematic mapping and graphing of model output. Modification and recompilation of computer codes of existing legacy programs like SWMM or EPANET to create integrated systems should be avoided. If recoding is necessary, developers should be extremely cautious to avoid coding errors and the associated liability. Users should remember that a difficult-to-use model is much better than an inaccurate model (Shamsi, 2001).

The development of a GIS for use in urban H&H modeling is an expensive investment (Heaney et al., 1999). Typically the most advanced tools are created for advanced applications, where a full-fledge GIS is in place. For some applications, a DOS-based model may still be the most appropriate. However, as more urban areas create GIS layers, the integration of modeling software and GIS software will become more useful and prevalent.

USEFUL WEB SITES

HEC software	www.hec.usace.army.mil/
GIS Hydro	www.crwr.utexas.edu/archive.shtml
BASINS software	www.epa.gov/OST/BASINS
WMS software	http://emrl.byu.edu/wms.htm

CHAPTER SUMMARY

This chapter described GIS applications that have revolutionized the way H&H modeling is performed. It showed how GIS can be used to develop H&H models, and also provided a software review of GIS application programs. From the material presented in this chapter, it can be concluded that GIS applications allow modelers to be more productive. Users devote more time to understanding the problem and less time to mechanical tasks of data input and checking, getting the program to run, and interpreting reams of output. More than just tabular outputs, models become an automated system evaluation tool. There are three modeling application methods:

interchange, interface, and integration. Interchange is the simplest method but offers limited benefits. Interchange systems have existed since 1980s in all domains. The interface method's benefits lie between those of the interchange and the integration methods. There are a few interface systems in the public domain and many in commercial and proprietary domains. The integration method offers the maximum productivity benefits. BASINS is the only public-domain modeling system that uses the integration method. A few integrated commercial systems are available starting at $5000. There are many in-house and proprietary integrated systems. The number of public-domain and commercial integration packages is expected to grow steadily in the near future.

CHAPTER QUESTIONS

1. What are the GIS applications in computer modeling of water, wastewater, and stormwater systems?
2. What are computer modeling application methods?
3. Give an example of each application method using programs other than those described in this chapter.
4. Which GIS application method is the easiest to use? Why?
5. Which GIS application method is the easiest to implement? Why?
6. List the typical steps used in the interface method.
7. Make a list of the H&H models used in your organization and the type of GIS application method used by each.

CHAPTER 12

Water Models

GIS saves time and money in developing water distribution system hydraulic models for simulating flows and pressures in the system. GIS also helps in presenting the model results to non-technical audiences.

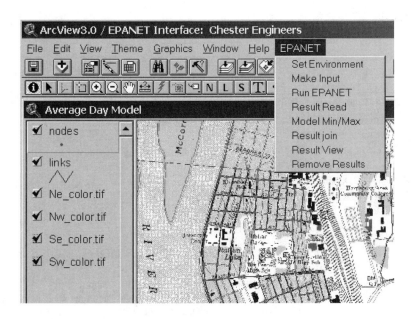

EPANET and ArcView GIS integration.

LEARNING OBJECTIVE

The learning objective of this chapter is to understand how GIS can be applied in developing water distribution system hydraulic models and presenting the results of the hydraulic models.

MAJOR TOPICS

- Development of hydraulic models
- Software examples
- EPANET model
- Network simplification (skeletonization)
- Estimation of node demands
- Estimation of node elevations
- Delineation of pressure zones
- Mapping the model output results
- GIS application examples and case studies

LIST OF CHAPTER ACRONYMS

AM/FM/GIS Automated Mapping/Facilities Management/Geographic Information System
CIS Customer Information System
COM Component Object Model
COTS Commercial Off-the-Shelf
DEM Digital Elevation Model
DHI Danish Hydraulic Institute
DLL Dynamic Link Library
DRG Digital Raster Graphic
GEMS Geographic Engineering Modeling Systems
MMS Maintenance Management System
ODBC Open Database Connectivity
PRV Pressure Regulating Valve
SCADA Supervisory Control and Data Acquisition
TIF Tag Image File

CITY OF GERMANTOWN'S WATER MODEL

Application	Water distribution system mapping and modeling
GIS software	ArcGIS, ArcGIS Water Utilities Data Model (formerly ArcFM Water Data Model), and MapObjects
Other software	WaterCAD
GIS layers	Digital orthophotos, pipes, hydrants, valves, manholes, pumps, meters, fittings, sampling stations, and monitoring wells
Hardware	Two Dell Precision 220 computers
Study area	City of Germantown, Tennessee
Organization	City of Germantown, Tennessee

The City of Germantown (Tennessee) water distribution system serves 40,000 people through 12,000 installations in a 17.5 mi^2 area. The system has approximately 95,000 ft of water pipes.

The City used ESRI's geodatabase, an object-oriented GIS data model, to design a database. The database design was done with minimal customization of ESRI's ArcGIS Water Utilities Data Model. This approach helped complete the database design in a few weeks instead of several months. ESRI's MapObjects was used to create a custom utility tool set for developing the water distribution network. Valve and hydrant locations from the digital orthophotos and planimetric mapping were used with the existing 1 in. = 500 ft-scale paper map of the water system as the basis for connecting water pipes.

The GIS layers (Figure 12.1) were initially developed using the ArcView Shapefiles. The Shapefiles were then migrated to a geodatabase. To ease the data migration, connectivity rules were stored in ArcCatalog rather than directly in the model. ArcMap "flag and solver" tracing tools were used to ensure topological integrity of the migrated data. ArcMap also provided the City with out-of-the-box network analysis functionality such as tracing paths and water main isolation.

For the development of the hydraulic model, the City did not have to create a modeling network from scratch. GIS was used to skeletonize the network for input to the Haestad Method's WaterCAD 4.1 hydraulic model. Spatial operations available in ArcGIS were used to assign customer-demand data — obtained from the City's billing database and stored in parcel centroids — to modeling nodes. Node elevations were extracted from the City's DEM.

The hydraulic model was run to assess the future system expansion and capital improvement needs of the City. The model indicated that the system needed a new elevated storage tank and a larger pipe to deliver acceptable pressure under peak demand conditions (ESRI, 2002b).

GIS APPLICATIONS FOR WATER DISTRIBUTION SYSTEMS

GIS has wide applicability for water distribution system studies. Representation and analysis of water-related phenomena by GIS facilitates their management. The GIS applications that are of particular importance for water utilities include mapping, modeling, facilities management, work-order management, and short- and long-term planning. Additional examples include (Shamsi, 2002):

- Conducting hydraulic modeling of water distribution systems.
- Estimating node demands from land use, census data, or billing records.
- Estimating node elevations from digital elevation model (DEM) data.
- Performing model simplification or skeletonization for reducing the number of nodes and links to be included in the hydraulic model.
- Conducting a water main isolation trace to identify valves that must be closed to isolate a broken water main for repair. Identifying dry pipes for locating customers or buildings that would not have any water due to a broken water main. This application is described in Chapter 15 (Maintenance Applications).
- Prepare work orders by clicking on features on a map. This application is described in Chapter 14 (AM/FM/GIS Applications).
- Identifying valves and pumps that should be closed to isolate a contaminated part of the system due to acts of terrorism. Recommending a flushing strategy to clean the contaminated parts of the system. This application is described in Chapter 16 (Security Planning Vulnerability Assessment).

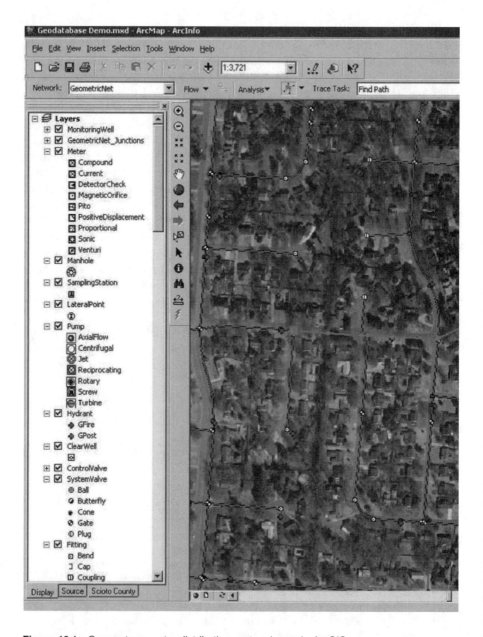

Figure 12.1 Germantown water distribution system layers in ArcGIS.

- Providing the basis for investigating the occurrence of regulated contaminants for estimating the compliance cost or evaluating human health impacts (Schock and Clement, 1995).
- Investigating process changes for a water utility to determine the effectiveness of treatment methods such as corrosion control or chlorination.
- Assessing the feasibility and impact of system expansion.
- Developing wellhead protection plans.

By using information obtained with these applications, a water system manager can develop a detailed capital improvement program or operations and maintenance plan (Morgan and Polcari, 1991). The planning activities of a water distribution system can be greatly improved through the integration of these applications.

In this chapter, we will focus on the applications related to hydraulic modeling of water distribution systems.

DEVELOPMENT OF HYDRAULIC MODELS

The most common use of a water distribution system hydraulic model is to determine pipe sizes for system improvement, expansion, and rehabilitation. Models are also used to assess water quality and age and investigate strategies for reducing detention time. Models also allow engineers to quickly assess the distribution network during critical periods such as treatment plant outages and major water main breaks.

Today, hydraulic models are also being used for vulnerability assessment and protection against terrorist attacks.

Hydraulic models that are created to tackle a specific problem gather dust on a shelf until they are needed again years later. This requires a comprehensive and often expensive update of the model to reflect what has changed within the water system. A direct connection to a GIS database allows the model to always remain live and updated.

Denver Water Department, Colorado, installed one of the first AM/FM/GIS systems in the U.S. in 1986. In the early 1990s, the department installed the $60,000 SWS water model from the Stoner Associates (now Advantica Stoner, Inc.). Around the same time, Genesee County Division of Water and Waste Services in Flint, Michigan, started developing a countywide GIS database. The division envisioned creating sewer and water layers on top of a base map and integrating the GIS layers with their water and sewer models (Lang, 1992).

GIS applications reduce modeling development and analysis time. GIS can be used to design optimal water distribution systems. An optimal design considers both cost-effectiveness and reliability (Quimpo and Shamsi, 1991; Shamsi, 2002a) simultaneously. It minimizes the cost of the system while satisfying hydraulic criteria (acceptable flow and pressures). Taher and Labadie (1996) developed a prototype model for optimal design of water distribution networks using GIS. They integrated GIS for spatial database management and analysis with optimization theory to develop a computer-aided decision support system called WADSOP (Water Distribution System Optimization Program). WADSOP employed a nonlinear programming technique as the network solver, which offers certain advantages over conventional methods such as Hardy-Cross, Newton Raphson, and linear system theory for balancing looped water supply systems. WADSOP created a linkage between a GIS and the optimization model, which provided the ability to capture model input data; build network topology; verify, modify, and update spatial data; perform spatial analysis; and provide both hard-copy reporting and graphical display of model results. The GIS analysis was conducted in PC ARC/INFO. The ROUTE and ALLOCATE programs of the NETWORK module of PC ARC/INFO were found to be very helpful. Computational time savings of 80 to 99% were observed over the other programs.

Generally, there are four main tasks in GIS applications to water system modeling:

1. Synchronize the model's network to match the GIS network.
2. Transfer the model input data from a GIS to the model.
3. Establish the model execution conditions and run the model.
4. Transfer the model output results to GIS.

As described in Chapter 11 (Modeling Applications), GIS is extremely useful in creating the input for hydraulic models and presenting the output. A hydraulic modeler who spends hundreds of hours extracting input data from paper maps can accomplish the same task with a few mouse clicks inside a GIS (provided that the required data layers are available). GIS can be used to present hydraulic modeling results in the form of thematic maps that are easy for nontechnical audiences to understand. Additional information about GIS integration methods is provided in Chapter 11, which provided a taxonomy developed by Shamsi (1998, 1999) to define the different ways a GIS can be linked to hydraulic models. According to Shamsi's classification, the three methods of developing GIS-based modeling applications are interchange, interface, and integration. For GIS integration of water distribution system hydraulic models, one needs a hydraulic modeling software package. The software can run either inside or outside the GIS. As explained in Chapter 11, when modeling software is run inside a GIS, it is considered "seamlessly integrated" into the GIS. Most modeling programs run in stand-alone mode outside the GIS, in which case the application software simply shares GIS data. This method of running applications is called a "GIS interface."

The interchange method offers the most basic application of GIS in hydraulic modeling of water distribution systems. In 1995, the Charlotte–Mecklenburg Utility Department (CMUD) used this method to develop a KYPIPE model for an area serving 140,000 customers from 2,500 mi of water pipes ranging in size from 2 in. to 54 in. (Stalford and Townsend, 1995). The modeling was started by creating an AutoCAD drawing of all pipes 12 in. and larger. The pipes were drawn as polylines and joined at the intersections. The AutoCAD drawing was converted to a facilities management database by placing nodes at the hydraulic intersections of the pipes, breaking the polylines as needed, adding extended attributes to the polylines from the defined nodes, extracting this information into external tables, and adding the nongraphical information to the tables. Then, by using ArcCAD inside AutoCAD, coverages were created for the pipes and nodes, which could be viewed inside ArcView. Although this method was considered state-of-the-art in 1995, it is not a very efficient GIS integration method by today's standards. The EPANET integration case study presented later in the chapter exemplifies application of the more efficient integration method.

In the simplest application of the integration method, it should be possible to use a GIS to modify the configuration of the water distribution network, compile model input files reflecting those changes, run the hydraulic model from within the GIS, use the GIS to map the model results, and graphically display the results of the simulation

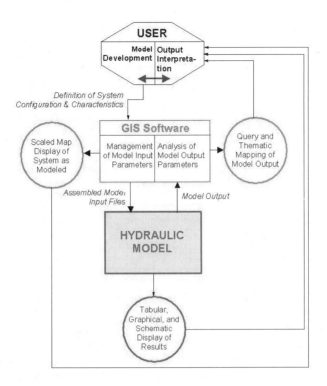

Figure 12.2 Integration method interrelationships.

on a georeferenced base map. The integration method interrelationships among the user, the hydraulic model, and the GIS software are illustrated in Figure 12.2.

SOFTWARE EXAMPLES

Table 12.1 lists representative water distribution system modeling packages and their GIS capabilities, vendors, and Web sites. The salient GIS features of these programs are described in the following subsections.

EPANET

EPANET was developed by the Water Supply and Water Resources Division (formerly the Drinking Water Research Division) of the U.S. Environmental Protection Agency's National Risk Management Research Laboratory (Rossman, 2000). It is a public-domain software that may be freely copied and distributed.

EPANET is a Windows 95/98/2K/NT program that performs extended-period simulation of hydraulic and water quality behavior within pressurized pipe networks. A network can consist of pipes, nodes (pipe junctions), pumps, valves, and storage tanks or reservoirs. EPANET tracks the flow of water in each pipe, the pressure at each node, the height of water in each tank, and the concentration of a chemical species

Table 12.1 Water Distribution System Modeling Software

Software	Method	Vendor	Web site
EPANET	Interchange	U.S. Environmental Protection Agency (EPA)	www.epa.gov/ord/NRMRL/ wswrd/epanet.html
AVWater	Integration	CEDRA Corporation	www.cedra.com
H₂ONET and H₂OMAP	Interface and Integration	MWH Soft	www.mwhsoft.com
InfoWorks WS and InfoNet	Interface and Integration	Wallingford Software	www.wallingfordsoftware.com
MIKE NET	Interface	DHI Water & Environment	www.dhisoftware.com
PIPE2000 Pro	Interface	University of Kentucky	www.kypipe.com
SynerGEE Water	Interface and Integration	Advantica Stoner	www.stoner.com
WaterCAD WaterGEMS	Interface Integration	Haestad Methods	www.haestad.com

throughout the network during a simulation period made up of multiple time steps. In addition to chemical species, water age and source tracing can also be simulated.

The Windows version of EPANET provides an integrated environment for editing network input data, running hydraulic and water quality simulations, and viewing the results in a variety of formats. These include color-coded network maps, data tables, time-series graphs, and contour plots. Figure 12.3 shows an EPANET screenshot displaying pipes color-coded by head-loss and nodes color-coded by pressure.

The EPANET Programmer's Toolkit is a dynamic link library (DLL) of functions that allow developers to customize EPANET's computational engine for their own specific needs. The functions can be incorporated into 32-bit Windows applications written in C/C++, Delphi Pascal, Visual Basic, or any other language that can call functions within a Windows DLL. There are over 50 functions that can be used to open a network description file, read and modify various network design and operating parameters, run multiple extended-period simulations accessing results as they are generated or saving them to file, and write selected results to file in a user-specified format. The toolkit should prove useful for developing specialized applications, such as optimization models or automated calibration models, that require running many network analyses while selected input parameters are iteratively modified. It also can simplify adding analysis capabilities to integrated network modeling environments based on CAD, GIS, and database management packages (Rossman, 2000a).

The current version of EPANET (Version 2.0) has neither a built-in GIS interface nor integration. Therefore, users must develop their own interfaces (or integration) or rely on the interchange method to manually extract model input parameters from GIS data layers.

H₂ONET™ and H₂OMAP™

MWH Soft, Inc. (Broomfield, Colorado), provides infrastructure software and professional solutions for utilities, cities, municipalities, industries, and engineering organizations. The company was founded in 1996 as a subsidiary of the global

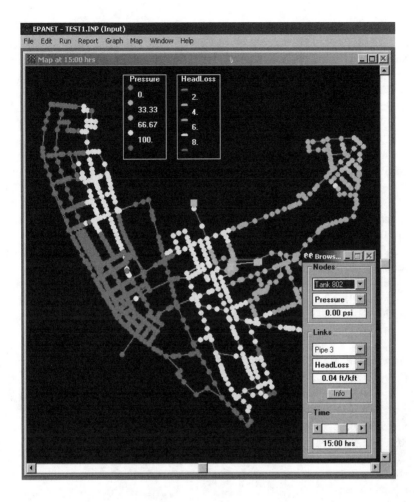

Figure 12.3 EPANET screenshot (Harrisburg, Pennsylvania, model).

environmental engineering, design, construction, technology, and investment firm
MWH Global, Inc. The company's CAD- and GIS-enabled products are designed
to manage, design, maintain, and operate efficient and reliable infrastructure systems.
MWH Soft, provides the following water distribution system modeling software:

- H_2ONET
- H_2OMAP Water
- InfoWater
- H_2OVIEW Water

H_2ONET is a fully integrated CAD framework for hydraulic modeling, network
optimization, graphical editing, results presentation, database management, and
enterprise-wide data sharing and exchange. It runs from inside AutoCAD and,
therefore, requires AutoCAD installation.

Built using the advanced object-oriented geospatial component model, H₂OMAP Water provides a powerful and practical GIS platform for water utility solutions. As a stand-alone GIS-based program, H₂OMAP Water combines spatial analysis tools and mapping functions with network modeling for infrastructure asset management and business planning. H₂OMAP Water supports geocoding and multiple mapping layers that can be imported from one of many data sources including CAD drawings (e.g., DWG, DGN, and DXF) and standard GIS formats (Shapefiles, Generate files, MID/MIF files, and ArcInfo coverages). The program also supports the new geodatabase standard of ArcGIS through an ArcSDE connection.

InfoWater is a GIS-integrated water distribution modeling and management software application. Built on top of ArcGIS using the latest Microsoft .NET and ESRI ArcObjects component technologies, InfoWater integrates water network modeling and optimization functionality with ArcGIS.

H₂OVIEW Water is an ArcIMS-based (Web-enabled) geospatial data viewer and data distribution software. It enables deployment and analysis of GIS data and modeling results over the Internet and intranets. Figure 12.4 shows a screenshot of H₂OVIEW Water.

Various GIS-related modules of H₂O products are described in the following subsections.

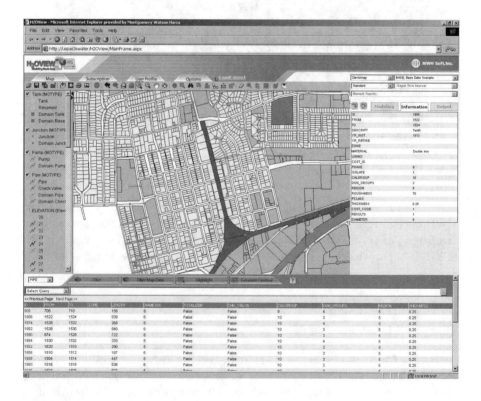

Figure 12.4 H₂OVIEW screenshot (Screenshot courtesy of MWH Soft).

Demand Allocator

Automatically computes and geographically allocates water demand data throughout the water distribution network model.

Skeletonizer

Automates GIS segmentation (data reduction, skeletonization, and trimming) to construct hydraulic network models and reallocates node demands directly from the GIS.

Tracer

H_2ONET Tracer is a network tracing tool that locates all valves which need to be closed to isolate a portion of the network, visually identify all customers affected by loss or reduction in service, and determine the best response strategy. It is a valuable tool for vulnerability analysis and security planning to protect systems against terrorist attacks.

WaterCAD™ and WaterGEMS™

Haestad Methods, Inc. (Waterbury, Connecticut), provides a wide range of H&H computer models, publishes books, and offers continuing education for the civil engineering community. Haestad Methods is a growing company with more than 100,000 users in more than 160 countries around the world. Haestad Methods was acquired by Bentley Systems (Exton, Pennsylvania) in August 2004. Haestad Methods provides the following water distribution system modeling software:

- WaterCAD Stand-alone
- WaterCAD for AutoCAD
- WaterGEMS for ArcGIS

WaterCAD is Haestad Method's water distribution analysis and design tool. It can analyze water quality, determine fire flow requirements, and calibrate large distribution networks. The stand-alone version does not require a CAD or GIS package to run. The AutoCAD version runs from inside AutoCAD and, therefore, requires AutoCAD.

Shapefile Wizard, the ArcView GIS interface module for WaterCAD, is sold separately. The wizard provides import and export capability to transfer data between GIS and WaterCAD models. For example, Import Shapefile Wizard imports Water-CAD model input parameters from ArcView GIS Shapefiles. Similarly, the Export Shapefile Wizard exports hydraulic model networks to ArcView. This feature lets users build and maintain their water and sewer networks directly inside a GIS.

Geographic Engineering Modeling Systems (GEMS) is the next generation of Haestad Methods' hydraulic network modeling products. GEMS is a new product family that combines water, wastewater, and stormwater system modeling with an open database architecture. It was developed using ESRI's ArcObjects technology and computing tools from Microsoft's .NET initiative. GEMS technology eliminates the need for a separate model database. It streamlines the process of accessing information on, for example, facilities management, zoning, topography, and gauging stations for use with the model. GEMS-based models can be run within ESRI's ArcGIS software using the geodatabase structure (Shamsi, 2001).

WaterGEMS uses a geodatabase, Water Data Model, for integrated WaterCAD modeling within the ArcGIS platform. Because WaterGEMS runs from inside ArcGIS, it requires an installation of ArcGIS software. Although WaterGEMS accommodates Shapefiles and ArcInfo coverages, it provides users with the tools to migrate to the new geodatabase format. Users can easily save an existing hydraulic model as a geodatabase.

In an interview by Haestad Methods, Jack Dangermond, president of ESRI, stated, "GEMS bridges the gap between engineering and planning departments and allows users to further leverage their GIS investment" (Haestad Methods, 2003).

MIKE NET™

DHI Water and Environment (formerly Danish Hydraulic Institute, Hørsholm, Denmark, www.dhi.dk) is a global provider of specialized consulting and numerical modeling software products for water, wastewater, river and coastal engineering, and water resources development. DHI has developed a large number of hydraulic, hydrologic, and hydrodynamic computer models for water, wastewater, and stormwater systems. The company also specializes in linking its computer models with GIS so that modelers can use both modeling and GIS technologies within a single product. Since 1998, DHI has embarked on an ambitious program to link its models with the ESRI family of GIS products. Many of DHI's modeling systems now support GIS data transfer. For example, DHI has developed ArcView GIS extensions and interface programs for a number of its products.

MIKE NET is a software package for the simulation of steady flow and pressure distribution and extended-period simulation of hydraulic and water quality behavior within drinking water distribution systems. MIKE NET was developed in cooperation with BOSS International (Madison, Wisconsin, www.bossintl.com). MIKE NET uses the EPANET 2.0 numerical engine.

Data can be shared with any standard Windows spreadsheet (e.g., Microsoft Excel) or relational database (e.g., Oracle, Microsoft SQL Server, Informix, and Sybase) either directly or using Open Database Connectivity (ODBC) links, or by simply cutting to and pasting from the Microsoft Windows clipboard.

Important GIS features of MIKE NET are described below:

- MIKE NET can read the network models directly from an ArcInfo, ArcFM, ArcView, or MapInfo GIS database. MIKE NET can build a link to a GIS database structure, using attribute mapping and geocoding. For example, a pump or valve can be represented as either a node or an arc in the originating GIS database, while still being linked with MIKE NET.
- MIKE NET can read the node elevation field from Shapefile polygons and assign it to each junction node within the polygon.

- MIKE NET can import a fire hydrants layer to automatically determine which pipe the hydrant belongs to. It temporarily inserts the hydrant into the pipe at the hydrant location, performs the fire flow simulation, and resets the network back to the original state without the hydrant.
- MIKE NET allows the user to assign nodal demands to the entire network system using the program's Distributed Demands capability. In addition, GIS layers can be used to geocode water consumption as well as multiple demand patterns from land-use and population-density aerial coverages. Multiple overlying layers can be aggregated and assigned to the appropriate junction nodes.
- It can perform automated and/or user-assisted skeletonization of the distribution network. Node demands are automatically accumulated during the skeletonization process so that the skeletal model continues to perform and respond closely to those of the original network model.

Other Programs

CEDRA AVWater from Cedra Corporation (Rochester, New York) provides an ArcView 3.x GIS extension interface to the University of Kentucky's KYPIPE and EPA's EPANET models.

The InfoWorks WS from Wallingford Software (Wallingford, United Kingdom) imports models from any data source using the InfoWorks CSV data gateway. It exchanges data with MapInfo Professional and ArcView GIS and exports data and simulation results to MapInfo Professional and ArcView. InfoNet has a multiuser database with built-in data models for water, wastewater, and stormwater infrastructure; a GIS user-interface; and a report generator. It can be integrated with GIS, Desktop Office systems, hydraulic modeling systems, maintenance management systems (MMS), field data systems, SCADA, and corporate databases.

PIPE2000 is the hydraulic modeling software from the developers of the legacy KYPIPE program from the University of Kentucky (Lexington, Kentucky). PIPE2000 uses the EPANET simulation engine for water quality simulation. The Pro version of PIPE2000 imports and exports PIPE2000 data to and from Shapefiles.

SynerGEE Water (formerly Stoner Model) is a hydraulic modeling software from Stoner Advatica, Inc. (Carlisle, Pennsylvania). Water Solver uses SynerGEE's simulation engine as a Microsoft Component Object Model (COM)-compliant component. This capability allows integration of hydraulic simulation capability into a GIS.

EPANET AND ARCVIEW INTEGRATION IN HARRISBURG

A representative example of model integration is presented in this section. Additional water system modeling case studies are provided in Chapter 17 (Applications Sampler).

Chester Engineers (Pittsburgh, Pennsylvania) developed a GIS integrated model by combining ESRI's ArcView GIS software and EPA's EPANET hydraulic modeling program (Maslanik and Smith, 1998). EPANET provides extensive capability to support decisions in the operation, management, planning, and expansion of water distribution systems. ArcView is a user-friendly desktop

Table 12.2 AVNET Functions

Function	Description
Menu Functions	
Set Environment	Enter model execution parameters (e.g., length of simulation)
Make Input	Makes EPANET input file
Run EPANET	Launches the EPANET program
Results Read	Translates the EPANET output file into GIS database files
Model Min/Max	Finds the minimum and maximum EPANET output parameters
Results Join	Joins the EPANET results for a selected hour or minimum/maximum values to a layer
Results View	Sets the legend to an EPANET output parameter
Remove Results	Removes the join of the EPANET results from a theme
Tool Functions	
N Tool	Adds a water system node
L Tool	Adds a water system main
S Tool	Splits an existing water system line and adds a node

mapping and GIS tool, which can be learned by most users without extensive training and experience. Due to its low cost, ease of use, and customization capability, ArcView was found to be a suitable tool for developing the EPA-NET–GIS integration.

The custom ArcView 3.x and EPANET 2.x integration is called AVNET. AVNET performs preprocessing of GIS data to create model input files. It also conducts postprocessing of model results to create GIS maps of the model output results. Both the GIS and the modeling functions are available in ArcView. There is no need to exit ArcView to edit or run the model. AVNET provides seamless model input creation, model data editing, model execution, and model output display capabilities from within ArcView GIS. Other important features of AVNET are:

- As shown in the figure on the first page of this chapter, AVNET adds a new EPANET menu to ArcView's standard interface. The AVNET menu provides eight functions listed in Table 12.2.
- As shown in Figure 12.5, AVNET adds three new editing tools to ArcView's standard interface. These tools provide three functions listed in Table 12.2. These CAD-like editing tools allow adding nodes and pipes and on-screen input of model attributes.
- AVNET uses ArcView to modify the configuration of the water distribution system.
- Water mains and links can be added interactively while maintaining the required network topology.
- AVNET compiles new EPANET model input files reflecting the edits made by the user.
- AVNET requires ArcView Shapefiles or ArcInfo coverages for nodes and pipes. The required attributes for these layers are shown in Table 12.3.

Table 12.3 AVNET Attributes

Attribute	Type	Description
Attributes for the Node Layer		
Node	Numeric	Node ID number
X_coord	Numeric	X coordinate value
Y_coord	Numeric	Y coordinate value
Type	Character	Node type (JUNCTION or TANK)
Elev	Numeric	Node elevation
Demand	Numeric	Node demand (water consumption)
Init_level	Numeric	Initial level of water system tank
Min_level	Numeric	Minimum level of water system tank
Max_level	Numeric	Maximum level of water system tank
Diam	Numeric	Diameter of water system tank
Attributes for the Pipe Layer		
Link	Numeric	Line ID
Start_node	Numeric	Node ID number of starting node of pipe
End_node	Numeric	Node ID number of ending node of pipe
Type	Character	Main type (PIPE or PUMP)
Length	Numeric	Pipe length
Diam	Numeric	Pipe diameter
Rough_coef	Numeric	Pipe roughness coefficient
Pump_H	Numeric	Pump design head
Pump_Q	Numeric	Pump design flow

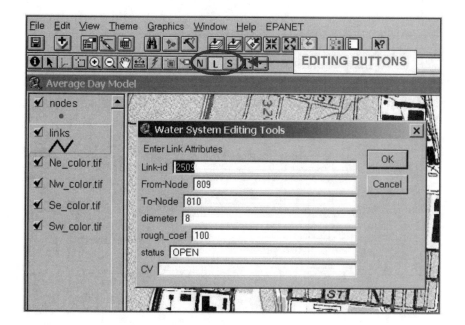

Figure 12.5 Editing pipe attributes in AVNET model.

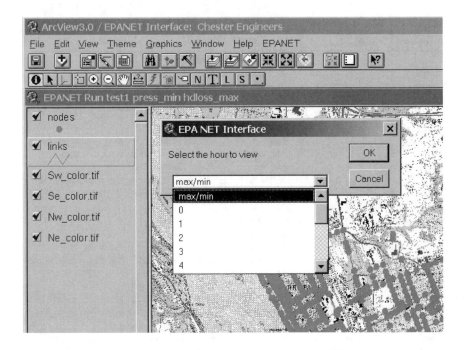

Figure 12.6   Reading the EPANET model output results in ArcView GIS.

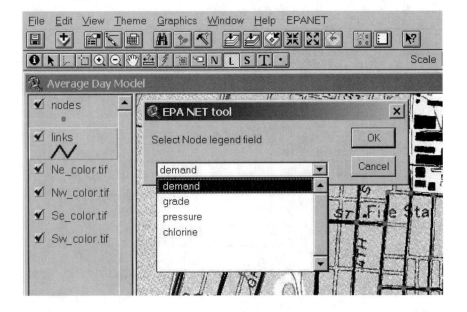

Figure 12.7   Classifying the legend to create model output thematic maps.

- AVNET creates a map file for EPANE model that contains a line for each node with the node ID and the x and y coordinates. The data are extracted from the water system node attribute table.
- AVNET runs the EPANET model from within ArcView.
- AVNET uses ArcView to map the water distribution system model and graphically display the results of the simulation on a georeferenced base map. The output file is linked to the GIS attribute files for pipes and nodes. As shown in Figure 12.6, AVNET reads the output file to extract the output parameters for minimum and maximum values or values at any user-specified simulation time step. Thematic maps of model output parameters (flows, pressures, residual chlorine, etc.) are created by classifying the ArcView theme legends (Figure 12.7).

Figure 12.8 shows the AVNET model for the city of Harrisburg, Pennsylvania. This model consists of 2500 pipes and 1650 nodes, 4 tanks, and 6 pumps. It shows layers for model nodes and links (pipes). The base-map layer consists of a USGS digital raster graphic (DRG) layer. A DRG is a TIF file of a scanned USGS 7.5-min paper topographic map.

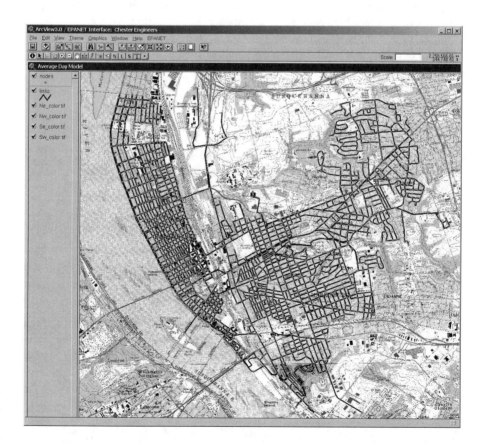

Figure 12.8 AVNET model for the City of Harrisburg, Pennsylvania.

Mapping the Model Output Results

The power of GIS gives a new meaning to the popular phrase "a picture is worth a thousand words." In fact, the thematic maps created by GIS are so effective in displaying large amounts of complex data that the popular phrase can be "plagiarized" to create a new cliché "a map is worth a thousand numbers."

The model output results in tabular or spreadsheet format are not very meaningful for nonmodelers. GIS technology is an effective means of bridging the gap between modeling information and its recipients. GIS can convert textual output results into easily interpretable thematic maps, for example, a map of pipes color-coded by flows and nodes color-coded by pressures. Figure 12.9 shows the Harrisburg model output results presented as a thematic map in ArcView GIS. The first layer (represented as dots) shows the minimum working pressures at the nodes. The second layer (represented as lines) shows the maximum head losses for the pipes. An overlay of the two layers clearly shows areas where pressures are low and head losses high. Such areas usually represent inadequate pipe sizes for delivering required flow rates and might be targeted for water main installation and replacement work. In fact, the GIS can automatically flag such hot spots for further study. The model output thematic maps are excellent for conveying modeling results to nontechnical audiences such as members of the city council or water utility board, or the general public.

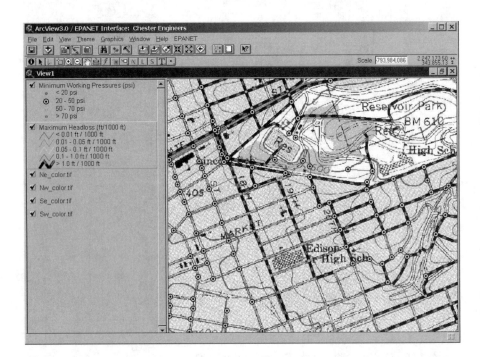

Figure 12.9 EPANET output results presented as a map in ArcView GIS.

NETWORK SKELETONIZATION

GIS and CAD layers of water distribution systems often have a level of detail that is unnecessary for analyzing system hydraulics. This can include hydrants, service connections, valves, fittings, and corner nodes. Although development of models that are an exact replica of the real system is now possible due to enormously powerful computers, the cost of purchasing, developing, and running very large models can be excessive. Attempting to include all the network elements could be a huge undertaking that may not have a significant impact on the model results (Haestad Methods, 2003). Modelers, therefore, resort to a network simplification process, which is also referred to as the "skeletonization" process.

Network simplification is needed to increase processing speed without compromising model accuracy. Manual skeletonization is a cumbersome process. Modelers, therefore, have resorted to rules-of-thumb, such as excluding all pipes smaller than a certain size. This approach requires a significant effort in locating and removing the candidate pipes. In addition, the network integrity and connectivity might be adversely impacted, with a potential to alter the output results significantly.

An example of how GIS performs skeletonization is shown in Figure 12.10 to Figure 12.16. Figure 12.10 shows a portion of the water distribution system of the city of Sycamore, Illinois. It shows the GIS layers for parcels, water mains, service lines, service valves, distribution system valves, and fire hydrants. If a modeling node is created at each pipe junction, valve, and hydrant, the resulting network model (shown in Figure 12.11) would have 93 nodes and 94 pipes (links). If this process is extended to the entire system, we could expect thousands of nodes and pipes. For large

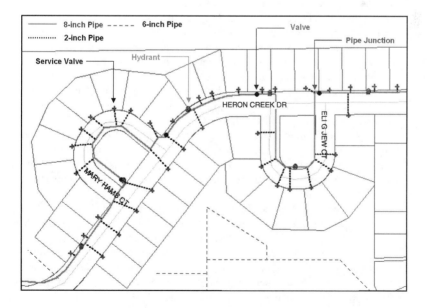

Figure 12.10 GIS layers for the Sycamore water distribution system.

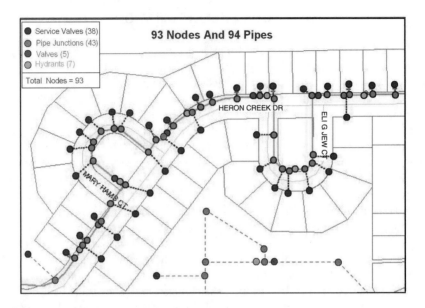

Figure 12.11 Network model without skeletonization.

systems, this process can generate several hundred thousand nodes and pipes. Skeletonization can be performed in a GIS using the following steps:

1. It is not necessary to create nodes at the fire hydrants because fire flow demand can be modeled at a nearby node. Thus, nodes corresponding to fire hydrants can be eliminated without compromising the modeling accuracy. Figure 12.12 shows the resulting skeletonized network with 86 nodes and 87 pipes. This represents an 8% reduction in the number of nodes and pipes.

2. It is not necessary to create nodes at all the distribution system valves. For example, the effect of an isolation valve (e.g., gate valves and butterfly valves) can be modeled as part of a pipe by using minor loss coefficients. In some models, closed valves and check valves can be modeled as links. Thus, nodes corresponding to most valves can be eliminated without compromising modeling accuracy. Figure 12.13 shows the skeletonized network after removing the valve nodes. It now has 81 nodes and 82 pipes representing a 13% reduction in the number of original nodes and pipes. This level of skeletonization is considered minimum.

3. It is not necessary to model corner nodes that describe actual layout of pipes. The pipe layout does not affect model output as long as the pipe length is accurate. Any potential head loss due to sharp turns and bends can be modeled using minor loss coefficients or friction factors. Thus, nodes corresponding to corner nodes can be eliminated without compromising modeling accuracy. Figure 12.14 shows the skeletonized network after removing the corner nodes. It now has 78 nodes and 79 pipes, representing a 16% reduction in the number of original nodes and pipes.

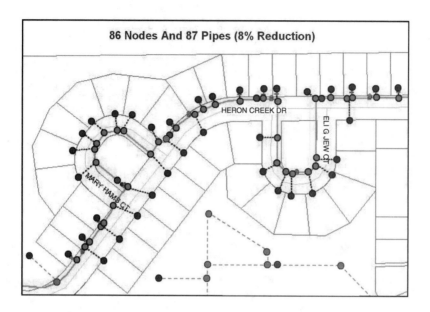

Figure 12.12 Skeletonized network model after removing nodes at fire hydrants.

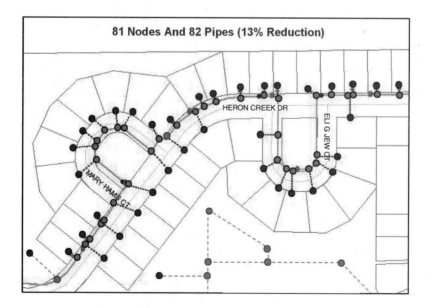

Figure 12.13 Skeletonized network model after removing nodes at valves.

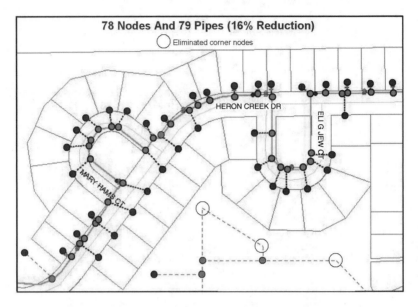

Figure 12.14   Skeletonized network model after removing corner nodes.

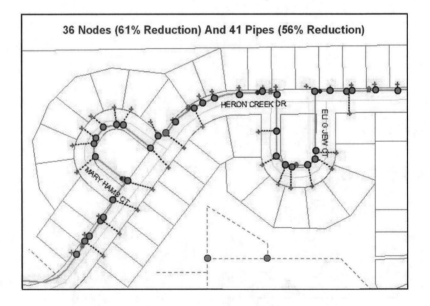

Figure 12.15   Skeletonized network model after removing pipes less than 8 in.

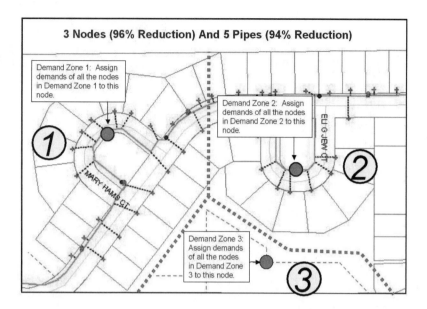

Figure 12.16 Skeletonized network model after aggregating demand nodes.

4. Small pipes with low flows can also be eliminated without significantly changing the modeling results. There are no absolute criteria for determining which pipes should be eliminated (Haestad et al., 2003). Elimination of pipes less than 6 in. or 8 in. is most common. Thus, small pipes and corresponding nodes can be eliminated for further network simplification. Figure 12.15 shows the skeletonized network after removing pipes of less than 8 in. width. The most dramatic effect of this step is the removal of all the service lines from the network model. The network now has 36 nodes and 41 pipes, representing a 56% reduction in the number of original nodes and pipes. This moderate level of skeletonization is acceptable in most cases.

5. Further reduction in the size of a network model is possible by aggregating the demands from several customers to a single node. Figure 12.16 shows the skeletonized network after dividing the network into three demand zones and assigning the total zone demand to a node inside that zone. The network now has only 3 nodes and 5 pipes, representing a 96% reduction in the number of original nodes and a 94% reduction in the number of original pipes. This high level of skeletonization can affect modeling results. For example, this model cannot model the flows and pressures at different houses within the subdivision.

Although the preceding steps can be performed manually in a GIS by editing the data layers, the process will be too cumbersome unless the system is very small. Automatic GIS-based skeletonization that can be completed quickly is now available in some software packages. Automatic skeletonization removes the complexity from the data, while maintaining the network integrity and hydraulic equivalency of the system. Reductions of 10 to 50% in the size of a network are possible. These programs allow the users to input the skeletonization criteria such as the smallest

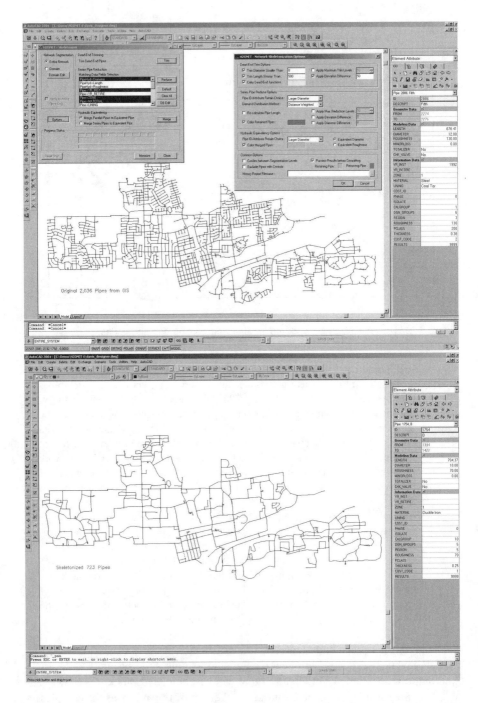

Figure 12.17   Skeletonizer's network simplification results. Top: original system with 2036 pipes. Bottom: skeletonized network with 723 pipes (65% reduction). (Maps courtesy of MWH Soft, Inc.)

pipe size that should be included in the model. The smallest-pipe-size criterion should be used carefully because both size and accuracy of the skeletonized model depend on it. A highly skeletonized model does not work well in areas where most of the water supply is through small pipes.

The Skeletonizer™ module of H₂OMap water system modeling software (MWH Soft, Inc., Broomfield, Colorado, www.mwhsoft.com) and Skelebrator™ module of WaterCAD and WaterGEMS water system modeling software (Haestad Methods, Waterbury, Connecticut, www.haestad.com) are examples of software tools that automate the skeletonization of CAD and GIS layers. Figure 12.17 shows an example of Skeletonizer skeletonization. The top screenshot shows the original network with 2036 pipes imported from a GIS. The bottom screenshot shows the skeletonized network with 723 pipes, indicating a 65% reduction in the size of the network.

Mohawk Valley Water (MVW) is a water utility serving more than 125,000 people in approximately 150 mi² area in central New York. In 2002, MVW created an all-pipe WaterCAD hydraulic model that was connected to a geodatabase. A reduce/skeletonize/trim (RST) process was used to eliminate hydrant laterals and private service connections to scale the hydraulic model to pipe segments 4-in. and greater. The demands, pipe attributes, and elevations were extracted from the geodatabase (DeGironimo and Schoenberg, 2002).

ESTIMATION OF NODE DEMANDS

Node demand is an important input parameter of a hydraulic model, representing the water consumption rate in units of volume/time (e.g., gallons per day or liters per second). Demands are grouped by customer type (e.g., residential, commercial, industrial, or institutional) or magnitude (e.g., average daily, maximum daily, peak hourly, or fire flow). Manual estimation of node demands from customer billing records is a cumbersome process. Modelers, therefore, have traditionally resorted to approximate methods, such as uniform distribution of the residential water use across all nodes followed by adding the nearest commercial and industrial use at various nodes. This approach requires a significant effort in fine-tuning the node demands during the model calibration phase of the project.

GIS allows implementation of more accurate demand estimation methods. One such method utilizes the GIS to geocode (or geolocate) all individual water users (defined by the address at which each meter is located) and to assign the metered water consumption rates for each meter location to the nearest node in the model. The geocoding process links a street address to a geographic location and finds the coordinates of a point location based on its street address. This method assumes that a computer file (spreadsheet or delimited text file) containing meter addresses and metered water use for specified durations (e.g., quarterly or monthly) can be produced from the utility's billing database. If this information is available, the following procedure can be used to assign demands to nodes:

1. Create a GIS layer (e.g., a Shapefile) representing all nodes in the model.
2. Obtain a street centerline layer of the system with address ranges. Such data can also be purchased from commercial GIS data vendors (Shamsi, 2002).
3. Obtain and preprocess customer meter location and consumption database.
4. Process the meter address and street address range data in the GIS to geocode the meter locations into a point layer registered to the model node layer.
5. Process the geocoded meter data to assign and aggregate individual demands to the nearest nodes.
6. Analyze the assigned metered demand to determine unaccounted-for water and metered demands that were not successfully geocoded. Uniformly distribute these remaining demands across nodes.
7. Load the node demand data into the hydraulic model.

The preceding process produces a highly accurate distribution of water use across the nodes. The intermediate product of point locations with associated demands can also be used to visualize the spatial distribution of individual demand points throughout the system. A metered demand database can also be created to include customer classes as defined in the billing system. This data can be used to assign differing demand patterns to residential, commercial, and industrial demand classes for use in extended-period simulations.

Haestad Methods' WaterGEMS hydraulic modeling software has a Loadbuilder module that allows modelers to automatically calculate and assign demands to junctions using the following three methods (Haestad Methods, 2002):

1. Customer meter data method: This method is used if geocoded customer billing data are available. It assigns the demands to nodes using any combination of the following three methods:
   - Nearest node method: This method assigns the meter demand to the nearest node.
   - Nearest pipe method: This method assigns the meter demand to the two nodes of the nearest pipe. Each node of the nearest pipe can receive equal demands or weighted demands in proportion to its distance from the meter location. Alternatively, the demand can be assigned to the pipe node closest to the meter location.
   - Meter aggregation method: This method creates *service areas* around the nodes of a hydraulic model. A service area polygon is associated with each model node, enclosing all service connections that consume their water from the node and whose consumption is aggregated to that node. Service areas can be drawn manually or using automatic methods such as the Theissen Polygon method. Programs are available for automatic delineation of Theissen Polygons in a GIS package. All the meters located within a service area are assigned to its associated node. Figure 12.18 shows the service areas created using the Thiessen Polygon method in WaterGEMS software.
2. Area water use data method: This method is used if billing meter data are not available. It assigns the demands to nodes using the following three methods:
   - Equal flow distribution: This method requires a polygon layer for various flow zones (e.g., meter routes or pressure zones) in the system. A GIS overlay of nodes and flow zone layers is conducted to divide each zone's flow equally among all the nodes within that zone.

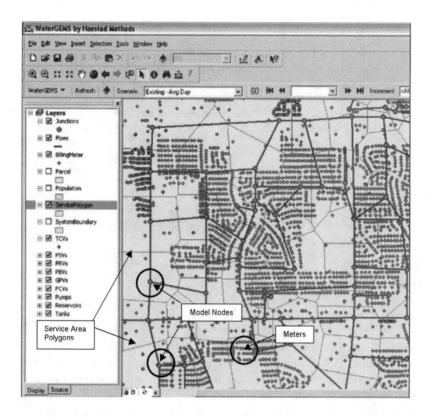

Figure 12.18 Example of meter aggregation method (WaterGEMS™ screenshot courtesy of Haestad Methods).

- Proportional distribution by area: In this method, a GIS overlay of service area (of nodes) and flow zone layers is conducted. Each zone's flow is divided among all the service area polygons within that zone in proportion to the area of service area polygons.
- Proportional distribution by population: In this method, a GIS overlay of service area (of nodes) and flow zone layers is conducted. Each zone's flow is divided among all the service area polygons within that zone in proportion to the population of service area polygons. The service area population can be calculated in GIS by conducting an overlay of service area polygons with census-block[*] (or census-tract) polygons.
3. Land-use/population data method: This method forecasts the future demands for nodes using the following two methods:
   - Land-use data method: This method uses future land-use polygons to forecast future demands. A GIS overlay of service area and land-use polygons is conducted to calculate the area of each land-use type within each service area. A demand factor (per unit area) is specified for each land-use type. Calculated

---

[*] Census blocks and census tracts from U.S. Bureau of Census are polygon layers of demographic data for the U.S. Additional information is available in Chapter 11.

land-use areas and demand factors are multiplied to estimate a future demand for each service area.

- Population data method: This method uses projected future population polygons to forecast future demands. A GIS overlay of service area and population polygons is conducted to calculate the future population of each service area. A demand density (gallons per capita per day) is specified by the user. Calculated future population and demand density are multiplied to estimate a future demand for each service area.

Detailed information on the preceding methods is available in *Advanced Water Distribution Modeling and Management* (Haestad et al., 2003). For future land-use mapping techniques and information on how to calculate population from U.S. Bureau of Census GIS data (e.g., Census blocks and tracts), please refer to *GIS Tools for Water, Wastewater, and Stormwater Systems* (Shamsi, 2002).

Demand-Estimation Case Studies

Newport News, Virginia

In 1996, Newport News Waterworks (Newport News, Virginia) developed a water system hydraulic model composed of 540 nodes representing approximately 110,000 service connections. The model contained all pipes 12 in. and larger and some 8 in. critical pipes. Model node demands were computed using the meter aggregation method described in the previous section. The service area polygons were drawn manually, which was found to be the largest engineering effort required by the modeling project. A GIS Demand Application was developed to create, modify, or delete model nodes and service area polygons. The application was developed using ArcTools, a collection of ArcInfo productivity tools implemented through an Arc Macro Language (AML) interface.

The demand data were extracted from a public utilities billing system. Average daily consumption of each customer was calculated by dividing the total consumption by the number of days in the billing cycle. The consumption data were loaded in a tabular GIS file. Account numbers were used to associate demands with service connections. Consumption for all service connections that geographically fell within a service area polygon were aggregated. The aggregated consumption was stored as an attribute of the model node (Sevier and Basford, 1996).

Round Rock, Texas

The City of Round Rock (Texas) used Haestad Methods' WaterCAD hydraulic modeling software, ESRI's ArcView GIS package, and Microsoft's Access billing database for estimating the node demands. Approximately 95% of the city's 22,000 water meters were assigned to precise geographic locations using geocoding. This approach provided accurate node demand data and reduced model calibration time (Haestad Methods, 2002). The following steps were undertaken:

1. The geocoding process was performed in ArcView by linking the addresses within an existing parcels Shapefile to the billing address in the water meter database.
2. Using WaterCAD's Shapefile connection wizard, the node data from WaterCAD was synchronized to ArcView.
3. The modelers then drew polygons around the nodes and corresponding meters.
4. ArcView was used to sum the demand data for the meters in each polygon.
5. The total demand was applied to appropriate nodes.
6. The junction file with the newly calculated node demands was synced back to WaterCAD.

Lower Colorado River Authority, Texas

Pimpler and Zhan (2003) used an alternative grid-based (raster GIS) approach that worked well for estimating water demand for Lower Colorado River Authority. The Authority serves more than 1 million customers in Texas in an area spanning more than 29,800 mi^2. Their method involved four steps:

1. Divide the service area into grid–cells.
2. Calculate a score for each grid–cell based on weighted variables (e.g., land use, proximity to employment centers, etc.) considered important for water demand analysis.
3. Distribute the service area population proportionally to each grid–cell based on the score assigned to each cell.
4. Estimate water demand associated with each cell based on its allocated population.

ESTIMATION OF NODE ELEVATIONS

Node elevation is also an important input parameter for a hydraulic model. It is usually defined by the ground elevation. Because 2.31 ft of elevation corresponds to 1 psi of water pressure, modeling water pressure to 1-psi precision requires elevation data accurate to approximately 2 ft. Table 12.4 provides the accuracy of common sources of ground surface elevation. USGS topographic maps are generally acceptable because an elevation accuracy of 5 ft is adequate for most nodes (Walski, 2001).

Surveying provides the most accurate ground surface elevations. Unfortunately, surveying is very costly and time consuming and is recommended for critical model nodes such as the centerline of a pump or a model calibration node (Walski, 2001). Manual estimation of node elevations from contour maps or as-built plans is also a

Table 12.4 Accuracy of Ground Elevation Data Sources

Source	Accuracy (ft)
USGS 7.5-min 1:24,000-scale topographic maps	10
USGS 30-m DEMs	5–10
GPS survey — code phase	10
GPS survey — carrier phase	1
Digital orthophotos	1

cumbersome process. A GIS allows automatic computation of node elevations from DEM data using the following procedure:

1. Create a GIS layer (e.g., a Shapefile) for the nodes in the model.
2. Obtain a DEM layer covering the water distribution service area.
3. Georeference (georegister) the node and DEM layers (i.e., assign both layers the same map projection system).
4. Overlay the node layer over the DEM layer to assign DEM elevations to nodes. Interpolate in between the DEM grid cells (pixels) to estimate the correct elevation at each node.
5. Load the node elevation data into the model.

Figure 12.19 shows an overlay of a water distribution system layer over a DEM layer for calculating node elevations.

WaterGEMS has a TRex Wizard that automates extraction of elevation data from a DEM layer and assigns the extracted data to model nodes. TRex uses the preceding steps to calculate node elevations. It interpolates in between the DEM pixels (grid–cells) rather than using a simple spatial join. A spatial join is less accurate

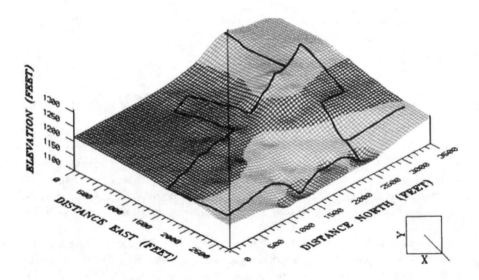

Figure 12.19 Water distribution system draped over a DEM

because nodes that fall in between the DEM pixels would be assigned the elevation of the nearest cell.

PRESSURE ZONE TRACE

A GIS also allows identification of various pressure zones in a water distribution network. This is done by selecting a network pipe or node that is part of a larger pressure zone. The GIS traces the water network in all possible directions until either sources of water or assets (e.g., pressure regulating valves — PRVs) that isolate water flow are found. All the pipes (or nodes) in the selected pressure zone are highlighted. A buffer can be created around the highlighted features to create a polygon for the pressure zone boundary. Once the trace is complete, the user can assign a pressure zone attribute to the selected pipe or node.

CHAPTER SUMMARY

This chapter shows that GIS offers many useful applications for water distribution systems. The applications described in this chapter indicate that GIS provides many powerful and efficient tools for performing various hydraulic modeling tasks such as network skeletonization, demand calculation and allocation, and elevation calculations. The case studies and examples presented in the chapter illustrate that many water utilities are successfully using GIS applications for hydraulic modeling of their water distribution systems. Various software examples presented in the chapter indicate that many commercial off-the-shelf software packages are available to help users benefit from hydraulic modeling applications of GIS.

CHAPTER QUESTIONS

1. Prepare a list of GIS applications for water distribution systems.
2. Which hydraulic modeling tasks can be performed using GIS?
3. What are the typical input parameters for a water distribution system hydraulic model? Which of these parameters can be obtained from a GIS?
4. List the GIS data layers that are useful for hydraulic modeling.
5. How does a GIS help in presenting model output results?
6. What is skeletonization? How does a GIS help this task?
7. List the hydraulic models you are familiar with. Which of these models use a GIS? Which method of GIS linkage (interchange, interface, or integration) do they use?

CHAPTER 13

Sewer Models

GIS saves time and money in developing sewer system models for simulating flows and depths in the collection system. GIS also helps to prepare maps of the model results, which can be easily understood by nonmodelers.

GIS-based sewer models bridge the gap between information and its recipients.

LEARNING OBJECTIVES

The learning objectives of this chapter is to understand how GIS can be applied in developing sewer system hydraulic models, and to understand how GIS can be used to present modeling results

MAJOR TOPICS

- Representative sewer models
- Storm Water Management Model (SWMM)
- Graphical user interface (GUI)
- SWMM applications
- Data preparation
- Interface development steps
- GIS application case studies

LIST OF CHAPTER ACRONYMS

CMOM   Capacity, Management, Operations, and Maintenance
CSO   Combined Sewer Overflow
DSS   Decision Support System
EPA   (U.S.) Environmental Protection Agency
GUI   Graphical User Interface
HEC   Hydrologic Engineering Center
HGL   Hydraulic Gradient Line
LTCP   Long-Term Control Plan (for CSOs)
NMC   Nine Minimum Controls (for CSOs)
NPDES   National Pollution Discharge Elimination System
ODBC   Open Database Connectivity
OLE   Object Linking and Embedding
SHC   System Hydraulic Characterization (for CSOs)
SSO   Sanitary Sewer Overflow
SWMM   Storm Water Management Model

MAPINFO™ AND SWMM INTERCHANGE

Application area	Basement flooding modeling using the Interchange GIS Linkage Method described in Chapter 11 (Modeling Applications)
GIS software	MapInfo
Modeling software	SWMM
Other software	Microsoft Access
GIS data	Sewer pipes, aerial photographs, buildings and parking lot polygons
GIS format	Raster and vector
Study area	City of Portland, Oregon
Organization	Bureau of Environmental Services, City of Portland, Oregon

In the late 1990s, Portland's Bureau of Environmental Services (BES) developed a suite of GIS and database tools to develop data for SWMM models. Model data were

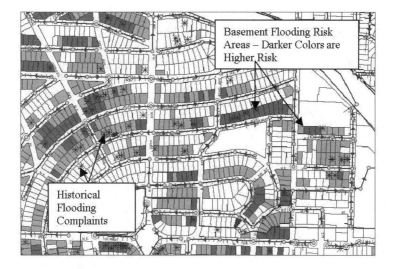

Figure 13.1 Thematic map showing flooding risk and historical complaints (map courtesy of Computational Hydraulic Int.).

developed from maps and databases with automation of many of the steps for data extraction, variable assignments, and model building. Detailed collection-system models of basins were developed to analyze local areas. These models contained essentially all of the pipes in the collection system and utilized two subcatchment definitions to place runoff into the system in the correct location. Surface water subcatchments were defined as the areas that flowed into the sewer through street inlets in the public right-of-way. Direct connection subcatchments were defined as the areas that contributed to sanitary and stormwater drainage through service laterals.

A GIS tool was developed to predict flooding of individual parcels by overlaying modeled hydraulic gradient lines (HGLs) with estimated finished floor elevations obtained from a DEM. Detailed SWMM RUNOFF and EXTRAN models were developed from facility maps and photogrammetric data with the use of a MapInfo GIS and a Microsoft Access database. The SWMM model showed that many of the system capacity problems were in the local areas at the upstream ends of the system and that the local collection system tended to hold water back from the downstream trunk sewers in the system. This better understanding of where flooding was likely to occur and where system capacity constraints actually occurred allowed BES a top-down view of the basin during alternatives analysis. Local inflow control alternatives were considered first, and pipe upsizing along with passing the problem downstream was the second alternative. Figure 13.1 shows a thematic map showing flooding risk and historical complaints (Hoffman and Crawford, 2000).

GIS APPLICATIONS FOR SEWER SYSTEMS

GIS applications for sewer systems include (Shamsi, 2002):

- Performing H&H modeling of collection systems, including:
  - Automatic delineation of watersheds, sewersheds, and tributary drainage areas

- Model simplification or skeletonization to reduce the number of manholes and sewers to be included in the H&H model
- Estimating surface elevation and slope from digital elevation model (DEM) data
- Estimating dry-weather sewage flow rates from land use, census data, and billing records
- Estimating wet-weather sewage flow rates from land use, soil, surface imperviousness, and slope
- Creating maps for wastewater National Pollution Discharge Elimination System (NPDES) permit requirements (as described in Chapter 9 [Mapping Applications]), such as:
  - U.S. Environmental Protection Agency's (EPA) combined sewer overflow (CSO) regulations, such as System Hydraulic Characterization (SHC), Nine Minimum Controls (NMC), and long-term control plan (LTCP)
  - U.S. EPA's sanitary sewer overflow (SSO) regulations, such as capacity, management, operations, and maintenance (CMOM)
- Documenting field inspection data (as described in Chapter 15 [Maintenance Applications]), including:
  - Work-order management by clicking on map features
  - Inspection and maintenance of overflow structures and manholes
  - Television (TV) inspection of sewers
  - Flow monitoring and sampling
  - Smoke-testing, dye-testing, and inflow/infiltration (I/I) investigations
- Performing and planning tasks such as assessment of the feasibility and impact of system expansion

In this chapter, we will focus on the sewer system H&H modeling applications of GIS.

SEWER SYSTEM MODELING INTEGRATION

Tight integration between a sewer model and a GIS is highly desirable.

The GIS applications in sewer system modeling are developed by using the three application methods (interchange, interface, and integration) described in Chapter 11 (Modeling Applications). These methods facilitate preparation of model-input data and mapping of model output results similar to their application for water system models described in Chapter 12 (Water Models).

Transfer of data between a GIS and a sewer model is handled differently by different software vendors. Some products offer nothing more than a manual cut-and-paste approach for transferring data between a GIS and the model. Others offer a truly integrated package. Obviously, tight integration between the sewer model and the GIS is desirable. However, it should not be assumed that a tightly integrated model is the best modeling tool. For example, the interchange or interface capability of a comprehensive sewer model might be more useful for some users compared with a tightly integrated model having limited modeling capability to simulate complex situations such as surcharged sewers.

Table 13.1 Representative Sewer System Modeling Software

Software	GIS Linkage Method	Vendor	Web Site
SWMM	Interchange	U.S. Environmental Protection Agency (EPA)	www.epa.gov/ceampubl/ swater/swmm/index.htm
CEDRA AVSand	Integration	CEDRA Corporation	www.cedra.com
H₂OMAP Sewer H₂OVIEW Sewer	Interface and Integration	MWH Soft	www.mwhsoft.com
InfoWorks CS and InfoNet	Interface and Integration	Wallingford Software	www.wallingfordsoftware.com
Mouse GM and MIKE SWMM	Interface	DHI Water & Environment	www.dhisoftware.com
PCSWMM GIS	Integration	Computational Hydraulics Int.	www.computationalhydraulics .com
StormCAD and SewerCAD	Interface	Haestad Methods	www.haestad.com
XP-SWMM	Interchange and Interface	XP-Software	www.xpsoftware.com

SOFTWARE EXAMPLES

Sewer system modeling software is used to compute dry- and wet-weather flow for hydraulic analysis and design of collection systems and wet-weather control facilities such as storage tanks and equalization basins.

Table 13.1 lists representative sewer system modeling packages and their GIS capabilities, vendors, and Web sites. The salient GIS features of some programs are described in the following subsections.

What is common among the recent software developments is a transferability of fundamental database information, also referred to as a decision support system (DSS). Under a DSS framework, neither the GIS nor the model is central to the process. Both perform satellite functions for a central master database. (Heaney et al., 1999).

SWMM

Numerous hydrologic models were created in the U.S. during the 1970s, including the U.S. EPA's legacy computer program SWMM and the U.S. Army Corps of Engineers' Hydrologic Engineering Center's HEC series of models (HEC-1 through 6). Two of the most popular models, HEC-1 and HEC-2, have been updated and renamed HEC-HMS and HEC-RAS.

Collection system hydraulics can be characterized by using an H&H computer model such as SWMM. SWMM is a comprehensive computer model for analysis of quantity and quality problems associated with urban runoff. Both continuous and single-event simulation can be performed on catchments having sanitary sewers, storm sewers, or combined sewers and natural drainage, for prediction of flows, stages, and pollutant concentrations.

SWMM is one of the most successful models produced by EPA for the wastewater environment. SWMM was developed from 1969 to 1971 as a mainframe computer program and has been continually maintained and updated. Versions 2, 3, and 4 of SWMM were distributed in 1975, 1981, and 1988, respectively. The first batch-mode microcomputer version of SWMM (Version 3.3) was released by EPA in 1983. The first conversational-mode user-friendly PC version of SWMM, known as PCSWMM, was commercially distributed in 1984 by Computational Hydraulics, Inc., of Guelph, Ontario (Computational Hydraulics, 1995). The first PC version of EPA's SWMM was Version 4, which was distributed in 1988 (Huber and Dickinson, 1988; Roesner et al., 1988). An excellent review of SWMM's development history can be found in the book by James (1993).

SWMM is regarded as the most widely used urban H&H model in the U.S. (Heaney et al., 1999). SWMM continues to be widely used throughout the world for analysis of quantity and quality problems related to stormwater runoff, combined sewers, sanitary sewers, and other drainage systems in urban areas, with many applications in nonurban areas as well. From 1988 through 1999, the EPA Center for Exposure Assessment Modeling in Athens, Georgia, distributed approximately 3600 copies of SWMM. The University of Florida distributed roughly 1000 copies of SWMM in the late 1980s. Third-party interfaces for SWMM, such as MIKE-SWMM, PCSWMM, and XP-SWMM, have several thousand users. The number of subscribers to the SWMM Users Group Internet discussion forum (Listserv) on the Internet is nearly 10,000 (EPA, 2002).

The current version (4.3) was released by EPA in May 1994 for the 16-bit MS-DOS operating system. Professor Wayne Huber of Oregon State University, one of the authors of SWMM, has developed several updated versions of SWMM 4. His latest version (4.4H) was updated in March 2002. All of these and previous SWMM versions were written in the FORTRAN programming language. None of them have a user interface or graphical capability. Users must provide ASCII text input and rely on ASCII text output. EPA is currently developing an updated version of SWMM referred to as SWMM 5. Written in the C++ programming language, SWMM 5 will incorporate modern software engineering methods as well as updated computational techniques. The modular, object-oriented computer code of SWMM 5 is designed to simplify its maintenance and updating as new and improved process submodels are developed. The new code will also make it easier for third parties to add GUIs and other enhancements to the SWMM engine. This approach follows the same approach used by the highly successful EPANET model from EPA (described in Chapter 12 [Water Models]), which analyzes hydraulic and water quality behavior in drinking water distribution systems (Rossman, 2000).

SWMM simulates dry- and wet-weather flows on the basis of land use, demographic conditions, hydrologic conditions in the drainage areas, meteorological inputs, and conveyance/treatment characteristics of the sewer system. The modeler can simulate all aspects of the urban hydrologic and quality cycles, including rainfall, snow melt, surface and subsurface runoff, flow routing through drainage

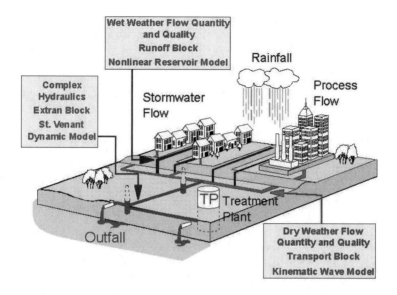

Figure 13.2 SWMM conceptual schematic.

network, storage, and treatment. Statistical analyses can be performed on long-term precipitation data and on output from continuous simulation. Figure 13.2 shows a conceptual schematic of SWMM. This figure shows that SWMM is a complex model capable of modeling various phases of the hydrologic cycle using different blocks (modules) such as RUNOFF, TRANSPORT, and EXTRAN.

SWMM can be used both for planning and design. Planning mode is used for an overall assessment of the urban runoff problem or proposed abatement options. SWMM is commonly used to perform detailed analyses of conveyance system performance under a wide range of dry- and wet-weather flow conditions. These days, SWMM is frequently used in the U.S. for modeling wet-weather overflows including CSO, SSO, and stormwater discharges from collection systems. As such, it is the model of choice for use in many collection-system modeling studies. For example, SWMM can be used to develop a CSO model to accomplish various tasks leading to the development of a CSO Plan of Actions mandated by EPA. These tasks include characterizing overflow events, developing the CSO vs. rainfall correlation, maximizing the collection-system storage, and maximizing the flow to the treatment plant. The CSO model can also be used to develop the EPA-mandated LTCPs to evaluate various CSO control options. Sizing of CSO control facilities such as a wet-weather equalization tank requires CSO volume and peak discharge, both of which can be modeled using SWMM.

The use of SWMM to model wet-weather overflows is particularly advantageous for the following reasons:

- SWMM produces estimates of present and future dry- and wet-weather flow rates. Flow estimates can be prepared based upon present and future land-use conditions, topography, sewer characteristics, and selected meteorological conditions. The model can be calibrated against measured flow rates.

- SWMM models the performance of the conveyance system under a range of dynamic flow conditions.
- Using SWMM, it is possible to assess hydraulic capacity in response to wet-weather input. This characteristic can be very useful for analyses related to abatement of overflows.

SWMM is flexible enough to allow for different modeling approaches to the same area. An approach that adequately describes the service area and accurately computes and routes the flows at reasonable computing time and effort should be adopted. The following modeling strategy is generally used for modeling wet-weather overflows:

- Delineate sewersheds (the drainage areas tributary to overflows, also referred to as subareas or subbasins).
- Use the TRANSPORT Block to generate sewershed dry-weather flows.
- Use the RUNOFF Block to generate sewershed wet-weather flows.
- Combine dry-weather and wet-weather flows to generate combined sewershed flows.
- Use the EXTRAN Block to route the flows through the collection and interceptor system.

USEFUL SWMM WEB SITES

SWMM4	www.epa.gov/ceampubl/swater/swmm/index.htm
SWMM5	www.epa.gov/ednnrmrl/swmm/
Oregon State University	ccee.oregonstate.edu/swmm/
SWMM-Users Internet Discussion Forum	www.computationalhydraulics.com/Community/ Listservers/swmm-users.html
SWMM-Online	www.swmm-online.com

SWMM Graphical User Interface

Most users have now become accustomed to modeling in a point-and-click computing environment that provides a user-friendly graphical user interface (GUI). A GUI is a computer program that acts as an interpreter between users and their computers. It is designed to minimize (but not eliminate) the need for human experts and to guide the modeler through the intricacies of a particular numerical model. A GUI provides a suite of tools to create a decision support system for the numerical model that has been adopted. It also stimulates user interest and facilitates interpretation of model results. A GUI improves productivity by increasing the efficiency of data entry, eliminating data errors through expert-checking, and the use of decision-support graphics and interpretation tools. It replaces difficult-to-remember text commands by interactive computer graphics consisting of menus, dialogue boxes, input and output windows, and icons. The main goal of using GUIs is to develop user-friendly computer applications or to add user-friendliness to existing command-line-driven applications. For example, Microsoft Internet Explorer is a GUI for user-friendly access to the Internet.

Modern software offer pull-down menus, toolbars, icons, buttons, dialogue boxes, hypertext, context-sensitive online help, etc., which are not available in legacy computer programs including SWMM 4. Many commercial modeling packages now offer on-screen point-and-click drawing and editing of drainage network maps, but unfortunately some legacy programs still rely on ASCII text input and output files. SWMM was developed in an era when input files were created on punched cards. After 30 years, SWMM 4 now runs on personal computers, but it is still a text-based, nongraphical DOS program. It reads ASCII input to produce ASCII output, which is most suitable for mainframe line printers. SWMM's ASCII format output is long, boring, difficult to interpret, and not very useful for nonmodelers. Creating computer models and reviewing the model results is often slowed by our inability to see the system being modeled. It is up to the modeler to review SWMM's voluminous output and construct a mental image of the physical system being modeled. Often, the limitation in understanding the model output has been the modeler's own comprehension of the output, not the model itself. Quite frequently, it is impossible for the modeler to absorb the large amount of information contained in the model output (TenBroek and Roesner, 1993).

A GUI and a GIS can be employed to overcome SWMM's input/output deficiencies. Two types of GUIs can be employed for GIS applications in SWMM modeling (Shamsi, 1997; Shamsi, 1998):

- An input interface (also called a front-end interface or preprocessor) extracts SWMM input from GIS layers and creates SWMM's traditional ASCII text input file. The input GUIs can also provide graphical tools to draw a network model that is subsequently converted to SWMM's ASCII input file. For example, an input interface may extract sewer-segment lengths and manhole coordinates from the sewer and manhole layers.
- An output interface (also called a back-end interface or postprocessor) converts text to graphics. It can convert SWMM's traditional ASCII text output file to graphs, charts, plots, and thematic maps that can be easily understood by everyone.

With the help of these GUI/GIS tools, everybody can understand the model output. These tools help one to see storm-surge progress through the sewer system and pinpoint areas of flooding and surcharging. More than just reams of computer paper, such models become automated system-evaluation tools. Other benefits of GUI/GIS tools are:

- Preparation of network schematics is not essential. Digitized plots of sewers and subarea boundaries can be used to create a drainage network diagram on a computer screen.
- Zoom and pan features make it possible to display even the largest networks conveniently on the screen.
- Connectivity data errors are easily detected and can be edited while still in the program. Instabilities in the model output, often the most difficult errors to find, are also easily located.

- Flow and depth data from SWMM output can be displayed in either plan or profile view, providing an animated display of the HGL during the simulation time steps.
- Flow and HGL time-series plots can be displayed for any conduit or node in the hydraulic network.
- Field-collected flow and depth data can be displayed along with the model output for model calibration and verification.
- Network graphics and modeled hydrographs can be exported to word processors to aid in report preparation.

XP-SWMM and XP-GIS

XP-SWMM by XP Software (Belconnen, Australia and Portland, Oregon) is a full-fledged 32-bit Microsoft Windows application. The program has been enhanced by the addition of a graphics database, and an adaptive dynamic wave solution algorithm that is more stable than the matrix method used in the original SWMM. The program is divided into a stormwater layer, which includes hydrology and water quality; a wastewater layer, which includes storage treatment and water quality routing for BMP analysis; and a hydrodynamic/hydraulics layer for simulation of open or closed conduits (Heaney et al., 1999).

XP-SWMM is also included in Visual SWMM from CaiCE Software Corp. (Tampa, Florida). Basically, XP-SWMM and Visual SWMM are GUI programs for SWMM. The user-friendly GUI is based upon a graphical representation of

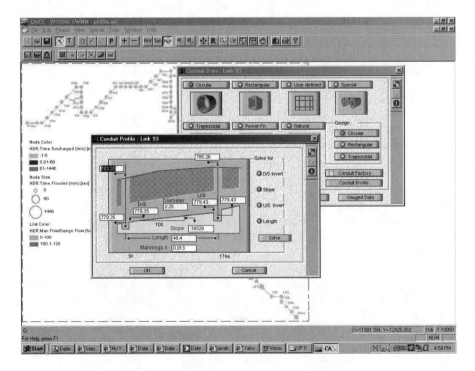

Figure 13.3   Visual SWMM screenshot.

the modeled system using a link-node architecture. It provides a shell that acts as an interpreter between the user and the model. It facilitates on-screen development of a SWMM model in the form of a link-node network. Figure 13.3 shows a Visual SWMM (Version 7) screenshot.

Because the links and nodes are set up on a coordinate-system basis, files can be translated between most CAD and GIS software systems using the interchange method. CAD or GIS files can also be used as a backdrop for the system being modeled. Although XP-SWMM is not linked or integrated to a GIS package, it provides an optional utility called XP-GIS to extract model input data from the existing GIS database tables. XP-GIS is a module for linking XP-SWMM to GIS databases. Its main purpose is to facilitate the import and export of modeling data from GIS and other data sources such as spreadsheets, asset management software, and OLE/ODBC compliant databases. It also allows for the inclusion of Shapefiles as background layers and for data to be viewed and manipulated in an XP-SWMM graphical environment.

GIS Data for SWMM

GIS layers for land use, elevation, slope, soils, and demographics can be processed to extract model parameters for input to various SWMM Blocks and other collection system models. The basic input parameters to be assembled into the RUNOFF Block input files include sewershed area, overland flow slope, overland flow width, percent imperviousness, roughness coefficients, and soil infiltration parameters. Area, slope, and percent imperviousness can be computed from the GIS layers. Table 13.2 shows sample sewershed RUNOFF data extracted from GIS layers using the interchange method for a study area in Pennsylvania.

GIS layers can also be used to estimate inputs for sewershed population, population density, number of houses, average family size, average market value of houses, average family income, and the predominant land-use type for the TRANSPORT Block. Table 13.3 shows sample sewershed TRANSPORT data extracted from GIS layers using the interchange method for a study area in Pennsylvania.

Estimating Green-Ampt Parameters Using STATSGO/SSURGO GIS Files

Based on their resolution, there are three types of U.S. Natural Resources Conservation Service (NRCS) (formerly the U.S. Soil Conservation Service or SCS) soils data that are useful in GIS applications:

- National Soil Geographic (NATSGO)
- State Soil Geographic (STATSGO)
- Soil Survey Geographic (SSURGO)

SSURGO provides the highest resolution soils data at scales ranging from 1:12,000 to 1:63,360. This resolution is appropriate for watersheds a few squares miles in area. STATSGO data are digitized at 1:250,000-scale, which is useful when

Table 13.2 Sample GIS Data for SWMM's RUNOFF Block

Sewershed	Area (Acres)	Mean PI	Mean % Slope	Land Use (Acres)					
				S.F.R.	M.F.R.	Comm.	Ind.	O.S.	S.W.
2	397.82	25.46	7.78	168.79	—	—	45.90	183.12	—
4	20.18	44.30	5.81	14.81	—	0.47	4.65	0.24	—
5	4.13	35.00	4.34	—	—	4.13	—	—	—
6	31.58	43.90	6.87	25.24	1.79	4.54	—	—	—
7	116.82	16.60	11.70	41.42	1.90	—	—	73.51	—
8	29.66	76.50	4.52	—	10.14	18.89	—	0.63	—
9	38.95	38.73	4.50	35.43	1.54	1.98	—	—	—
10	13.93	40.50	6.39	11.55	2.06	0.31	—	—	—
11	16.69	31.30	6.86	13.46	0.24	0.25	—	2.74	—
12	29.47	27.60	9.13	22.18	—	—	—	7.29	—
13	13.86	21.60	9.76	7.68	—	—	—	6.18	—
14	18.49	26.20	7.56	13.07	—	—	—	5.42	—
15	64.49	17.40	13.63	26.64	—	—	—	37.85	—
16	84.24	23.19	8.49	46.04	2.52	—	—	35.68	—
17	29.13	50.40	9.15	5.10	—	—	17.47	6.56	—
18	84.93	21.50	7.10	46.67	—	—	—	38.26	—
19	31.76	33.20	7.12	22.84	0.01	—	3.13	5.77	—
20	159.91	38.41	6.47	118.24	10.71	4.57	11.75	14.63	—
21	58.43	52.10	5.87	30.98	12.92	5.79	8.75	—	—
23	78.67	39.20	6.72	36.66	—	—	23.78	18.23	—
24	77.83	33.00	6.79	72.52	—	0.01	—	5.31	—
25	297.06	22.60	5.52	166.87	3.77	—	—	126.41	0.02
26	5.05	35.00	18.55	5.05	—	—	—	—	—
27	45.16	33.30	6.94	28.96	1.56	—	4.70	9.95	—
28	21.82	45.70	5.70	13.84	—	—	7.05	0.93	—
29	28.79	49.50	5.11	16.90	3.10	0.02	8.75	—	0.01
30	9.71	61.10	3.47	3.07	3.92	2.68	—	0.04	—
32	33.46	46.90	4.64	24.47	2.47	6.51	—	—	—
35	24.97	41.90	5.24	18.84	—	4.41	—	1.70	0.04
37	8.82	40.20	6.70	7.60	—	—	1.23	—	—
38	30.77	54.20	4.93	14.83	7.78	4.35	3.80	—	—
S21	13.43	25.33	8.88	9.10	—	—	—	4.33	—
Total/Mean	1920.01	37.24	7.26	1068.85	66.43	58.91	140.96	584.78	0.07

Notes: P.I. = Percent Imperviousness; S.F.R. = Single-Family Residential; M.F.R. = Multifamily Residential; Comm. = Commercial; Ind. = Industrial; O.S. = Open Space; and S.W. = Surface Water.

Table 13.3 Sample GIS Data for SWMM's TRANSPORT Block

Sewershed	Population	Population Density	Houses	Family Size	Family Income	Market Value
2	944	2.37	406	2.3	27,709	58,491
4	165	8.18	73	2.3	19,777	36,823
5	15	3.63	13	1.2	19,777	34,180
6	359	11.37	177	2.0	20,461	39,054
7	363	3.11	150	2.4	27,708	50,258
8	223	7.52	156	1.4	17,227	42,920
9	504	12.94	242	2.1	22,184	46,114
10	195	14.00	102	1.9	16,375	38,077
11	191	11.44	88	2.2	16,639	37,734
12	241	8.18	106	2.3	19,540	35,694
13	38	2.74	17	2.2	21,439	33,337
14	98	5.30	43	2.3	21,439	38,712
15	282	4.37	132	2.1	23,260	52,256
16	330	3.92	148	2.2	21,439	48,634
17	47	1.61	20	2.3	21,439	56,297
18	717	8.44	321	2.2	19,192	36,142
19	501	15.77	234	2.1	16,610	33,060
20	2,282	14.27	1,127	2.0	19,769	36,348
21	774	13.25	437	1.8	19,237	31,908
23	659	8.38	282	2.3	17,454	33,436
24	761	9.78	374	2.0	22,598	29,289
25	1,301	4.38	576	2.3	22,642	46,708
26	23	4.55	11	2.0	22,646	26,955
27	403	8.92	178	2.3	16,480	31,922
28	260	11.92	119	2.2	16,889	29,544
29	410	14.24	186	2.2	16,214	26,441
30	94	9.68	56	1.7	14,814	28,387
32	247	7.38	110	2.2	12,783	33,810
35	102	4.08	48	2.1	13,074	38,263
37	26	2.95	10	2.6	21,439	56,620
38	421	13.68	220	1.9	18,369	39,140
Total/Mean	12,976	8.14	6,162	2.10	19,568	38,921

analyzing large regional watersheds. NATSGO data describe variations in soil type at the multistate to regional scale, which is not suitable for wastewater and storm-water modeling applications (Moglen, 2000).

The steps for estimating Green-Ampt parameters using STATSGO GIS data are given in the following text:

1. Download the STATSGO database files from the NRCS Web site (www.ftw.nrcs. usda.gov/stat_data.html).
   - comp.dbf — Attribute HYDGRP has Hydrologic Soil Groups (A, B, C, etc.)
   - layer.dbf — Attribute TEXTURE2 has soil texture (S [sand], SIL [silt], etc.)
2. Join the soils layer to the preceding tables using the MUID attribute that provides a common link between the soils layer and the STATSGO database files.

Table 13.4   Sample EXTRAN Input Data

KS Look-Up Table	
HYDGRP	**KS**
A	0.375
B	0.225
C	0.100
D	0.025

MD Look-Up Table		
Soil	**TEXTURE2**	**KS**
Sand	S	0.31
Sandy loam	SL	0.28
Silt loam	IL	0.26
Loam	L	0.23
Sandy clay loam	SCL	0.18
Clay loam	CL	0.16
Clay	C	0.13

SU Look-Up Table		
Soil	**TEXTURE2**	**SU**
Sand	S	4
Sandy loam	SL	8
Silt loam	SIL	12
Loam	L	8
Clay loam	CL	10
Clay	C	7

3. Perform an overlay of soils with subareas layer to estimate:
    - Percentage of HYDGRP in each subarea
    - Percentage of TEXTURE2 in each subarea
4. Calculate the Green-Ampt parameters (saturated hydraulic conductivity KS, initial moisture deficit IMD, and average capillary suction SU) for each subarea using the look-up tables shown in Table 13.4. These look-up tables are based on the data provided in SWMM's users manual (Huber and Dickinson, 1988).

GIS APPLICATIONS FOR SWMM

Representative GIS applications in SWMM modeling are given in the following subsections.

AVSWMM

Shamsi (1997) developed an ArcView GIS interface called AVSWMM for collection-system and wet-weather overflow modeling. Both RUNOFF and EXT-RAN Blocks were included. The interface was developed by customizing ArcView 3.2, using Avenue. Avenue is ArcView's native scripting language and is built into ArcView 3.x. Avenue's full integration with ArcView benefits the user in two ways: (1) by eliminating the need to learn a new interface and (2) by letting the user work

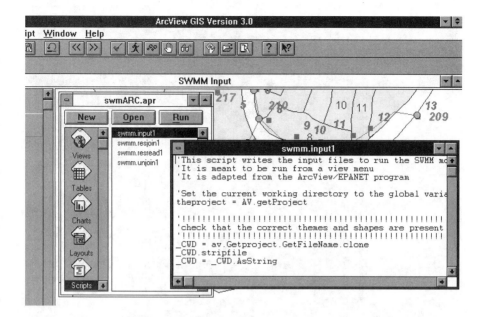

Figure 13.4 ArcView 3.x integrated programming interface for writing avenue scripts.

with Avenue without exiting ArcView (ESRI Educational Services, 1995). Figure 13.4 shows ArcView's integrated programming interface, using Avenue.

The AVSWMM interface was developed as two separate extensions, one each for the RUNOFF and the EXTRAN Blocks of SWMM.

AVSWMM RUNOFF Extension

When the extension is loaded, a menu named SWMM is added to the ArcView interface as shown in Figure 13.5. The SWMM menu has seven calculators and two functions that calculate various RUNOFF Block parameters for subareas (sewersheds). A polygon layer for subareas is required to use this extension. ArcView's Spatial Analyst Extension is required for the analysis of raster data layers. Each calculator prompts the user for the layer containing the subareas and the layer containing the data from which the model parameter is to be extracted. Each calculator adds the calculated input parameters to fields in the subarea layer. The fields are created if they do not exist. Various functions and calculators of this extension are described in the following text:

1. Average Slope Calculator

Function	Assigns average slope to subareas
Input Layers	Subarea polygon layer
	Slope grid
Output	Adds or updates MEANSLOPE field in subarea layer

2. Census Parameter Calculator

Function	Assigns total population and housing counts to subareas

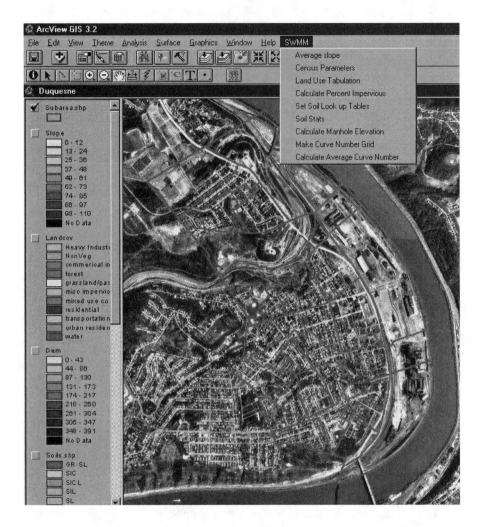

Figure 13.5   AVSWMM RUNOFF Extension.

Input Layers	Census block polygon layer
	Subarea polygon layer
Output	Adds or updates total population and housing fields to subareas
	Pop100	Total Population
	Hu100	Total Households
	PopDens	Population density per acre
	HouseDens	Households density per acre
	MnFamSize	Mean family size

3. Land Use Calculator

Function	Calculates area of each land use class for subareas
Input Layers	Landuse grid layer
	Subarea polygon layer
Output	Adds fields for each land use type to subarea layer
	Calculates area for each land use type

4. Percent Imperviousness Calculator

Function	Assigns percentage of percent imperviousness for each subarea. The calculator will add a percent impervious field to the landuse polygon layer and will prompt for the percent impervious value for each land use type.
Input Layers	Landuse grid layer
	Subarea polygon layer
Output	Adds or updates field containing percent impervious, PERIMP, to subarea layer

5. Set Soil Look-up Tables Function

Function	Joins the look-up tables for soil infiltration parameters for the Green-Ampt method (Huber and Dickinson, 1988)
	IMD Initial moisture deficit
	KS Saturated hydraulic conductivity or permeability
	SU Capillary suction
	Additional information is presented below
Input Layers	Soils polygon layer
	IMD.dbf look-up table
	KS look-up table
	SU look-up table
Output	Joins the look-up tables to the soils layer

6. Soil Parameter Calculator

Function	Calculates Green-Ampt infiltration parameters, IMD, KS, and SU for subareas. The user may also select any numerical fields of the Soils layer to average.
InputLayers	Soil polygon layer
	Subarea polygon layer
Output	Adds or updates average soil parameters to subarea layer
	MNIMD Mean IMD value
	MNKS Mean KS value
	MNSU Mean SU value

7. Manhole Elevation Calculator

Function	Assigns elevation to manholes based from a DEM
Input Layers	DEM grid
	Manhole point layer
Output	Adds or updates ELEV field containing elevation to Manhole layer

8. Make Curve Number Grid Function

Function	Prepares a grid layer for runoff curve numbers
Input Layers	Land use layer
	Soil polygon layer
	Subarea polygon layer
Output	It overlays the land use and soil layers to make a curve number grid. The values of the grid are based on each unique combination of land use and hydrologic soil group.

9. Curve Number Calculator

Function	Calculates subarea runoff curve numbers. Although EPA's SWMM release does not use runoff curve numbers for input, many hydro-logic models (e.g., HEC-HMS) and some SWMM GUI programs, such as XP-SWMM (described above) can use this parameter.

Input Layers   Curve number grid
                    Subarea polygon layer
Output   Calculates the average curve number for each subarea. The values
                    are added to the CURVNUM field of the Subarea layer.

AVSWMM EXTRAN Extension

Development of the AVSWMM EXTRAN Extension required the two main tasks
described below:

Task 1: Create EXTRAN input file

This task requires that ArcView-compatible GIS layers (themes) exist with
attribute fields for all model input parameters. The interface fetches the model input
parameters from the GIS layers and exports (copies) them to an ASCII text input
file specified by the user. The user should create a header file containing the model
control parameters. The control parameters are entered on the A1 line (title) and
on B0 to B9 lines (solution and printing instructions). The interface automatically
appends this header file to the input file. A simplified example is given in the
following text:

Given Data:

1. ArcView GIS layer for manholes
   Manholes layer attributes:
   - ID
   - Bottom elevation
   - Top elevation
   - Inflow rate
2. ArcView GIS layer for sewers
   Sewers layer attributes:
   - ID
   - Upstream manhole
   - Downstream manhole
   - Type (circular, rectangular, etc.)
   - Depth (diameter for circular pipes)
   - Width (not required for circular pipes)
   - Length
   - Material
   - Date installed (material and date installed can be used to automatically
     calculate Manning's roughness coefficient)
   - Upstream invert elevation
   - Downstream invert elevation

Data To Be Created: An ASCII text input file containing the following parameters:

- Junction (usually manhole) data:
  - Data type (D1 for junctions)
  - Manhole number
  - Top elevation (GR-EL)

- Bottom elevation (INV-EL)
- Inflow rate
- Initial depth (Y0, usually 0.0)
- Conduit (usually sewer pipe) data:
  - Data type (C1 for conduits)
  - Pipe number
  - Upstream junction (FR)
  - Downstream junction (TO)
  - Initial flow (Q0, usually 0.0)
  - Type (1 for circular, 2 for rectangular, etc.)
  - Area (0.0 for circular and rectangular)
  - Depth (DEEP = diameter for Type 1, depth for Type 2)
  - Width (WIDE = width for Type 2)
  - Length (LEN)
  - ZP1 (upstream invert elevation minus upstream node bottom elevation)
  - ZP2 (downstream invert elevation minus downstream node bottom elevation)
  - N (Manning's roughness coefficient, a function of pipe material and age)
  - STHETA and SPHI (usually zero, required for open channels only)

Note that ZP1, ZP2, and N are not part of the given sewer's data, but can be estimated in ArcView from the given data. For example, if the material is concrete and the age of the pipe is less than 10 years (fair condition), then ArcView can use a look-up table to estimate N as 0.015. ZP1 and ZP2 can be similarly estimated from the upstream and downstream manhole data. Sample SWMM input lines for junctions and conduits are shown in Table 13.5.

Table 13.5   Sample EXTRAN Input Data

Sample EXTRAN Input Data Lines for Junctions

*	**JUNCTION DATA**				
*	ID	GR-EL	INV-EL	INFLOW	Y0
D1	10	554.7	522.0	0.0	0.0
D1	20	551.1	522.3	0.0	0.0
D1	30	554.7	522.7	0.0	0.0
D1	40	553.5	523.0	0.0	0.0
D1	50	550.6	524.6	0.0	0.0

Sample EXTRAN Input Data Lines for Conduits

*	**CONDUIT DATA**												
*	ID	FR	TO	Q(0)	TYPE	AREA	DEEP	WIDE	LEN	ZP1	ZP2	N	STHETA	SPHI
C1	330	50	40	0.0	1	0.0	6.0	0.0	820	0.0	0.0	0.015	0.0	0.0
C1	331	40	30	0.0	1	0.0	6.0	0.0	173	0.0	0.0	0.015	0.0	0.0
C1	332	30	20	0.0	1	0.0	6.0	0.0	200	0.0	0.0	0.015	0.0	0.0
C1	333	20	10	0.0	1	0.0	6.0	0.0	140	0.0	0.0	0.015	0.0	0.0

Table 13.6 Sample EXTRAN Output Data

Junction Summary Statistics

Sample EXTRAN Output Data for Junctions

Junction Number	Ground Elevation (FT)	Uppermost Pipe Crown Elevation (FT)	Mean Junction Elevation (FT)	Junction Average % Change	Maximum Junction Elev. (FT)	Time of Occurence Hr.	Min.	Feet of Surcharge At Max Elevation	Feet Max. Depth is Below Ground Elevation	Length of Surcharge (Min)	Length of Flooding (Min)	Maximum Junction Area (FT²)
10	554.70	528.00	527.74	3.1377	554.59	8	30	26.59	0.11	69.5	0.0	8.7171
20	551.10	528.30	528.14	2.2338	551.10	8	30	22.80	0.00	60.8	0.5	9.8781
30	554.70	528.70	528.69	1.9667	551.55	8	30	22.85	3.15	60.8	0.0	1.0841
40	553.50	529.00	529.03	1.7722	551.40	8	30	22.40	2.10	60.7	0.0	2.8831
50	550.60	530.60	530.72	1.2461	548.47	8	30	17.87	2.13	60.3	0.0	5.3431

Sample EXTRAN Output Data for Conduits

Conduit Summary Statistics

Conduit Number	Design Flow (CFS)	Design Velocity (FPS)	Conduit Vertical Depth (IN)	Maximum Computed Flow (CFS)	Time of Occurence Hr.	Min.	Maximum Computed Velocity (FPS)	Time of Occurence Hr.	Min.	Ratio of Max. to Design Flow	Maximum Inv. at Upstream (FT)	Depth Above Conduit Ends Downstream (FT)	Length of Norm Flow (Min)	Conduit Slope (FT/FT)
330	1.62E+02	5.73	72.00	2.10E+02	9	4	7.43	9	4	1.30	23.87	28.40	46.3	0.00195
331	1.53E+02	5.41	72.00	2.11E+02	9	2	7.45	9	2	1.38	28.40	28.85	22.2	0.00173
332	1.64E+02	5.81	72.00	3.21E+02	8	35	11.35	8	35	1.96	28.85	28.80	4.8	0.00200
333	1.76E+02	6.22	72.00	3.52E+02	8	35	12.45	8	35	2.00	28.80	32.46	10.3	0.00214

Task 2: Create SWMM EXTRAN output layers in ArcView GIS

This task creates model output layers in ArcView by reading EXTRAN's ASCII text output file. A database table of the relevant output results is created from the output file. The output database file is then temporarily joined to ArcView layers (usually the pipes layer). The linked output results can then be queried or used to classify a legend. A simplified example is given in the following text:

Given Data: EXTRAN's ASCII text output file. Sample SWMM output lines for junctions and conduits are shown in Table 13.6.

Data to Be Created

- An ArcView layer for manhole SWMM output
- An ArcView layer for pipe SWMM output

As shown in Figure 13.6, the AVSWMM EXTRAN interface adds a SWMM menu to ArcView's main menu. This menu has four functions: Make Input File, Read Output, Join Results, and Unjoin Results. The interface allows a user to create an EXTRAN input file, translate an output result file, join the output result file to the themes, and remove the results joined to the themes. All model runs that have been translated are eligible to be viewed. Model run data are stored in separate folders. AVSWMM EXTRAN functions are described in the following text.

Make Input File: This menu choice creates an EXTRAN input file. As shown in Figure 13.7, the user is prompted for a model input name. The input file is placed in a folder called *name*, which is under the project directory. If the folder does not exist, it is created. The function prompts for a description of the model run. It is a string that is inserted in the input file as a comment line. The function then prompts for an optional header file to attach to the input file. The header files have an extension of .hdr and are created separately by the user in a text editor. The header file contains EXTRAN input file data other than C1 and D1 records, which are already stored in the "name.inp" file.

Read Output: This menu choice translates an EXTRAN output file into GIS database files. The function prompts for an output file to translate. The program assumes the user has run the SWMM model from the input file created previously. The output file should be located in the same folder as the input file. The program translates the Conduit Summary and Junction Summary Statistics tables of the EXTRAN output into GIS databases named condsum.dbf and junctsum.dbf and stores them in the "name" folder. The function also writes a file "name.res" in the project directory. The .res file contains the absolute path to the directory where the model results are stored. This file indicates that a model has been translated. It is used to find all model runs available for linking.

Result Join: This menu choice temporarily joins the translated EXTRAN output to the ArcView attribute tables of manholes and sewers. The tool prompts the user to select a model run to link. The routine temporarily links the result databases to the attribute data. The result data can now be queried, listed, or used to classify a legend. Figure 13.8 shows ArcView database tables showing imported EXTRAN results.

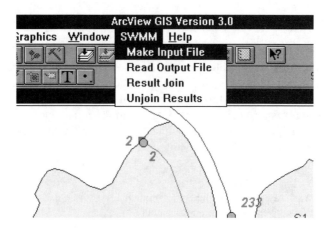

Figure 13.6   AVSWMM EXTRAN Interface.

Figure 13.9 shows a thematic map of maximum sewer flows based on classifying the Q(MAX) output variable from the EXTRAN output.

Unjoin Results: This menu choice reverses (unjoins) the link between the SWMM output and the ArcView attribute tables established by the "Results Join" menu choice. The purpose of this menu choice is to *undo* the last import of SWMM output to ArcView if the user does not like the SWMM results and wants to rerun SWMM and import the output again.

SWMMTools

SWMMTools is an ArcView extension developed by Heineman (2000). Inspired by Shamsi's work described in the preceding subsections, it allows users of SWMM 4.4 to visualize a SWMM model in conjunction with existing GIS data. SWM-MTools links SWMM input and output files with ArcView GIS. It allows viewing of model input and output summary data within ArcView. The extension adds a

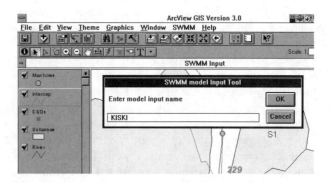

Figure 13.7   AVSWMM EXTRAN: Make Input File feature.

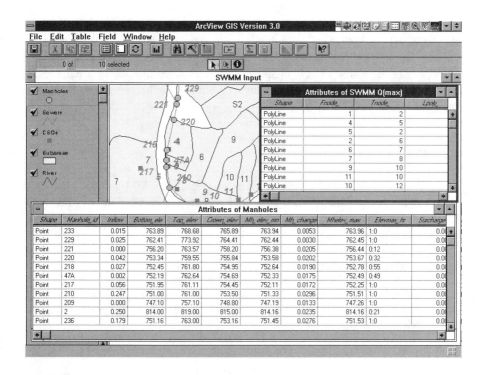

Figure 13.8 EXTRAN results in ArcView.

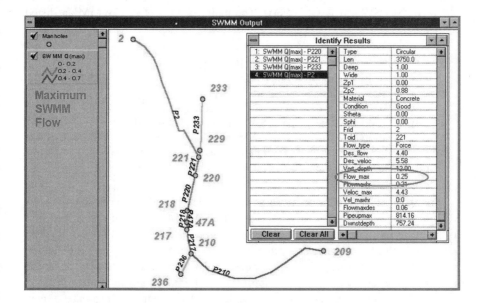

Figure 13.9 Thematic map of maximum sewer flows.

single pull-down menu to ArcView. Three scripts load plan views of RUNOFF, TRANSPORT, or EXTRAN conduit networks. Three scripts link SWMM output with existing RUNOFF, TRANSPORT, and EXTRAN GIS themes respectively. Two scripts work with a stormwater catchment theme to facilitate catchment parameterization. One estimates RUNOFF catchment widths. Another estimates RUNOFF impervious area fractions (parameter DCIA) from GIS land-use data.

AGSWMM

Existing SWMM4-ArcView3 interfaces such as AVSWMM and SWMMTools described in the preceding subsections will continue to work with SWMM4 and ArcView3 Unfortunately, these interfaces cannot work with SWMM5 or ArcView/ArcGIS8 and higher versions because of the following reasons:

1. SWMM4 and SWMM5 input file formats are different
2. SWMM4 and SWMM5 output file formats are different
3. SWMM4 and SWMM5 have different data structures (e.g., three different representations of the same pipe are not required in Runoff, Transport, and Extran Blocks)
4. ArcView3 interfaces are written in Avenue, which is not compatible with ArcView8

To benefit from new features of the new software versions, new interfaces should be developed. Shamsi and Smith (2004) developed a SWMM5-ArcView8 interface called ArcGIS SWMM (AGSWMM). The main features of AGSWMM are listed below:

- AGSWMM links SWMM5 with ArcGIS
- AGSWMM was developed using VBA
- AGSWMM setup installs a DLL file
- AGSWMM adds a new "ArcSWMM" toolbar in ArcGIS
- AGSWMM allows creation of a SWMM5 input file from GIS layers
- AGSWMM allows creation of a SWMM5 backdrop map (.emf file) in ArcGIS
- AGSWMM allows on-screen editing of input parameters in ArcGIS
- AGSWMM allows thematic mapping of SWMM5 output in ArcGIS

Figure 13.10 shows the AGSWMM toolbar. The toolbar consists of two tools: SWMM Input and SWMM Results View. The SWMM Input tool opens the input form shown in Figure 13.10. It allows users to create the SWMM5 input file. The left side of the form is used for creating input data for nodes (e.g., manholes). The right side of the form is used for creating input data for links (or conduits, e.g., pipes). Users can assign the appropriate GIS layer attributes to SWMM5 input parameters using dropdown pick lists. Users can either create a new SWMM5 input file using the "Make SWMM Input File" button at the bottom of the input form, or they can update the input parameters in an existing SWMM5 input file by selecting the "Synchronize SWMM File" button. The synchronization

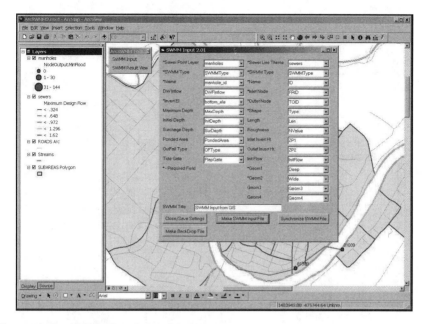

Figure 13.10 AGSWMM toolbar and input form.

capability allows users to edit the input file attributes in ArcGIS (e.g., change a sewer pipe diameter), and the changes are automatically reflected in the SWMM5 input file. This feature keeps the SWMM5 input files current. The "Make Backdrop File" button creates a backdrop map file for SWMM5 in the required .emf file format. This button automatically updates the SWMM5 input file with the appropriate scaling factor and offset distance so that it will properly align with the nodes and links in the SWMM5 map window. Figure 13.11 shows a SWMM5 map screenshot with a backdrop file creating by AGSWMM.

PCSWMM GIS™

PCSWMM GIS (Computational Hydraulics International, Guelph, Canada) is a good example of the model-based integration method described in Chapter 11 (Modeling Applications). PCSWMM GIS is a preprocessor for EPA's SWMM, and it also facilitates output visualization through a variety of plug-in tools. Model input parameters (node, conduit, and subcatchment) are extracted from an ODBC-compliant database such as Microsoft Access using SQL queries. Model input data can also be imported from an underlying dBASE-compatible GIS database. Extracted data are saved in an intermediate database (MS Access) for preprocessing into a useful model. Processed data are exported to a SWMM input file (Runoff, Transport, or Extran). The cost of PCSWMM GIS 2000 is

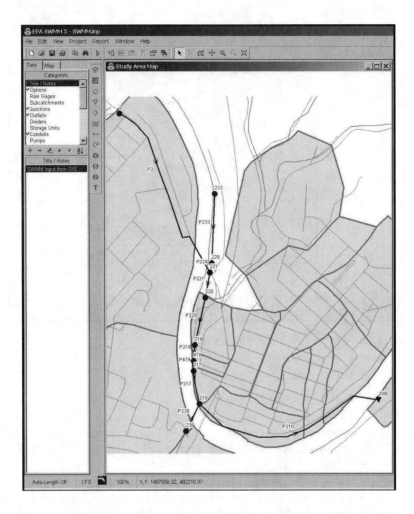

Figure 13.11   SWMM5 map window with a backdrop file created in ArcGIS.

$400. Figure 13.12 shows the PCSWMM GIS interface that has been developed using Visual Basic and resembles the ArcView GIS interface.

SWMM and BASINS

EPA has shown an interest in incorporating SWMM into its BASINS package, which is already integrated with the ArcView GIS software. EPA is now developing a SWMM linkage to its BASINS package. This link would be a valuable tool for urban planners and watershed managers to evaluate the urban pollution load in relation to other loads in the watershed and provide a more holistic view of watershed management. This project is being done through an EPA Office of Administration and Resources Management (OARM) contract with Lockheed

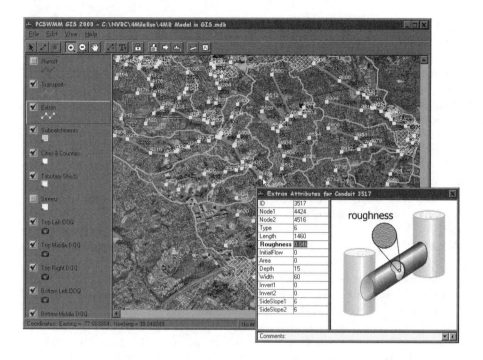

Figure 13.12 PCSWMM GIS: an example of a model-based integration method.

Martin Information Services (LMIS). When completed, this project will bundle SWMM and BASINS together with HSPF, TOXIROUTE, and QUAL2E programs that are currently part of the BASINS modeling environment. It should also allow extraction of SWMM input data from the GIS files that are part of the BASINS CD-ROMs. Detailed BASINS information is presented in Chapter 11 (Modeling Applications).

SWMMDUET

The Delaware Department of Natural Resources and Environmental Control, Division of Soil and Water, developed this program in 1994. SWMMDUET is an example of SWMM and GIS integration. This program integrates SWMM 4.2 (1993) with ArcInfo (ARC/INFO) 6.0. For this reason, it does not function with the latest versions of SWMM and ArcInfo (Schell et al., 1996). It has been written in ArcInfo's native Arc Macro Language (AML). It creates a computing environment that does not require arcane knowledge of SWMM and ArcInfo. SWMMDUET incorporates a preprocessor and a postprocessor, and therefore, provides interfaces both for creating SWMM's input file and for graphing SWMM's output file. SWMMDUET takes advantage of the graphical paradigm around which ArcInfo was constructed and utilizes relational database capabilities to organize data. Beyond data management, the program uses expert-system logic to assemble data that define the modeling process, prepares SWMM input, executes the SWMM program, and converts the output into meaningful graphical displays. SWMMDUET, therefore, uses the integration method of GIS applications described in Chapter 11 (Modeling Applications). Data-entry sheets and forms eliminate the need for the modeler to know detailed ArcInfo processing techniques. Similarly, feature selection, spatial joins, and processing commands and options of ArcInfo are specified and executed for the user. Hyetographs are related to rain gauges where rainfall data were recorded. The storms are selected simply by a georeference to the gauges. Time-series data are stored as sequential records in the database files. SWMMDUET simplifies the management of vast amounts of hydrologic data, and lets the hydrologist concentrate on hydrologic matters rather than data input and output interpretation. The SWMMDUET software is in the public domain and can be obtained for the cost of distribution (Curtis, 1994).

AVsand™

AVsand from Cedra Corporation (Rochester, New York) is an example of ArcView and SWMM integration. This program offers only limited SWMM modeling capability because not all of the SWMM Blocks and capabilities are available. AVsand can be used for modeling sanitary sewers using a simplified implementation of SWMM's TRANSPORT Block. Storm and combined sewers are not modeled using SWMM; the Rational Method or surface hydrographs are used instead. AVsand accepts user model input via a series of dialogue boxes. AVsand also provides model development and editing tools, such as a Service Area Tool to define service areas, a Node Definition Tool to define nodes, and a Pipe Definition Tool to define pipes between nodes. A Wastewater Loads Tool enables the user to introduce wastewater loads at the selected nodes. A "Query" menu option helps the user to create GIS maps from the model output. It can also create a profile plot of the HGL, which is very effective to help pinpoint the areas of hydraulic overloads.

OTHER SEWER MODELS

The preceding section focused on SWMM models. Other sewer system models with GIS application capabilities are described in the following subsections.

DHI Models

DHI Water and Environment (formerly Danish Hydraulic Institute, Hørsholm, Denmark) and DHI, Inc. (Newton, Pennsylvania), specialize in developing GIS-based H&H computer models. For example, DHI has developed the following sewer system models for the ArcView GIS software package.

MOUSE™

Modeling of Urban SEwers (MOUSE) is a package for the simulation of surface runoff, flow, water quality, and sediment transport in urban catchments and sewer systems. MOUSE combines complex hydrology, hydraulics, water quality modeling, and sediment transport modeling capabilities in a graphical and easy-to-use interface. MOUSE is a 32-bit Windows application specifically designed to operate within Microsoft Windows and Windows NT operating systems, and is optimized for fast simulations and graphics. Both metric (SI) and imperial units are supported. The MOUSE system is organized in several modules:

- MOUSE Runoff: Surface-runoff models for urban catchment applications
- MOUSE HD: Hydrodynamic network model with some limited RTC capabilities
- MOUSE RDII: Advanced hydrological model for continuous simulation
- MOUSE RTC: Advanced reactive RTC capabilities for MOUSE pipe models
- MOUSE LTS: Long-term hydraulic simulations with statistics
- MOUSE SRQ: Pollutant buildup and transport on catchments surfaces
- MOUSE AD: Pollutant advection-dispersion in drainage networks
- MOUSE WQ: Water quality processes in drainage networks
- MOUSE ST: Sediment transport in drainage networks

Some of these modules are marketed as optional add-on modules to the standard configuration. The various modules together with the user interface part, including the time-series and cross-section editors, advanced graphics both on the input and output sides, and the online HELP system, make the MOUSE system a desirable sewer system modeling package. MOUSE is a sophisticated hydraulic model that is commonly compared to SWMM (Heaney, 1999).

MIKE SWMM™

MIKE SWMM is DHI's implementation of SWMM. Resulting from a collaborative effort between DHI and Camp, Dresser, and McKee (CDM), MIKE SWMM combines the power of SWMM with a user-friendly interface. Data can be digitized and viewed with background images and edited with the same graphical editors that are also part of the MOUSE system. Data can also be entered and edited through database forms, which include scrollable spreadsheet-like sections for efficient editing of tabulated data. These different editors or views of the same information are dynamically linked, so that changes introduced through one editor are automatically and instantly updated in the other views. MIKE SWMM's database tool for storage of network data gives access to a range of query options.

MOUSE GIS™

MOUSE GIS is an application linking the MOUSE numerical sewer modeling system with ArcView. MOUSE GIS is an interface between ArcView and the MOUSE model. The MOUSE GIS interface is used for preprocessing external database information into MOUSE data sets. Once the link to an external database is established, manhole, link, weir, pump, and catchment data can be uploaded. The entire database can be used for the model networks, or a subset can be

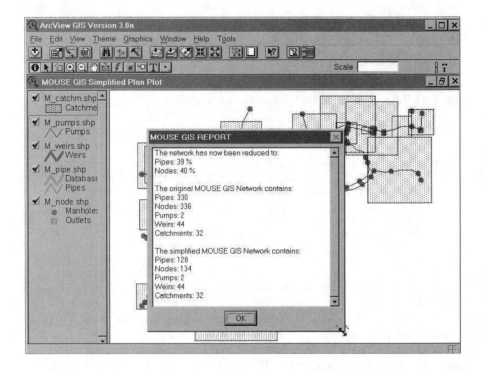

Figure 13.13   MOUSE GIS screenshot showing network skeletonization results.

selected. The selected set can subsequently be simplified (nodes excluded) to reduce the model size. MOUSE GIS also acts as a postprocessor for simulation results. Pipe flow results (maximum flow, velocity, depth and surcharge, flooding, etc.) can be displayed and overlayed with other GIS data. Mouse GIS also provides an interface to MIKE SWMM. Figure 13.13 shows a MOUSE GIS screenshot of the sewer system model for the city of Ljubljana (Slovenia) developed by DHI. The figure also shows the results of network skeletonization performed by MOUSE GIS. The automatic skeletonization has reduced the network size to 40%.

MOUSE GM™

MOUSE GM (in which MOUSE GIS is embedded), an ArcView-based module for MIKE SWMM and MOUSE users, promotes spatial and visual representation of the model and provides a direct link to GIS databases. MOUSE GM includes model-simplification tools that enable the user to create simplified (skeletonized) models from detailed data sets. This productivity tool may save days or weeks of tedious and error-prone work in the early phases of modeling projects. MOUSE GM reads and writes the SWMM file formats. It can be used as the ArcView GIS link and model-simplification tool for all SWMM modeling packages that apply the standard SWMM data formats. With MOUSE GM, MIKE SWMM users can link to GIS databases and with any asset management system that supports ODBC connectivity.

InfoWorks™

Wallingford Software Ltd. (Wallingford, U.K.) develops and supplies data management, operation, and simulation software for the water industry worldwide. It offers two main software products for sewer system modeling:

- InfoWorks CS (formerly HydroWorks): Integrates asset and business planning with urban drainage network modeling.
- InfoNet: Provides tools, reports, displays, and data management capabilities. It integrates with GIS, desktop office systems, hydraulic modeling systems, maintenance management systems (MMS), field data systems, SCADA, and corporate databases. Data exchange interfaces have been designed not only for import and export purposes but also to connect directly with GIS (ArcView and MapInfo), hydraulic modeling systems (InfoWorks), SCADA, and logger systems. InfoNet has been designed to ensure that data exchange with Microsoft Windows applications and geographic information systems is a seamless operation.

InfoWorks CS is recognized for its fast, efficient, and stable algorithms for wastewater modeling. It models systems containing complex pipe connections, complex ancillary structures and static or variable control devices, open channels and trunk sewers, full backwater effects, and reverse flow. It can be upgraded to include one or more of the following optional modules:

- Water Quality module: Predicts quality components, volume of spillage and floods, and helps develop cost-effective solutions to pollution and sedimentation problems.
- Real Time Control (RTC) module: Allows central management of flow throughout an entire network and local flow control at individual ancillary structures. RTC helps users maximize wastewater system storage, optimize flow routing, optimize operation of ancillary structures, and design cost-effective systems.

InfoWorks products are designed to link H&H models with GIS and information management systems. InfoWorks CS combines a relational database with spatial analysis. It can be used to transfer data to and from third-party applications. One can also import asset data directly from Microsoft Access or .CSV files into the InfoWorks database. For instance, InfoWorks CS allows importing of subcatchment areas, area breakdown, and population data directly from MapInfo Professional, ArcView, or GeoMedia Professional GIS. Figure 13.14 shows the InfoWorks CS output results imported in MapInfo Professional (top) and ArcView (bottom).

In Figure 13.14, nodes are symbolized according to modeled flood volume in cubic meters, which represents the water overflowing from a manhole. Smaller (green) nodes are OK (no overflow), those in light blue and slightly larger are loosing a small amount

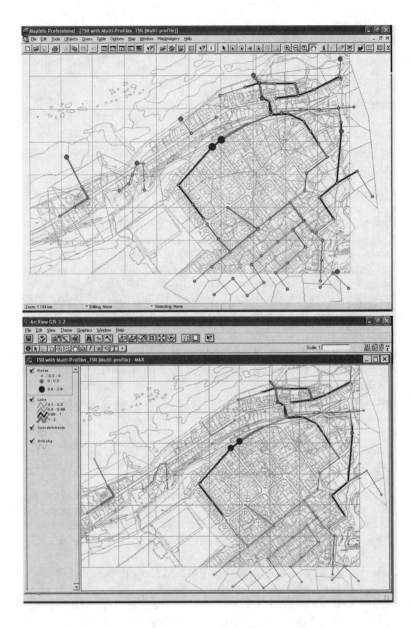

Figure 13.14   InfoWorks CS results in MapInfo Professional (top) and ArcView 3.2 (bottom).

of water, those in dark blue and larger are loosing lots of water. Link theme is showing surcharge depth. This represents 0.1 to 0.5 of pipe height (up to half full), 0.5 to 0.99 of pipe height (between half full and completely full), 1.0 of pipe height, or 100% full, and then 2.0, which is pressurized flow. The associated colors are black, light blue, dark blue and purple.

SewerCAD™ and StormCAD™

Haestad Methods, Inc.[*] (Waterbury, Connecticut), provides the following sewer system modeling:

- Sanitary sewers
  - SewerCAD Stand-alone
  - SewerCAD for AutoCAD
  - SewerGEMS for ArcGIS
- Storm sewers
  - StormCAD Stand-alone
  - StormCAD for AutoCAD

Stand-alone versions do not require a CAD package to run. The AutoCAD versions run from inside AutoCAD and, therefore, require AutoCAD. An ArcGIS version of SewerCAD called SewerGEMS (Geographic Engineering Modeling System) may be released in 2004.

SewerCAD is a design, analysis, and planning tool that handles both pressurized force mains and gravity hydraulics. SewerCAD performs steady-state analysis using various standard peaking factors, extended-period simulations, and automatic system design. StormCAD provides the same functionality for storm sewer systems.

GIS/database connection modules for these programs are marketed as optional modules that are sold separately. These programs establish a connection between a sewer model and relational and nonrelational database management systems, spreadsheets, and ESRI Shapefiles. For example, the Shapefile Connection Wizard provides import and export capabilities to transfer data between GIS and sewer models.

SEWER MODELING CASE STUDIES

Representative case studies of GIS applications in sewer system modeling are given in the following subsections.

XP-SWMM and ArcInfo Application for CSO Modeling

Application	Hydrodynamic sewer system modeling
Study period	1995–1998
GIS software	ArcInfo
Modeling software	XP-SWMM (sewer system), RMA-2V (river hydraulics and geometry), HEC-RAS (river hydraulics), and WASP5 (river water quality)
Other software	FoxPro and IBM Data Explorer
GIS data	Sewer components, hydrologic soil groups, land use, and curve number
Study area	Fort Wright, Kentucky
Organization	Sanitation District No. 1

---

[*] Haestad Methods was acquired by Bentley Systems (Exton, Pennsylvania) in August 2004.

The Sanitation District No. 1 (the District) located in northern Kentucky used GIS, GPS, RDBMS, and computer modeling technologies to simulate virtual flows of sewage through pipes, CSOs, and into and through local rivers (Ohio and Licking Rivers). With the concurrent tasks of updating the GIS and starting up of the facility plan, a common database was developed. The objective of the database was to provide a central location where data could be stored and used for the benefit of both tasks. An XP-SWMM Management System (XMS) was developed within FoxPro, which acts as the translator between different software packages. By including all five data sources (roads and streams, rain-gauge data, XP-SWMM output, CSO loadings, and fecal coliform response of receiving water), a more complete cause–effect relationship was developed. The visualization of the system using IBM Data Explorer provided an overall picture of the system that was not available before. This case study is a good example of GIS being used as an integrator technology or integrating platform (Martin et al., 1998).

AM/FM/GIS and SWMM Integration

Application	Sewer system modeling using the interface method
GIS software	ArcInfo
Modeling software	SWMM
Other software	CASS WORKS (RJN Environmental Associates)
Study area	Boston, Massachusetts, metropolitan area
Organization	Massachusetts Water Resources Authority (MWRA)

MWRA serves the sewerage needs of over two million people in 43 communities in the Boston metropolitan area. It was estimated that nearly half the flow in their system is a direct result of rainfall-induced inflow and infiltration (I/I) due to deteriorated sewers and manholes. In order to evaluate both the short-term and long-term needs of their sanitary sewer interceptor system, MWRA initiated and completed a sewage analysis and management system (SAMS) project, which involved linking of ArcInfo, CASS WORKS, and SWMM using the interface method of model integration. ArcInfo was used to create a link between SWMM data and computerized maps of the MWRA interceptor system and communities served. The maps were used as both a tool to input different planning scenarios into the CASS WORKS system and as a visual medium to geographically show the model results. The integrated AM/FM/GIS system was installed on a multiuser DEC VAX 4000-200. The integrated model allowed the MWRA planners to predict the I/I reduction from elimination of groundwater and rainwater from the interceptor system. Elimination of these flows resulted in significant operational cost savings to the MWRA and its ratepayers (ESRI, 1994).

SWMM and ArcInfo™ Interface

This application was developed as part of a watershed approach to assessing the ecological health of Bayou Chico, which is a subestuary of Pensacola Bay. Bayou Chico is the receiving water body of small, mostly urbanized watersheds located in southern Escambia County, Florida. A GIS database and the use of remotely sensed satellite images were combined to determine surface characteristics, storm drainage systems, area, and slope of the watershed. These data layers were in turn linked to SWMM, which mathematically represents these physical characteristics and uses this information to determine both runoff and pollutant loading. The SWMM RUNOFF Block was broken into a hydrological component and a pollutant loading component. Both have parameters that may be retrieved from spatially based data. Pollutant loading is partially determined either by land-use class or gutter length. Gutter length can be derived from digitized road layers or are available for some metropolitan areas in the U.S. Census TIGER data files. ArcInfo tools allowed for querying of polygons for areal extent, perimeter, length, width, and other associated attributes. AML programming was used to query ArcInfo layers for SWMM parameter data and to write the information to an ASCII text file in the SWMM input file format (Schell et al., 1996).

Hydra™ and ArcInfo™ Interface

The City of Stockton, California, developed an interface between their ArcInfo GIS and Pizer's HYDRA/HydraGraphics sewer and stormwater modeling software. The GIS contained sanitary sewers and land-use layers, and was utilized as the raw data source for developing a sewer model for master planning purposes. To facilitate this process, a customized GIS data management interface was developed using ArcView GIS as the graphical interface between the City's GIS and the sewer model. This interface included several project-specific Avenue routines to import the City's GIS layers, extract the system model and land-use data, and export the model components to the sewer model (Prins et al., 1998).

USEFUL WEB SITES

Danish Hydraulic Institute	www.dhisoftware.com
Haestad Methods	www.haestad.com
MWH Soft	www.mwhsoft.com
SWMM	www.epa.gov/ceampubl/swater/swmm/index.hm
	www.epa.gov/ednnrmrl/swmm/
SWMM-Online	www.swmm-online.com
Wallingford Software	www.wallingfordsoftware.com

CHAPTER SUMMARY

This chapter shows that GIS offers many useful applications in sewer system H&H modeling. GIS applications are developed using the three methods of interchange, interface, and integration as described in Chapter 11 (Modeling Applications). The interface method seems to be the most prevalent method at the present time. The case studies and examples presented in the chapter illustrate that many sewer utilities are successfully using GIS applications for modeling their sewer systems. The software examples presented in the chapter indicate that many public-domain (e.g., SWMM) and commercial-off-the-shelf software packages (e.g., MOUSE) are available to help users benefit from sewer system modeling applications of GIS.

CHAPTER QUESTIONS

1. Prepare a list of GIS applications for sewer systems.
2. What is a GUI? How does it help modelers?
3. Which sewer system modeling tasks can be performed using GIS?
4. What are the typical input parameters for a sewer system hydraulic model? Which of these parameters can be obtained from a GIS?
5. List the GIS data layers that are useful for sewer system modeling.
6. How does GIS help in presenting the output results of a sewer system model?
7. List the sewer system models you are familiar with. Which of these models use GIS? Which method of GIS linkage (interchange, interface, or integration) do they use?
8. Is a GIS-integrated sewer model always better than a stand-alone model? Provide reasons for your answer.

CHAPTER 14

AM/FM/GIS Applications

At present our water and wastewater infrastructure, especially in the older cities, is in critical stages of deterioration and has started to crumble. AM/FM/GIS technology provides a spatial infrastructure management approach that is very effective in prioritizing your infrastructure improvement and maintenance needs.

Field inspection crews inspecting a combined sewer overflow (CSO) diversion chamber located near Pittsburgh, Pennsylvania.

LEARNING OBJECTIVE

The learning objective of this chapter is to find out how GIS is integrated with automated mapping/facilities management (AM/FM) systems for efficient inspection and maintenance of water industry infrastructure.

MAJOR TOPICS

- Infrastructure maintenance issues
- AM/FM/GIS basics
- AM/FM/GIS software
- AM/FM/GIS application examples

LIST OF CHAPTER ACRONYMS

AM/FM/GIS Automated Mapping/Facilities Management/Geographic Information System

CAD Computer-Aided Drafting/Computer-Aided Design

CAFM Computer-Aided Facility Management

CCTV Closed-Circuit Television

CIFM Computer Integrated Facility Management

CMMS Computerized Maintenance Management System

GASB Government Accounting Standards Board

RBE Rule Base Engine

RDBMS Relational Database Management Systems

SDE Spatial Database Engine

HAMPTON'S WASTEWATER MAINTENANCE MANAGEMENT

GIS software	ArcView, GBA GIS Toolkit Extension for ArcView, MapObjects Internet Map Server (IMS)
Other software	GBA Master Series and WinCan TV system
GIS data	Sewers, manholes, water mains, valves, hydrants, roads, street centerlines, buildings, and digital orthophotos
Study area	City of Hampton, Virginia
Project duration	1998–2001
Organization	Public Works Department, Hampton, Virginia

Hampton, Virginia, has a population of 150,000 and has 12,000 manholes and valves. The City conducted a GPS survey to collect coordinates of more than 8000 manholes located in alleys, buried in yards, or covered with asphalt. The City utilizes more than 35 licenses of ArcView GIS. The City stored attribute data in an external infrastructure database from GBA Master Series and linked it to ArcView GIS. The GBA software performs infrastructure inventory of manholes and pipes, work-order management, and parts inventory management. It can also interface with the City's WinCan system used for CCTV inspection of sewers. The GBA GIS Toolkit, an extension of ArcView GIS, is used to synchronize data consistency between ESRI Shapefiles, CAD, and the infrastructure management database. This application provides a GIS map linked to a comprehensive inventory, inspection, and maintenance management software program. The system also provides a graphical display of maintenance history and profile plots of user-specified sewer segments (ArcNews, 2001).

Although GIS applications in water, wastewater, and stormwater systems are not new, getting beyond the basic inventory and mapping functions is challenging. After a utility GIS database has been set up, GIS applications must be developed or else the GIS would be nothing but a pretty map. For instance, a GIS-based computerized maintenance management system (CMMS) can be implemented for more efficient maintenance of a utility because it can track problems within the utility network more accurately. The GIS and CMMS integration can facilitate proactive (preventive) maintenance, compliance with regulatory mandates, and systemwide planning activities.

According to the U.S. Government Accounting Standards Board (GASB), an infrastructure asset management system is a requirement for reporting capital asset activity. Electronic asset management systems extend the life of existing infrastructure, thereby optimizing maintenance schedules and deferring major capital expenditures until they become necessary and economically justified. Such systems can also identify risky assets and schedule rehabilitation projects that can save the cost of total replacement (Irrinki, 2000).

INFRASTRUCTURE PROBLEM

A civilization's rise and fall is linked to its ability to feed and shelter its people and defend itself. These capabilities depend on infrastructure — the underlying, often hidden foundation of a society's wealth and quality of life. Infrastructure is a Latin word meaning the structure underneath. According to a 1992 report from the Civil Infrastructure Systems Task Group of the U.S. National Science Foundation, "*A society that neglects its infrastructure loses the ability to transport people and food, provide clean air and water, control disease, and conduct commerce.*" (Tiewater, 2001)

The water, wastewater, and stormwater infrastructure is aging fast throughout the world, including in the U.S. For example, according to the 2001 counts, the U.S. has 54,000 drinking water and 16,000 wastewater systems. America's water and wastewater systems incur an expenditure of $23 billion per year on infrastructure projects. Much of the existing sewerage system in the U.S. was constructed in the 1950s and 1960s — some date back to the early 1900s and late 1800s. According to 2001 estimates, U.S. sewer systems are approaching 100 years in age. Even the newest U.S. water infrastructure systems are over 50 years old. What is worse, because of their material, they are more likely to be deteriorating than some systems dating back to 100 years. Many water systems are structurally obsolete and are now serving two to three times as many people as their design capacity. Many systems have not received the essential maintenance and repairs necessary to keep them working properly. The American Society of Civil Engineers (ASCE) projects $1.3 trillion in infrastructure needs in the U.S. during 1999 to 2004, of which approximately one third is needed for water and wastewater improvements.

Unfortunately, adequate government funds are not available for this expensive fix. This problem can be partially blamed on the "out-of-sight, out-of-mind" mentality of the decision makers. The lack of funding for the badly needed infrastructure improvements is making people worried. This is evident in the recently conducted public opinion polls. According to a recent Water Environment Federation (WEF) survey, 31% of Americans would not swim in their own rivers. A 1998 ASCE poll showed that more than 75% of the U.S. voters were concerned about the quality of roads, drinking water, and school buildings. In Boston, more voters were concerned about roads, bridges, and drinking water than Social Security and taxes. In New York City, concerns about school buildings and drinking water outscored drug abuse worries (ASCE, 1998). Many indicated that they would vote for a candidate whose election campaign addressed waste disposal (78%), drinking water (73%), and roads and bridges (62%). These surveys indicate that water industry infrastructure is particularly important because it is so closely related to public health and safety.

In the U.S., it has been recently observed that a combination of reduced federal spending and increased federal mandates is taking its toll on the infrastructure of the country. For instance, according to American Water Works Association (AWWA), annual pipe replacement costs will jump to $3 billion by 2020 and $6 billion by 2030. Replacement expenses projected into the future indicate that sewer main replacements are more urgent than water mains and will require more money.

As of 2001, ASCE estimates that the U.S. must invest nearly $277 billion in drinking water and wastewater infrastructure repairs over the next two decades. U.S. water systems have an $11-billion annual investment shortfall and need $151 billion by 2018. U.S. sewer systems have a $12-billion annual shortfall in investment and need $126 billion by 2016. The largest need, $45 billion, is for projects to control combined sewer overflows (CSOs). The second-largest category of needs, at $27 billion, is for new or improved secondary treatment.

In a report "Water Infrastructure Now" published in February 2001, a water conservation organization called Water Infrastructure Network (WIN) estimated a $23-billion per year gap between infrastructure needs and current spending. The report indicated that U.S. water and wastewater systems faced infrastructure funding needs of nearly $1 trillion over the next 20 years and a shortfall of half a trillion dollars. The report called for a 5-year, $57-billion federal investment in water infrastructure to replace aging pipes and upgrade treatment systems. WIN estimates that household water bills must double or triple in most U.S. communities if utilities are forced to absorb the entire infrastructure bill. For instance, the City of Gloucester, Massachusetts, is charging $20,000 per home (to be paid over the next 20 years) to finance the city's conversion from septic to sewer systems (CE News, 2001).

According to a 2001 statement by ASCE President Robert W. Bein, "Something is terribly wrong ... America has been seriously underinvesting in its infrastructure."

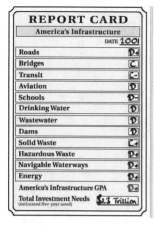

Figure 14.1   ASCE Infrastructure Report Card for the U.S.

In March 2001, ASCE released its 2001 Report Card (Figure 14.1) for America's infrastructure in which the nation's infrastructure received a cumulative grade of D+ for 12 infrastructure areas. Causes for such a dismal grade include: explosive population growth and school enrollment, which outpace the rate and impact of current investment and maintenance efforts; local political opposition and red tape, which outpace the development of effective solutions; and the growing obsolescence of an aging system — evident in the breakdown of California's electrical generation system and the nation's decaying water infrastructure. The 2001 Report Card follows the one released in 1998, at which time the ten infrastructure categories rated were given an average grade of D. The 2001 wastewater grade declined from a D+ to a D, while drinking water remained a D. Solid waste and schools scored the highest and the lowest grades, respectively (ASCE, 2001; EWRI, 2001). Despite the disturbing findings contained in the 2001 Report Card, the nation failed to heed the call to improve its deteriorating infrastructure, according to an update to the assessment released by ASCE in September 2003 (ASCE, 2003).

When the infrastructure condition is bad and funding is scarce, we need infrastructure management tools that can:

- Accurately define (or map) our infrastructure
- Identify the worst portions of the infrastructure
- Determine how to most cost-effectively improve (replace or repair) the worst parts of the infrastructure

AM/FM/GIS products provide an efficient infrastructure management system.

AM/FM/GIS BASICS

Many solutions for managing facilities have spawned from CAD-based architectural applications. Database solutions have evolved from the financial sector to

Table 14.1 Comparison of AM, FM, AM/FM, GIS, and AF/FM/GIS

Feature	AM	FM	AM/FM	GIS	AM/FM/GIS
Layers	Y	N	Y	Y	Y
Topology	N	N	N	Y	Y
Network definition	N	Y/N	Y	Y	Y
Lines	Y	N	Y	Y	Y
Nodes	N	Y/N	Y	Y	Y
Polygons (areas)	N	N	N	Y	Y
Attributes	N	Y	Y	Y	Y
Actual locations	Y/N	N	Y	Y	Y
Map intelligence	N	N	N	Y	Y
Facilities management	N	Y	Y	N	Y

Notes: Y = Yes; N = No; and Y/N = Both Yes and No.

include scheduling, inventory, and purchasing functions. Implementation of facility management automation can occur at any scale, from a small spreadsheet recording preventative maintenance items to a full corporate system with CAD capabilities managing the complete maintenance process with integrated contracting, accounting, inventory, asset, and personnel records. CAD-based products have normally been classified as computer-aided facility management (CAFM), whereas the database solutions, dealing with work-order management without a CAD component, have been termed computerized maintenance management systems (CMMS). The FM user community reflects this polarization, because the majority of current implementations are oriented either one way or the other. Computer integrated facility management (CIFM) brings both CMMS and CAFM capabilities into synergy, thereby addressing the implementation of a true facility management automation program. The core consideration is that, in practice, CMMS and CAFM capabilities augment and are essential to each other. For example, maintenance activities keep the data required for planning accurate and up-to-date (GEOTELEC, 2004).

AM/FM/GIS is a combination of four separate systems:

- Automated Mapping (AM)
- Facilities Management (FM)
- Automated Mapping/Facilities Management (AM/FM)
- Geographic Information System (GIS)

The differences between the three systems are summarized in Table 14.1 and illustrated in Figure 14.2.

Automated Mapping (AM)

Automated mapping (AM), also known as computer-aided mapping (CAM), is a CAD application for producing maps. It can be considered an alternative to the traditional manual cartographic maps. Data are organized in layers that are conceptually like registered film overlays. Layers organize data by theme (streams vs. roads) and type (linework vs. text).

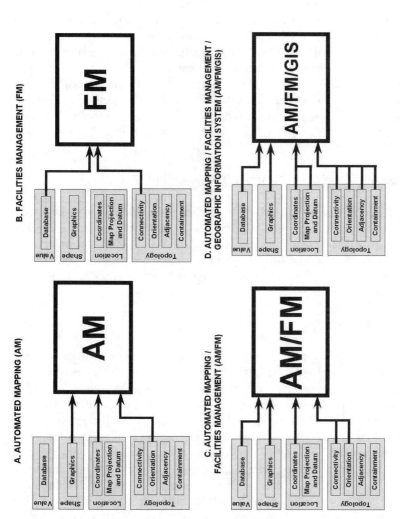

Figure 14.2 Comparison of AM, FM, AM/FM, and AM/FM/GIS.

There are no spatial relations (topology) among data elements except orientation. Figure 14.2(A) shows AM capabilities.

Facilities Management (FM)

Facilities management (FM), also referred to as asset management, is a CAD technology for managing utility system data. FM consists of such activities as inventory, inspection, and maintenance performed by cities, utilities, and government agencies. Organizations incur considerable expenses and resources as these functions are performed on a routine basis. FM includes an infrastructure management database. Compared with AM, there is less emphasis on graphical detail or precision and more emphasis on data storage, analysis, and reporting. Relationships among utility system components are defined as networks. Because FM systems have network definitions, an FM system "knows" the pipes connected upstream or downstream of a given pipe. As shown in Figure 14.2(B), FM systems generally do not have full topology; they offer connectivity and orientation only.

Automated Mapping (AM)/Facilities Management (FM)

AM/FM is a combination of AM and FM technologies. AM/FM software is used to automate maintenance. It allows the integration and automation of maintenance management. AM/FM systems have both orientation and network definitions as shown in Figure 14.2(C). The benefits of AM/FM are improved work-order management, better integrated inventory control, capture of maintenance data and costs, and allocation of costs.

AM/FM/GIS Systems

GIS and AM/FM are different systems but each has its own advantages and applications (Shamsi and Fletcher, 2000). A GIS can help locate the worst pipe. An AM/FM can help users prioritize the work required to bring their worst pipes up to a minimum operating standard. For many years, people have used the GIS and AM/FM systems separately. Developing and maintaining two different systems is expensive and inefficient. Thanks to the latest advances in computer hardware and software, integrated AM/FM and GIS systems called AM/FM/GIS systems are now available. These systems are especially useful for asset inventory, inspection and maintenance, and work management. They are especially popular among visual learners who prefer maps over tables and databases. Visual learners like to click on a manhole in a GIS map to initiate a work order rather than locate it by querying a database. As shown in Figure 14.2(D), AM/FM/GIS systems have more capabilities than AM/FM systems, such as orientation, network, database, and topology.

AM/FM/GIS SOFTWARE

There are currently a variety of computer applications on the market, which were developed for specific components of facility management. In applying automation

Table 14.2   AM/FM/GIS Software Examples

Software	Company	Web Site
AM Focus		
ArcFM	Miner and Miner	www.miner.com
ArcFM Water	ESRI	www.esri.com
FRAMME	Intergraph	www.intergraph.com
GeoWater and GeoWasteWater	MicroStation	www.bentley.com
FM Focus		
Cityworks	Azteca Systems	www.azteca.com
GBA Water Master and Sewer Master	GBA Master Series	www.gbamasterseries.com
GeoPlan	Regional Planning Technologies	www.rpt.com
WATERview, SEWERview, STORMview	CarteGraph Systems	www.cartegraph.com
Proprietary Systems		
CASS WORKS	RJN Group, Inc.	www.rjn.com
IMS-AV and IMS-ARC	Hansen Information Technologies	www.hansen.com

to these diverse functions, vendors have taken approaches that originate from two different technical directions: CAD or relational database management systems (RDBMS) (GEOTELEC, 2004).

GIS-based facilities management requires AM/FM software extensions that can be run from within the GIS software. These add-on programs are also referred to as AM/FM/GIS software. Representative AM/FM/GIS software are listed in Table 14.2. According to this table, there are two types of AM/FM/GIS software:

1. AM focus: These software provide more mapping (AM) capabilities. ArcFM is an example of this type of software.
2. FM focus: These software provide more database (FM) capabilities. Cityworks is an example of this type of software.

Software with an AM focus provide better data editing and mapping capabilities and generally require an FM add-on to provide work-order management functions. Software with an FM focus provide maintenance management functions (e.g., work-order management) but may lack the map maintenance functions. Projects that require both strong AM and FM capabilities may have to use two software products. For example, ArcFM's strong suite of CAD-like map-editing capabilities can be supplemented by Cityworks' work-order management functions.

All the packages listed in Table 14.2 are available for purchase from their respective companies except CASS WORKS, IMS-AV, and IMS-ARC, which are proprietary software programs that are installed by the vendors in their turnkey projects. Figure 14.3 shows Hansen's Infrastructure Management System CCTV-inspection database. Note that each defect has been linked to its location on the sewer pipe and a picture of the defect captured from the inspection videotape.

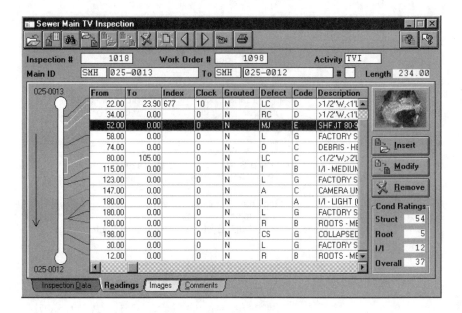

Figure 14.3   Hansen's Infrastructure Management System CCTV-Inspection Database (Photo courtesy of Hansen Information Technologies).

ArcFM

Developed jointly by ESRI and Miner & Miner, ArcFM is an off-the-shelf AM/FM/GIS application for the utility industry and for enterprise implementations. It is designed for editing, maintenance, modeling, and data management of utility information. The first version was written in Visual Basic 5.0 for the Windows NT platform. ArcFM has two components: (1) ArcFM, the main program for creating and editing maps and (2) ArcFM Viewer, the companion program for querying and viewing maps created in ArcFM. ArcFM is suitable for data conversion work by GIS technicians. ArcFM Viewer is a simple program that enables managers to review the finished maps. The relationship between ArcFM and ArcFM Viewer is similar to the relationship between ArcInfo 7.x and ArcView 3.x packages.

In ArcFM, features are modeled as objects with drag-and-drop capabilities. For example, as shown in Figure 14.4, to draw a butterfly valve, users can simply select the valve symbol from the objects library and drag-and-drop it to its correct location. In ArcFM software, GIS tools deal with features that are not simply points and lines, but valves, pipes, manholes, inlets, drainage basins, and more. ArcFM has a rule base engine (RBE) that determines how features are drawn (symbology, placement, rotation, etc.). The RBE can be individualized for different applications. For example, the stormwater RBE may define pipes, their associated features, and the methods that pertain to them. RBE can also validate user-entered data to ensure valid attribute values. With the help of these capabilities, ArcFM can perform routine editing functions, such as adding a new catch basin or service connection or splitting a pipe, with remarkable simplicity.

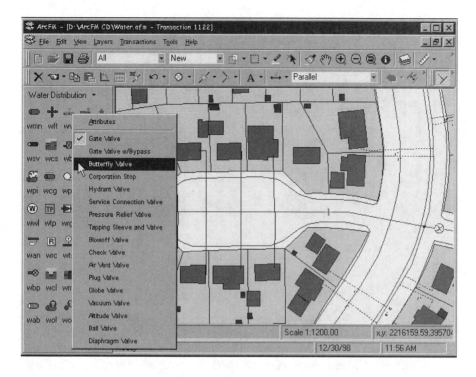

Figure 14.4 ArcFM screenshot showing water objects.

ArcFM has powerful and user-friendly data editing capabilities for routine map maintenance functions, e.g., for adding a new valve or hydrant, splitting pipes, or adding a service connection. ArcFM provides sample data models and business-rule templates for water, wastewater, and stormwater systems. These templates, called *modules,* can be modified for project-specific applications. ArcFM does not have maintenance management capability but it can be integrated with other work management or customer information systems. ArcFM is not a stand-alone program. It requires ArcInfo and Spatial Database Engine (SDE) software and an RDBMS, such as Oracle or Microsoft Access (Shamsi and Fletcher, 2000).

Greenville Utilities was chartered by the North Carolina Assembly in 1905. It is organized as an independent agency of the City of Greenville and operates electric, water, sewer, and gas utilities for the City of Greenville and a portion of Pitt County. Greenville Utilities water operations serve 25,000 customers with 475 mi of main, and sewer operations serve 20,000 customers with 333 mi of main. Greenville Utilities used ArcFM, SDE for its SQL server, and ArcInfo software for implementing an enterprise-wide GIS project for its electric, water, sewer, and gas services. ArcFM Viewer was used by casual users, customer-service representatives, field crews, and others to easily query, plot, and analyze spatial and related attribute data. ArcFM was used as an engineering tool (ESRI, 1999).

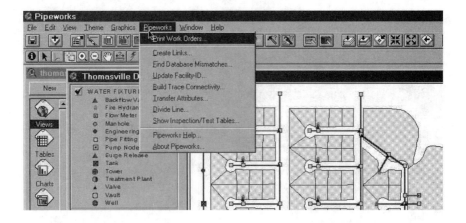

Figure 14.5   Pipeworks menu screenshot.

Cityworks

Cityworks (formerly known as Pipeworks) is available as ArcView 3.x or ArcGIS 8.x and 9.x Extension, helps users integrate their GIS and facilities management. Cityworks' water, wastewater, and stormwater module which for ArcView 3.x costs approximately $4000 per license and works with ESRI coverages and Shapefiles. Cityworks can store project data in any SQL database, such as SQL Anywhere, Oracle, Sybase, etc. Cityworks' capabilities include data inventory, data editing, work-order management, work-order scheduling, network tracing, maintenance histories, inspections, and condition ratings. Cityworks can also be used for managing and recording CCTV inspection programs for sewers. Figure 14.5 shows the Pipeworks menu in ArcView 3.x.

The Cityworks extension provides additional capabilities to make ArcView 3.x or ArcGIS 8.x a complete work-management system. Cityworks work-order modules are used to create and manage work orders, inspections and tests, and detailed inventory for water and sewer systems. The attribute data managed by Cityworks are referred to as extended attributes because these attributes are stored in the extended-coverage database or RDBMS. Examples of extended attributes are inventory data, work orders, work histories, inspections, customer complaints, etc. Cityworks for ArcView 3.x requires that all inventoried features be created and maintained as coverage data types, i.e., line, node, or point feature. The difference between Cityworks and other work-management solutions is its GIS-centric approach. Cityworks does not treat the GIS as subordinate to the work-management system. Instead of integrating the work-management system with GIS, it uses the GIS as a work-management system (Azteca, 1999). General functions of Cityworks are listed below:

- Assign work activity
- Link to service request
- Track tasks
- Track resources: labor, material, and equipment (estimated and actual)
- Link multiple inspections and tests

Table 14.3 Cityworks Inspections and Tests

Water	Wastewater	Storm
Meter inspection	Manhole inspection	Manhole inspection
Hydrant inspection	CCTV inspection	CCTV inspection
Valve inspection	Inlet inspection	Inlet inspection
Hydrant flow test	Dye test	Dye test
Meter changeout	Smoke test	Smoke test
	General test	General test

- Maintenance:
  - Corrective or reactive
  - Preventive or routine
  - Scheduled or cyclical
- Track permits
- Key dates and personnel
- Attach images
- Geocode and geolocate
- Print using Microsoft Word and ArcView templates and automated printing functions
- Search on any combination of work-order ID, work activity, selected map features, feature, type, key dates, key personnel, tasks, status, address, or geographical constraints.

Cityworks' inspections and tests are listed in Table 14.3. Another Azteca product called the Cityworks ArcGIS Interface simplifies the ArcGIS interface by providing a subset of key ArcGIS functionality (presented using large icons) commonly used by maintenance and operations staff. Cityworks Desktop ArcGIS Extension is an extension of the standard ArcGIS interface. It is accessible from a palette of tools that can be positioned anywhere on the map document.

CHAPTER SUMMARY

Deteriorating infrastructure and limited funds require that maximum system improvement be achieved at the lowest possible cost. AM/FM/GIS systems can be used to determine optimal infrastructure improvement strategies to accomplish this goal. Today, AM/FM/GIS systems are helping the water industry with their planning, design, and operational needs. AM/FM/GIS software is engineered to organize asset information, optimize maintenance activities, and prioritize necessary improvements. AM/FM/GIS software offers the integrated strength of AM/FM and GIS programs. AM/FM/GIS software is available in AM-centric and FM-centric flavors. Software selection should be made based on user-specific needs. Applications of AM/FM/GIS technology are presented in the next chapter.

CHAPTER QUESTIONS

1. What is AM/FM/GIS, and how does it differ from AM/FM and GIS?
2. What is work-order management, and how is it done using AM/FM/GIS software?
3. List ten applications of AM/FM/GIS technology.

CHAPTER 15

Maintenance Applications

GIS can be used to prepare inspection or maintenance work orders simply by clicking on a sewer pipe or manhole. This approach simply takes just a few minutes compared to the conventional method of locating and copying maps and typing the work order forms, which usually takes several hours.

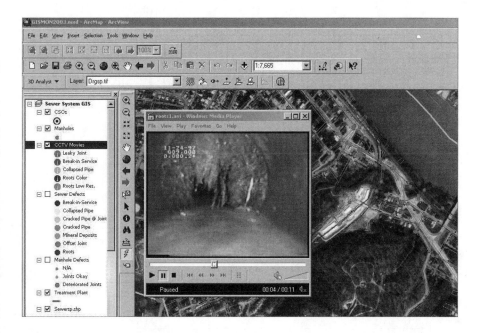

Displaying sewer CCTV inspection video in GIS. The movie shows a sewer blocked by heavy root growth.

LEARNING OBJECTIVE

The learning objective of this chapter is to understand GIS applications in inspection and maintenance of water, wastewater, and stormwater systems.

MAJOR TOPICS

- Field inspections
- Asset management
- GASB 34 applications
- Wet-weather overflow management applications
- CCTV inspections
- Video mapping
- Thematic mapping of inspection data
- Work-order management
- Identifying valves to be closed for repairing or replacing broken water mains
- System rehabilitation and repair
- Case studies

LIST OF CHAPTER ACRONYMS

AM/FM/GIS Automated Mapping/Facilities Management/Geographic Information System
AVI Audio Video Interleaved (digital movie format)
CCTV Closed-Circuit Television
CD Compact Disc
CMMS Computerized Maintenance Management System
DVD Digital Versatile Disc
GAAP Generally Accepted Accounting Principles
GASB Government Accounting Standards Board
HTML Hyper Text Markup Language (a file extension)
MPEG Moving Picture Experts Group (digital movievideo format)
O&M Operation and Maintenance
PDA Personal Digital Assistant (electronic handheld information device)
PDF Portable Document Format (Adobe Acrobat)
ROM Read Only Memory
VCR Video Cassette Recorder
VHS Video Home System (video cassette format)

This book focuses on the four main applications of GIS, which are mapping, monitoring, modeling, and maintenance and are referred to as the "4M applications." In this chapter we will learn about the applications of the last *M* (maintenance).

To fully appreciate the benefits of GIS-based inspections, consider the following hypothetical scenario. On March 10, 2004, following a heavy storm event, a sewer customer calls the Clearwater Sewer Authority to report a minor basement flooding event without any property damage. An Authority operator immediately starts the GIS and enters the customer address. The GIS zooms to the resident property and shows all the sewers and manholes in the area. The operator queries the inspection data for a sewer segment adjacent to the customer property and finds that a mini movie of the closed-circuit television (CCTV) inspection dated July 10, 1998, is available. The operator plays the movie and sees light root growth in the segment. A query of the maintenance history for that segment shows that it has not been cleaned since April 5, 1997. This information indicates that the roots were never cleaned and have probably grown to "heavy" status. The operator highlights the sewer segment, launches the work-order module, and completes a work-order form for CCTV inspection and root removal, if necessary. The export button saves the work-order form and a map of the property and adjacent sewers in a PDF file. The operator immediately sends the PDF file by e-mail to the Authority's sewer cleaning contractor. The entire session from the time the customer called the Authority office took about 30 min. The operator does not forget to call the customer to tell him that a work order has been issued to investigate the problem.

BUNCOMBE COUNTY'S SEWER SYSTEM INSPECTION AND MAINTENANCE

During the 1990s, the Metropolitan Sewerage District (MSD) of Buncombe County (North Carolina) spent more than $111 million rehabilitating the most problematic sewer lines in its system. MSD's aggressive sewer rehabilitation program has successfully implemented numerous rehabilitation technologies. In 1999, MSD began assessing a method of comprehensive basinwide rehabilitation. In this approach, every pipe in a drainage basin was evaluated using CCTV footage and engineering analysis. After evaluating this comprehensive basinwide rehabilitation method for 4 years, MSD determined that it is overly time- and capital-intensive. Seeking a more efficient and quicker method to fix the system, MSD developed and implemented a new GIS-based rehabilitation method called "pipe rating."

The pipe rating method has five main elements: (1) CCTV information, (2) a defect scoring system, (3) GIS database software, (4) sanitary sewer overflow (SSO) history, and (5) engineering analysis. These components are combined to generate specific projects for problem lines. Each structural defect noted in the video inspection is given a defect score, in accordance with MSD's standardized scoring system. For example, a circumferential crack is given a score of 20, and a collapsed pipe is assigned a score of 100. Defects are then embedded within the GIS database, with appropriate scores attached. Upon quantifying all structural defects within a pipe segment, three defect ratings are generated in the GIS: (1) peak defect rating, (2) mean defect rating, and (3) mean pipeline rating. These ratings allow users to visualize the severity of pipe defects using GIS. Figure 15.1 shows an ArcGIS screenshot of mean pipeline ratings. Such maps are used in prioritizing pipe segments for repair work. They are also used to determine when several point repairs should be made to

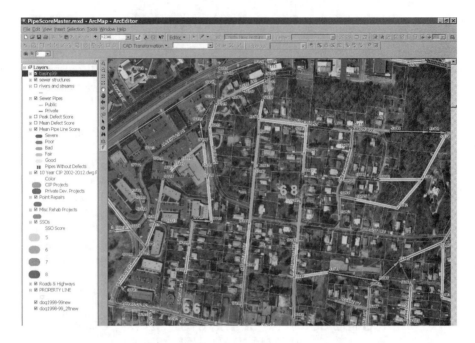

Figure 15.1   ArcGIS screenshot of mean pipeline ratings for the Metropolitan Sewerage District (MSD) of Buncombe County, North Carolina.

a line as apposed to excavating the entire line (Bradford et al., 2004). Additional information about this application is provided in Chapter 17 (Applications Sampler).

ASSET MANAGEMENT

As discussed in Chapter 14 (AM/FM/GIS Applications), at present our water and wastewater infrastructure, especially in the older cities, is in critical stages of deterioration and has started to crumble. Nationally and internationally, aging water and wastewater infrastructure is imposing enormous costs on local communities (Booth and Rogers, 2001). In the U.S., cities and utilities are demanding billions of dollars of government grants and funds for renovating their water infrastructure. Due to an "out-of-sight, out-of-mind" philosophy and the lack of funds to follow a preventive maintenance practice, the replacement is mostly performed on a react-to-crisis basis. A crisis maintenance program only corrects infrastructure problems after they have happened. Notwithstanding the conventional wisdom, this reactive approach may not be the best strategy as substantial expenditure and inconvenience can be avoided by replacing a deteriorated pipeline before it actually breaks. A preventive maintenance program is proactive because it strives to correct a problem before it occurs.

For water and wastewater systems, asset management can be defined as managing infrastructure capital assets to minimize the total cost of owning and operating them while delivering the service levels customers desire (Booth and Rogers, 2001). A typical asset management system has five components (Doyle and Rose, 2001):

1. Facilities inventory: Describes each system element in an asset group. GIS can be very useful in completing this task.
2. Condition assessment: Classifies each asset according to its capability to perform the intended function.
3. Valuation: Assigns a financial value to inventoried assets consistent with Generally Accepted Accounting Principles (GAAP).
4. Operations, maintenance, repair, and replacement management: Arguably the heart of a management system, this component tracks and records data about work orders and customer complaints, issues and tracks preventive and predictive maintenance schedules, and generates crew assignments and work-site maps. GIS has extensive capabilities to fulfill this part.
5. Analysis and evaluation: Considered as the brains of an asset management system, this component prioritizes work effort, analyzes cost-effectiveness, and optimizes asset performance.

An asset management system helps predict the future condition of assets and major rehabilitation costs for planning purposes. An effective asset management system can reduce the cost of system operation and maintenance (O&M). Every successful maintenance program should begin with an accurate system map because it is difficult to maintain a system if the users do not know where the water or sewer lines are. A well-constructed GIS should be used to create the system map. Historical maintenance data should also be linked with the GIS because it is difficult to schedule maintenance when you do not know the last time it was done (Gaines, 2001). A GIS-based asset management system can be used as a decision support system for capital improvement planning (CIP). For example, CH2M HILL (Atlanta, Georgia) used ESRI's MapObjects GIS software to develop an infrastructure capital asset management (ICAM) toolkit that can be used in a Web-based browser/server or stand-alone computing environment (Booth and Rogers, 2001).

In many cases, through more effective planning and management of infrastructure improvements and system operations, organizations can realize annual savings of 20 to 40% (Stern and Kendall, 2001). Boston Water and Sewer Commission (BWSC) has approximately 27,500 catch basins. In 2000, BWSC started enhancing its asset management program for locating, inspecting, and maintaining its catch basins. BWSC improved the productivity and efficiency of its catch basin preventive maintenance program by integrating it with GIS and a computerized maintenance management system (CMMS). In lieu of using global positioning system (GPS), the crews determined the precise location of catch basins by using a measuring wheel to measure distance from known points to the basin's center. BWSC utilized handheld touch-screen computers in the field to collect geographic location information and more than 25 attributes. GIS data in Shapefile format and a GIS interface provided the backbone of this application. The traditional paper-based data collection methods typically averaged 24 catch basin inspections per day, but the GIS-based method boosted this rate to approximately 42 per day.

Boston Water and Sewer Commission's GIS-based field-data-collection approach increased the catch basin mapping productivity by more than 40% compared with traditional, paper-based data collection methods (Lopes et al., 2002).

In the U.S., two key drivers are motivating improved asset management practices in water and wastewater utilities: (1) GASB 34 requirements, and (2) wet-weather overflow requirements.

GASB 34 APPLICATIONS

In 1999, the Government Accounting Standards Board (GASB) issued Rule 34 (known as GASB34) to govern financial reporting requirements of approximately 85,000 state and local governments in the U.S. Considered the most significant change in the history of government financing reporting in the U.S., GASB 34 requires the cities to adequately account for and report their capital asset inventory in a complete, accurate, and detailed manner (Booth and Rogers, 2001). The capital assets include infrastructure networks such as roads, bridges, and water, wastewater, and stormwater systems.

The literature indicates that a GIS-based approach saves time spent in locating, organizing, and confirming the accuracy of field inspection information (Criss, 2000). Industry experts believe that utilities can cut their maintenance costs in half by implementing GIS-based preventive maintenance programs. Thus, integration of GIS and maintenance management software is a natural progression for GIS applications in the water and wastewater industry. For instance, the Wastewater Collection Division of Fairfax County, Virginia, linked the county's sewer maps with the sanitary sewer maintenance management system database, making it easier to access maps during field activities (Fillmore et al., 2001). The City of Denver, Colorado, linked the city's sewer system GIS to an information management system to efficiently track and manage maintenance operations (Gaines, 2001).

WET WEATHER OVERFLOW MANAGEMENT APPLICATIONS

Management of wet weather overflows is a fertile field for GIS technology. By using geographic information in mapping, facilities management, and work order management, a wastewater system manager can develop a detailed capital improvement program or operations and maintenance plan for the collection system.

Broken and damaged sewers, laterals, and manholes usually contribute significant amounts of wet weather inflow and infiltration (I/I) to a wastewater collection system. This contribution often results in combined sewer overflows (CSO) from combined sewer systems and sanitary sewer overflows (SSO) from sanitary sewer systems.

In the U.S., CSO discharges are regulated by U.S. EPA's National CSO Policy. The policy requires a System Inventory and Physical Characterization report. Major portions of this report can be completed using GIS. The CSO policy's Nine Minimum Controls (NMC) mandate proper operation and regular maintenance programs for the sewer system and CSO outfalls that can also benefit from inspection and maintenance applications of GIS. SSO discharges are being regulated by U.S. EPA's SSO rule that requires implementation of a Capacity, Management, Operations, and

Maintenance (CMOM) program. CMOM requires the system owners/operators to identify and prioritize structural deficiencies and rehabilitation actions for each deficiency. CMOM requirements offer a dynamic system management framework that encourages evaluating and prioritizing efforts to identify and correct performance-limiting situations in a wastewater collection system. CMOM is a combination of planning tools and physical activities that help communities optimize the performance of their sewer systems. CMOM requirements mandate that the system owner/operator properly manage, operate, and maintain, at all times, all parts of the collection system. The owner/operator must provide adequate capacity to convey base flows and peak flows for all parts of the collection system. CMOM requirements include "maintaining a map" (Davis and Prelewicz, 2001), which is the simplest application of GIS as described in Chapter 9 (Mapping Applications).

Some wet weather overflow management experts believe that use of GIS is a must for CMOM compliance.

AutoCAD Map GIS Application for CMOM

The Stege Sanitary District (SSD) located near San Francisco Bay serves a population of approximately 40,000 in a 5.5 mi^2 area. The goal of a CMOM program is the ultimate elimination of any type of overflow from the sanitary sewer system. SSD was challenged with the goal of "no overflows" in 1996. To meet this goal, a new maintenance model was implemented for sanitary sewers that used AutoCAD Map GIS and Microsoft Access database software to identify system conditions causing overflows, and prompted immediate action to correct problems via repair or replacement. Those lines showing the greatest damage and whose repair or replacement were within the budget constraints established by the District, would be immediately set right. The ability to make effective and economic decisions regarding the capital replacement needs of the system based on the actual degradation of the system provided a degree of asset management that had not been readily available to the management. The SSD experience indicated that proactive maintenance provides the most cost-effective means of managing the system to achieve the goal of "no overflows."

During 1992 to 1994, SSD created the sewer and manhole layers in AutoCAD format and linked the capacity model output to these layers to identify line segments that were overloaded. Complete GIS capability was added later with the use of AutoCAD Map, an extension of the common AutoCAD drafting program that adds the feature of linking information from external databases. During 1994 to 2001, information was collected in a database on the physical system characteristics, system hydraulic performance under selected flow conditions, system overflows, routine maintenance activities, CCTV inspections, and repairs and replacements. All database information was linked to the mapped line segments, thereby allowing an easy evaluation of problem distribution throughout the system with a textual and graphical response from the database query. The relational database program used was Microsoft Access, which is compatible with AutoCAD Map. All of the work was completed on a desktop PC (Rugaard, 2001).

CCTV INSPECTION OF SEWERS

Traditionally, municipalities have relied on analog video for internal inspection of sewers. In this technique, remote-controlled and self-propelled cameras move through sewer lines to record the pipe's internal conditions on a video tape. In a parallel operation, technicians create paper logs identifying the location, size, and other key information about the pipe. Referred to as CCTV inspection, this method has been considered to be the most effective and economical method of pinpointing the sources of I/I contribution. For instance, by the late 1990s the City of Boston, Massachusetts, had conducted 12 million ft of CCTV inspection on video tapes. Using an inspection cost of one dollar per foot, the cost of collecting this information could be at least $12 million. CCTV inspection videotapes contain a wealth of useful information about the state of a collection system, yet they have been often treated as single-use items. A lot of money is spent on video inspections, yet inspection data are frequently underutilized mainly because accessing the information from the conventional video tapes and paper inspection log sheets has been difficult and time consuming (Criss, 2000).

GIS can be used as a document management system for CCTV inspection data. For example, users can click on a rehabilitated sewer pipe to see the "before" and "after" movies on their computer screen. This application, however, requires converting VHS video tapes to computer files (digital movies). Once converted and stored on computers, the valuable information once hidden in videotapes can be retrieved with the click of a mouse, eliminating the need for other equipment (TV, VCR, and cables) and office space to house the equipment. CDs take up 70% less shelf space than VHS tapes. As a side benefit, digital movies can be used to create multimedia presentations for utility management when requesting maintenance and rehabilitation funds. After all, a video is worth a thousand pictures. Most video tapes have a short shelf life of approximately 10 years and their video quality and performance are inferior to digital media. Unlike VHS tapes, digital videos do not lose picture quality when copied. Digital technology also allows high-resolution snapshots of defective pipe segments taken from the video. Dynamic segmentation and image integration features allow storage and display of these images with footage reading and a description of the pipe defect. For benefits like these, experts predict that by 2008, approximately 90% of wastewater utilities will be using digital video technology (Bufe, 2003).

As stated in the preceding text, integrating CCTV videos with GIS requires migrating from video tapes to digital movies. Four migration methods are possible:

1. Convert existing video tapes to digital files
2. Digitize existing video tapes
3. Retrofit tape systems with digital systems
4. Record directly in digital format

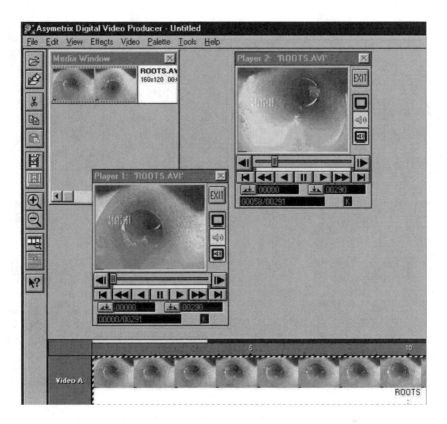

Figure 15.2   Screenshot of Digital Video Producer's video capture capability.

Convert Existing Video Tapes to Digital Files

In this method (which is the simplest), the existing video tapes are converted to multimedia computer files in the AVI or MPEG format. The hardware required for this method includes TV, VCR, multimedia computer, CD or DVD burner, and a video capture card. Video capture cards allow display of TV or VCR output on a computer screen. Some multimedia computers have built-in (internal) video capture cards. If an internal card is not installed, an external card that typically costs a few hundred dollars should be used. A video capture and editing software, which typically comes with the card, is also required. Figure 15.2 shows a screenshot of Digital Video Producer, a video capture software from Asymetrix Corporation.

The TV is connected to the VCR, and the VCR is connected to the video capture card (if external) or the computer (if internal) using AV cables. When a tape is played in the VCR, the user sees the video on the TV and on the computer screen. The video capture software has a VCR-like console with play, stop, and record buttons. When the user clicks on the software's record button, usually at the beginning of a sewer defect (e.g., collapsed pipe, root growth, break-in lateral, etc.), the computer starts to record the video in a computer file, usually in AVI or MPEG format. The

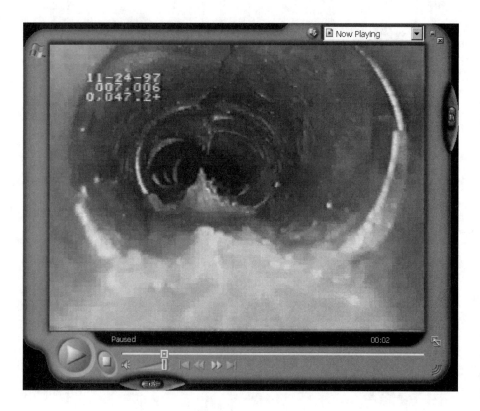

Figure 15.3   Sample mini movie in AVI format showing collapsed sewer pipe.

file recording stops when the STOP button is clicked, usually at the end of the sewer defect. Instead of converting the entire video tape, generally 5 to 30 sec mini movies showing problem areas are captured because the resulting files can be very large. For example, a 5 to 15 sec video segment recorded in color with sound can be 30 to 60 MB in size. Sound can be turned off to reduce the file size. Finally, in order to free up the computer hard disk space, the recorded digital movies are copied to digital media (e.g., CD-ROM or DVD-ROM). Raw video files can be compressed to optimize disk space. A CD cannot store more than 15 to 30 min of uncompressed data. However, in compressed format such as MPEG-1 or MPEG-2, 1 to 1.5 h of video can be stored on a CD and 6 to 12 h on a DVD. Figure 15.3 shows a sample mini movie in AVI format showing a collapsed pipe.

Digitize Existing VHS Tapes

This is basically a more efficient and automatic implementation of the previous method. Special video software is used for compressing and indexing video tapes to digital media. For example, the Tape-to-CD module of flexidata pipe survey reporting software from Pearpoint, Inc. (Thousand Palms, California), uses advanced optical character recognition (OCR) technology to read the footage count displayed on the monitor screen. This product was designed using skills and expertise gained in the development of automatic recognition of vehicle license plates. It allows the Pearpoint system to provide indexing of sewer records against the footage shown on the original videotape. The resulting CD will not only contain the original real time video but also a report.

WinCan

WinCan from WinCan America, Inc. (Durango, Colorado), is another example of video software that allows the user to capture either videos or pictures. Pictures may be captured in several different formats such as JPG or BMP. Mini movie files can also be created in increments of 5 to 35 sec. Videos may be captured in several different formats, including the most common, MPEG and AVI. WinCan can integrate stored pictures, live video from the camera or VCR, and stored movies in the pipe inspection database. Figure 15.4 shows a screenshot of the WinCan interface. The left window lists the inspected pipes in the WinCan project database. The right windows shows the mini movies for the selected pipes. The bottom window displays the observed defects for the selected pipe. WinCan also allows personal digital assistant (PDA)-based manhole reporting. The PDA (e.g., an iPAQ or PalmPilot) is used in the field to gather data with the user standing above the structure. A customized WinCan software is added to the user's PDA. Pull-down menus and easy fill-in fields make gathering the data as easy as using the full WinCan software. Back in the office, the user HotSyncs the PDA with the computer and the information is simply added to WinCan (WinCan, 2001).

WinCan interacts with GIS packages in two ways: (1) interfaces with the GIS database and (2) updates the GIS database. In the first method, the WinCan database interfaces with the GIS database by using a metadatabase. This metadatabase contains information from WinCan projects, such as unique manhole identifiers, line section numbers, street names, and pipe sizes. When a GIS user attempts to access a particular feature (e.g., a manhole or line segment), the software searches the metadatabase to determine if the feature is available. If an associated WinCan report is available, the software opens the report and highlights the specific item. The user can then display associated pictures, videos, or inspection data. This method requires toggling between WinCan and the GIS software. In the second method, an external utility appends the WinCan data to the GIS database. GIS users can then select the locations in their maps and view WinCan database and associated graphics without starting the WinCan program (Criss, 2000).

Retrofit Tape Systems with Digital Systems

The typical cameras on board the CCTV inspection vehicles are adequate for use in digital applications. Most simply need the addition of a compression unit to convert their analog signals to a digital format. In this method, digital video processing hardware and software are installed in the CCTV inspection vehicle and linked to an existing tape system.

Pearpoint equipment users can interface the Pearpoint hardware directly with WinCan. This eliminates the need for a separate overlay system with encoder.

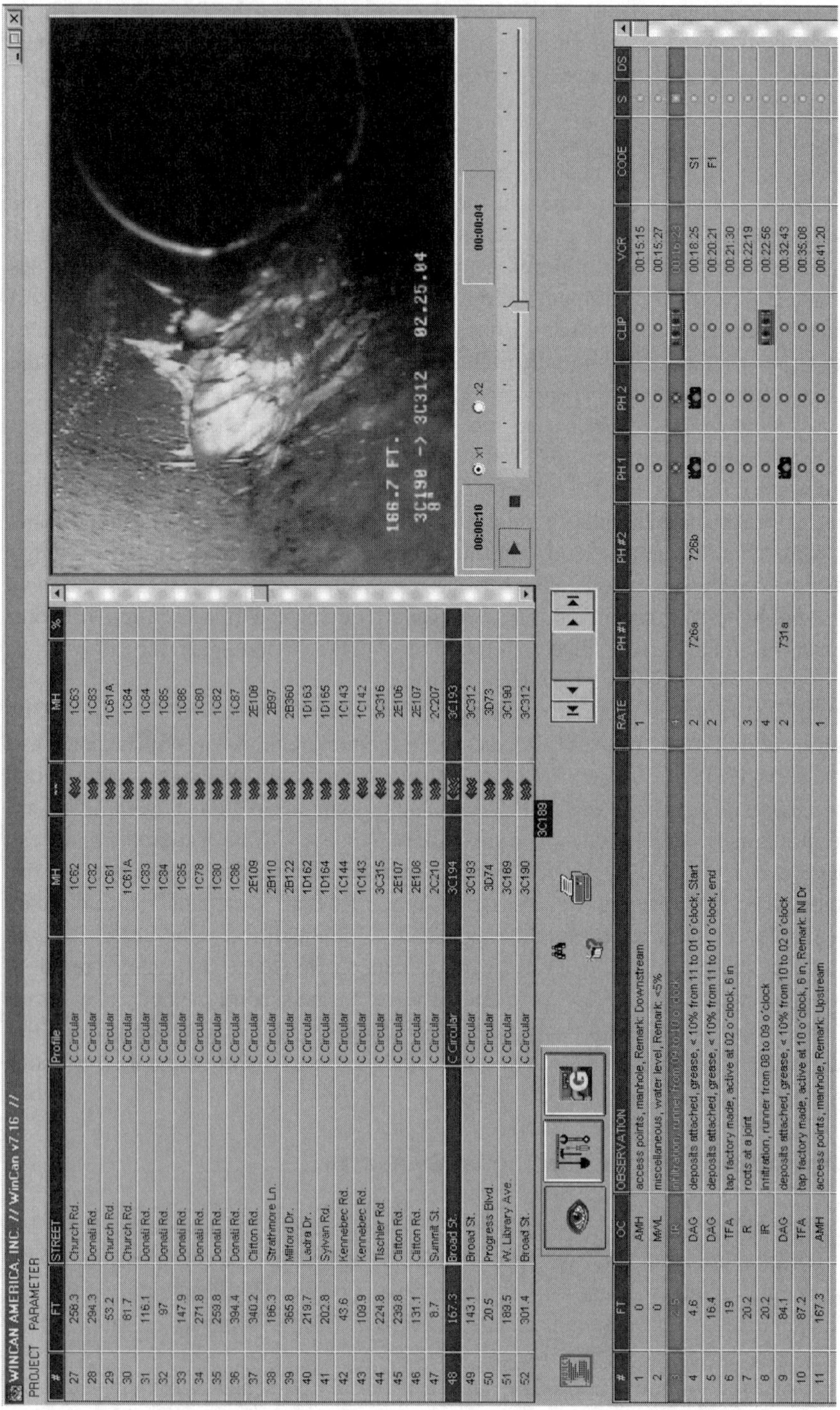

Figure 15.4   WinCan software screenshot.

According to Pearpoint, the installation is much easier with no hardware integration and wiring/soldering necessary. Camera inclination and other Pearpoint features function instantly with appropriate setup.

Another example of this approach is the PipeTech™ Software Suite from Peninsular Technologies, Inc. (Ada, Michigan). PipeTech is an innovative solution for simplifying sewer inspection projects. It allows collection, editing, analysis, and archiving of all inspection data digitally, eliminating the need for VHS tapes and the stacks of paper attachments that normally accompany an inspection project. PipeTech's Scan product is an on-site solution for video recording and incident reporting. Scan captures and compresses video in real time and facilitates the recording of incident locations. Incidents are entered on an electronic log sheet and are automatically cross-referenced with their occurrences in the video. Scan eliminates the need for a TV, VCR, and VHS tapes in the video vehicle. Videos captured using Scan are clearer than those captured through a VCR (PipeTech, 2000).

Video software such as WinCan and PipeTech also provide an on-screen pipeline or a video viewer for retrieval of stored images and movies. Consisting of a line segment graphic, these utilities allow the user to scroll along a line segment and display observations, photos, or movies by clicking on any desired location. An example is shown in Figure 14.3 for Hansen's Infrastructure Management System. This is possible because the defect location is cross-referenced with the video. In addition to the typical stop, play, and pause controls, video step and loop controls help the user to direct video progression. Seeking to any instant of video is also possible.

Record Directly in Digital Format

In this method, no VHS tapes are involved at all. The sewer inspection video is recorded directly in onboard computers as digital movies, using digital cameras. This method requires installation of appropriate digital recording hardware in the CCTV vehicle. For example, Pearpoint manufactures several video pipeline inspection systems with digital color video cameras.

This system offers complete integration of data and video. It enables a technician to watch the video as it is recorded and flag the video for defects in real time. Users can then jump to the flagged video frames without searching through the entire video (Bufe, 2003).

Linking Digital Movies to GIS

The digital movies created by using one of the preceding methods should be linked to the sewers layer in the GIS database. This will enable users to click on a sewer line in the GIS map to display the digital movie of the sewer defect. The following steps should be performed to accomplish the linkage in ArcView 3.x:

1. Add a new field to the sewer layer table to store the movie filename. Figure 15.5a shows an ArcView 3.2 database table for sanitary sewers containing a field named "Movie" for holding the movie filenames.
2. Enter the video filename for each sewer segment for which a movie is available.

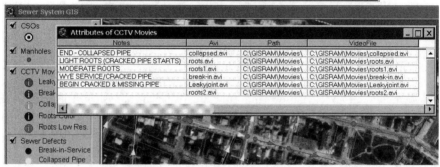

Figure 15.5 Linking digital movies to ArcView GIS. Above (a) linkage to sewers; below (b) linkage to points.

3. Write an Avenue script (shown in Figure 15.6 and described in the following text) for displaying the movie.
4. In the "Theme Properties" dialogue box for the sewers theme, click on the "Hot Link" tab as shown in Figure 15.6. Set "Field" to the field created in Step 1, set "Predefined Action" to "Link to User Script," and set "Script" to the script developed in Step 3.
5. Pick the Hot Link tool (the lightning bolt) shown in Figure 15.6 from the tool bar and click on a pipe with a video file to launch the digital movie.

Figure 15.6 shows a sample script called "Movie&ImageHotLink" for displaying movie files. The code of the script is included in Chapter 1 (GIS Applications, Applications Programming section).

The successful linkage of movies to sewers requires that both the CCTV inspection data and the sewer database table in the GIS have a common ID. This would essentially require planning ahead before conducting CCTV inspections to make sure that the inspection crews use the GIS IDs in their inspection reports. As an alternative to linking videos to sewers, a point layer can be created for video locations. Figure 15.5b shows an ArcView 3.2 point database table for CCTV movie locations. It has three fields for entering movie filenames. The users can enter the video filename and path separately in the AVI and Path fields. Or, an absolute pathname can be

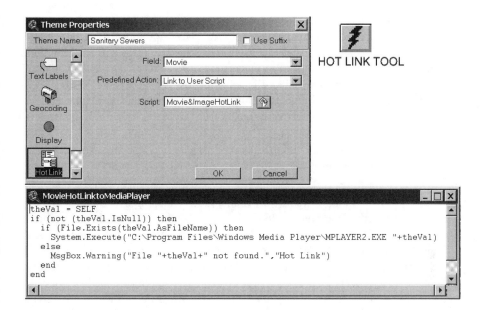

HOT LINK TOOL

Figure 15.6   ArcView GIS Hot Link Tool for linking sewer movies to sewer pipes.

entered in the VideoFile field. The disadvantage of absolute pathnames is that the application will not work if the pathname becomes invalid when video files are moved to another location (e.g., a CD or network drive). Here the users can click on a video point to see the digital movie.

In ArcGIS, the video files can be displayed without a script using the Hyperlink tool, which also looks like a lightning bolt. As in ArcView 3.x, the movie filename is entered in a database field. When the end of the bolt is near a defective sewer, the hyperlinked filename is displayed if the file exists. Clicking on the sewer will launch the movie. The hyperlinking is configured by right-clicking on the sewer layer, selecting properties, and then the display tab. The figure on the first page of this chapter shows a digital movie being played in ArcGIS 8.2.

VIDEO MAPPING

Today, GIS map features can be linked to any kind of file to document inspection of a location through pictures, field notes, audio, streaming video, etc. When digital video, GPS, and GIS maps are combined, the possibilities are limited only by the user's imagination.

Traditionally, GIS users have employed a low-tech and cumbersome approach to associate an inspection picture (e.g., inside of a sewer manhole) or video with its location. This method involves taking pictures (or recording video), writing the pictures' (or video's) description and frame numbers, scanning the pictures (or converting the video from analog to digital format), collecting GPS coordinates of picture locations, and entering all this information in the GIS database. In today's information age, this method is error prone and inefficient. The high-tech alternative known as video mapping (or media mapping) refers to georeferencing video footage by associating GPS data with video (Pfister et al., 2001). A GIS software can then read the GPS locations referenced to video time codes into its database. A map of video locations is created that can be clicked by the users to access the desired video footage. Video mapping technology can be applied to document sewer inspections and smoke-testing in GIS maps (Murphy, 2000).

Video mapping integrates GPS data and GIS maps to create interactive multimedia maps without manual scanning of photos or data entry into a computer.

Multimedia Mapping System (MMS) and Video Mapping System (VMS) software programs from Red Hen Systems, Inc. (Fort Collins, Colorado), are typical examples of video mapping products. VMS is a self-contained GPS encoder that records all data types — video, audio, and GPS — on a single tape. Users collect video, location, and audio information at the same time. It contains an integrated GPS receiver that receives location coordinates from satellites and then converts them to an audio signal for storage on video tape (Morrison and Ruiz, 2001). The user attaches the product to a video camera and records the field inspections. VMS sends GPS data from its internal receiver to the camera's microphone input. After recording the video, the user connects the system to a computer with Red Hen's MediaMapper software installed. VMS then sends the data to the computer to create a video map in TAB, SHP, or HTML format. These maps allow the users to click and view still images or video clips, play audio files, or display documents or data sets. MMS integrates the VMS unit and its built-in GPS receiver with a GPS antenna, headphone, and microphone.

MediaMapper software provides basic GIS functions and enables users to link georeference audio and video clips to GIS features. This functionality is essentially similar to ArcView's Hot Link tool described in the previous section but it is enabled by MediaMapper automatically. Software extensions are available for accessing MediaMapper data in MapInfo and ArcView GIS packages.

THEMATIC MAPPING OF INSPECTION DATA

Some applications use information from several different databases or tables. For example, a smoke-testing application might require information from the manhole and customer account databases. It is inefficient and cumbersome to enter the data from different tables in one table. Relational database tables allow accessing information from different tables without physically joining them.

A relational database stores information in records (rows) and fields (columns) and conducts searches by using data in specified fields of one table to find additional data in another field. This capability allows linking two tables that have a common field. The common field can be used to link the GIS feature tables to virtually any external database table. This linkage capability allows the GIS to make effective use of existing databases without requiring additional data entry in the GIS database.

Once the GIS and external database tables have been linked, the external data can be queried or mapped from within the GIS.

Typical manhole inspection reports have valuable attributes, such as:

- Structure number
- Street name
- Inspection date
- Offset from curb
- Structure type
- Cover or grate diameter
- Number of holes in the cover
- Frame adjustment type
- Corbel type
- Manhole wall type
- Manhole area
- Depth of structure
- Bricks missing (Yes/No)
- Ladder rungs (Yes/No)
- Ladder rungs condition
- Apron leaking (Yes/No)
- Pipe seals leaking (Yes/No)
- Manhole wall leaking (Yes/No)
- Catch basin connected (Yes/No)
- Root intrusion (Yes/No)
- Joint deterioration (Yes/No)

A manhole theme table can be linked to a manhole inspection table to create thematic maps of manhole conditions, which are excellent decision support tools for scheduling maintenance and inspection activities. The linkage can be created easily as long as both the inspection reports and the GIS database use the same manhole IDs.

Figure 15.7 shows the linkage of a manhole theme table in ArcView with the external manhole inspection table. The linkage is accomplished using the fields "Mhid" of the manhole theme table and "Mh_id" of the manhole inspection table. When the manhole "MH9-2" is selected in the GIS theme table, corresponding records in the manhole inspection table are also selected. Figure 15.8 shows a thematic map of manhole joint defects based on the "Joint Deterioration" attribute of the manhole inspection report created by using the table-linkage approach.

Similarly, typical CCTV inspection logs also have valuable attributes, such as:

- Date televised
- Type of pipe
- Begin and end manhole
- Total footage televised
- Overall pipe condition
- Manhole condition
- Joint spacing
- Joint alignment
- Number of houses
- Remarks

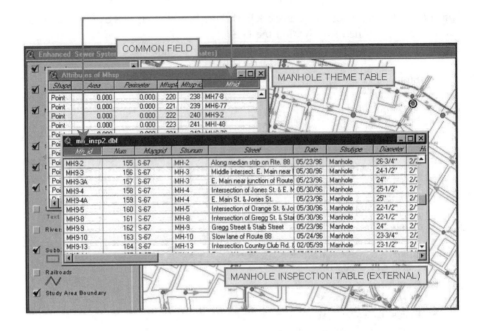

Figure 15.7   Linking manhole theme table and external manhole inspection table in ArcView.

Using the method described in the preceding text, the CCTV inspection logs can be integrated with the GIS database. This allows creation of thematic maps of sewer conditions. As described earlier, the successful linking of GIS features and inspection data requires that both the GIS database and the inspection data have a common ID.

Figure 15.8   Thematic map of manhole joint defects.

Figure 15.9 Thematic map of pipe defects.

Figure 15.9 shows a thematic map of pipe defects based on the "Remarks" attribute of the CCTV inspection logs, which contained information about sewer defects such as root growth, collapsed pipes, leaks, break-in service, etc.

WORK ORDER MANAGEMENT

Although this section focuses on wastewater system work orders, the methodology is equally applicable to water and stormwater systems.

Conventional work order management requires searching and printing the appropriate sewer system maps and manual completion of paper forms. GIS can be used to prepare work orders simply by clicking on the desired sewer or manhole in the GIS map. When the user clicks on an object, a work order inspection form is displayed that the user can fill by entering work information or by selecting predefined actions from drop-down lists. The completed work-order form and corresponding map area can be printed for the inspection and maintenance crews. The entire process takes just a few minutes compared to the conventional method of locating and copying maps and typing the work order forms, which usually takes several hours.

GIS-based work-order management requires automated mapping/facilities management (AM/FM) software extensions that can be run from within the GIS software. These add-on programs, referred to as AM/FM/GIS software, are described in Chapter 14 (AM/FM/GIS Applications). AM/FM/GIS systems are especially useful for asset inventory, inspection and maintenance, and work management (Shamsi and Fletcher, 2000). The Cityworks (formerly Pipeworks, Azteca Systems, Sandy, Utah) AM/FM/GIS software suite includes ArcView and ArcGIS extensions, which help users integrate their GIS and work management. Its Wastewater Module can be used for inspection and maintenance of sewer systems and for managing and recording a CCTV inspection program for sewers. Cityworks Work Order modules are used to create and manage work orders for inspections and tests, and for detailed inventory for water and sewer systems. The Cityworks work management solutions are based on a GIS-centric approach. Cityworks does not treat the GIS as subordinate to the work management system. Instead of integrating the work management system with the GIS, it uses the GIS as a work management system (Azteca, 1999). Figure 15.10 shows the Cityworks work order management screen for CCTV inspection of sewers.

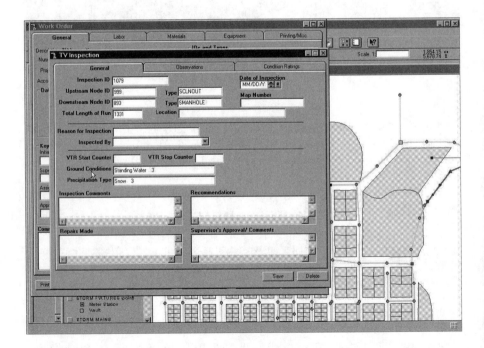

Figure 15.10 Cityworks work order management screen for CCTV inspections.

WATER MAIN ISOLATION TRACE

In many cases, the labor and material costs for fixing a broken water main are smaller than the hidden costs resulting from the pipe break. The hidden costs might be incurred due to human injury, property damage, litigation, customer inconvenience, traffic delays, police and emergency services, lost economic opportunities, and diminished fire-fighting capacity in other parts of the system. GIS offers many applications in the inspection and maintenance of water distribution systems. Water main isolation trace is one such application.

When a water main breaks, it should be isolated from the rest of the water distribution network for repairing the break or replacing the pipe. Water main isolation trace is defined as identifying those water distribution system valves that must be closed to repair or replace a broken water main. A water distribution system network layer that "knows" which links are connected to which nodes (i.e., has topology) can be used to traverse networks (Haestad et al., 2003) for solving the water main isolation problem. In addition to valve identification, GIS also allows an advance analysis of various water main shutdown scenarios, such as:

- Refining the valve search based on inoperable valves
- Identifying dry pipes (i.e., pipes without water)
- Identifying the affected customers served by dry pipes
- Identifying critical water service customers (e.g., hospitals), who need uninterrupted service
- Determining the extent of a service interruption to avoid unintentional shutdowns
- Capturing leak history data that are useful in pipeline replacement modeling

A GIS must have *topology* to benefit from the isolation trace application. Topology is defined as the spatial relationship between features. Spatial relationships between connecting or adjacent features, such as the pipes connected to a hydrant, which are so obvious to the human eye, must be explicitly defined to make the digitized maps "intelligent." For example, our eyes can instantly identify the pipe serving a house just by looking at a paper map. A GIS, on the other hand, must mathematically compare the pipe and parcel layers to complete this task. Topology is, therefore, defined as a mathematical procedure for explicitly defining spatial relationships between features. The subject of topology is so sophisticated that it is considered a branch of mathematics (Shamsi, 2002). The GIS layers that have topology can be used to trace transportation or utility networks. A water distribution system network layer that knows which links are connected to which nodes can be used to traverse networks (Haestad et al., 2003).

ESRI's ArcMap software provides many network analysis functions. ArcMap's standard interface can be customized to add network analysis functions such as network tracing. For example, a sample toolbar called the Utility Network Analyst (UNA) was developed by ESRI to help ArcGIS users develop custom applications (ESRI, 2002c). The UNA consists of a toolbar, a pull-down menu for saving and loading flags and barriers and setting options, a set of tools for specifying flags and barriers, and a combo box containing the list of available trace tasks. The UNA toolbar has a set of trace tasks (Find Connected, Trace Upstream, etc.) that can be used to solve many problems. The Isolation Trace Task of the UNA toolbar takes a single specified edge flag and determines the valves that would need to be turned off to isolate the area. The UNA toolbar can be downloaded from http://arcobjectsonline.esri.com. The steps required to use the Isolation Trace Task are:

1. Register the trace task with Categories.exe as an ESRI Utility Network Task.
2. Close and restart ArcMap.
3. Add your network to ArcMap. Use the Add Edge Flag Tool on the Utility Network Analyst toolbar to place an edge flag at the location you wish to isolate.
4. Choose "Isolation Trace" from the list of available trace tasks and perform the trace. The result will be a display of the isolated area and a selection set of the valves to turn off.

CASE STUDIES

Representative case studies of GIS applications in maintenance are given in the following subsections.

Isolation Trace Case Studies

Figure 15.11 shows an isolation trace example from the PipeWorks example data. Pipeworks (now known as CityWorks) is an AM/FM/GIS software package from Azteca Systems (Sandy, Utah, www.azteca.com), that runs as an ArcView Extension. The trace is conducted by selecting the broken water main (midway on the pipe in the center of the map) and clicking on the "Isolation Trace" button. The top screenshot shows that four valves must be closed to isolate the broken water main for repair or replacement. Next, it was assumed that one of these four valves located at the bottom of the screen is "open inoperable," which would prevent this valve from being closed down. This scenario was modeled by changing the valve status from "open operable" to "open inoperable." The actual valve status can be made available in the GIS database if field inspections data are integrated with it. The bottom screenshot shows that due to the inoperable status of one of the valves, two additional (i.e., a total of six) valves must be closed.

In 2001, San Jose Water Company, which services the City of San Jose, California, created an ArcView 8.1 application called "Main Break" to accomplish the preceding functions using Visual Basic and ArcObjects running against a personal geodatabase. The application was installed at SJWC's main office and on seven laptop computers. The laptop computers empowered the field personnel with a mobile GIS that allowed on-site identification of the valves to be closed (Coates, 2003).

Sewer System Inspections in Washington County

A sewer system inspection GIS was implemented for a small community located in the Washington County of Pennsylvania with a population of 4900 people and a

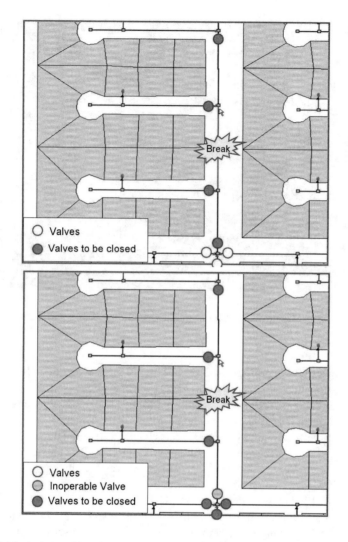

Figure 15.11 Isolation trace example. Above: all valves operable; below: one valve inoperable.

1.5 MGD (million gallons per day) sewage treatment plant. This GIS project consisted of the following tasks:

1. Develop manhole and sewer line coverages
2. Add manhole attributes from corrective action plan (CAP) manhole inspection reports
3. Add sewer attributes from CCTV inspection logs
4. Develop service laterals coverage from CCTV inspection logs
5. Develop video coverage from CCTV inspection tapes
6. Add street names to the existing roads coverage
7. Align image and feature themes
8. Package the information for delivery

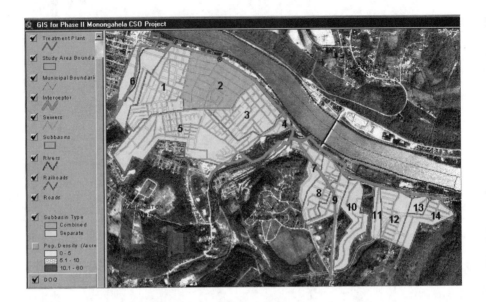

Figure 15.12   ArcView layers created from sewer system inspections data.

The project was completed in 2000 in 6 months for a cost of $15,000. Various free public-domain GIS data were used to keep the project cost low. These data included USGS digital orthophoto quarter quadrangles (DOQQ), digital raster graphics (DRG), and digital elevation models (DEM). Tasks 3, 4, and 5 were accomplished using the various techniques described in this chapter. Fifty mini movies were created from the existing VHS tapes and stored on five CD-ROMs. The movie files consisted of uncompressed AVI files of 40 to 60 MB size and resulted in approximately 3 GB of digital data.

Figure 15.12 shows various ArcView GIS themes of the project overlayed on the DOQQ image. The figure on the first page of this chapter shows an ArcView project screen displaying a mini movie of root growth for a user-specified location. Figure 15.8 and Figure 15.9 presented earlier show thematic maps of manhole and sewer defects, respectively.

Sewer Rehabilitation in Baldwin

Facing stringent regulatory requirements at the turn of the century, many communities in the Allegheny County (Pennsylvania) started undertaking comprehensive sewer rehabilitation projects to reduce systemwide inflow and infiltration. Rehabilitation activities included internal televising, pipe lining, pipe bursting, pipe replacement, manhole repair, and manhole replacement.

Baldwin Borough, a 5.5 mi^2 municipality located in the Greater Pittsburgh area with a population of 22,000, undertook a similar project. Baldwin's consultant,

(Chester Engineers, Pittsburgh, Pennsylvania) developed a sewer rehabilitation data management system (SRDMS) to assist in the planning and design phases of Baldwin's sewer rehabilitation projects. Initially, the system was successfully tested in the bidding phase for internal inspection of over 200,000 ft of sanitary sewers and 3,000 manholes. The system utilized digital map data compiled during a sewer-mapping effort initiated in 1996. The project team field-verified the sewer network data based on CCTV inspections conducted during a 3-month period in 2001. Concurrent with this effort, a Microsoft Access™ database was developed to capture conditional information for each surveyed sewer feature, apply unit costs for all necessary repair work, and generate a variety of reports for use in bid document submittals. Subsequent rehabilitation projects proved the effectiveness of SRDMS, leading to its implementation in several Allegheny County communities.

SRDMS is a GIS-based system because it reads and writes the information in a GIS database. It utilizes a relational database (e.g., Microsoft Access) to manage disparate data related to rehabilitation work, including manhole inspections, televised sewer line inspections, and flow data. This relational database is linked with the GIS database to facilitate mapping of the extent and severity of system damage and the recommended remedial activities. The required GIS layers can be created from a variety of sources, such as ArcInfo™ coverages, ArcView™ Shapefiles, and CAD files (e.g., Integraph™ or AutoCAD™). Preprocessing of CAD data to ensure the connectivity of sewer features may be necessary as the system requires that connectivity be established between all lines and manholes. SRDMS can use field inspection data from the conventional paper forms or the latest digital data collectors (e.g., PDAs, or tablet PCs). The GIS interface was initially developed as an ArcView 3.2 Extension. An ArcGIS version was created subsequently. SRDMS links inspection and repair information contained in the Access database to the corresponding GIS features. Applications of this system have resulted in a 20 to 30% reduction in project management labor effort over traditional methods. SRDMS provides the end user with the following functionalities:

- Provides the ability to visualize the overall condition of the sewage collection system using thematic maps of problematic areas.
- Identifies the precise locations of specific rehabilitation activities based upon an engineering interpretation of CCTV and manhole inspection results.
- Develops budgetary pricing for system repairs and rehabilitation work. It generates cost estimates for manhole rehabilitation and line repair work by applying multipliers to unit repair costs. Unit costs are entered by the project manager. The unit cost estimates may be modified at any time. The system will automatically update all aggregate costs based on the updated figures.
- Provides a series of reports detailing system damage, recommended remedial action, and cost estimates for the repair of damaged structures.
- Develops plans for distribution to prospective contractors during the bidding process.
- Provides a mechanism for tracking and reporting the progress of the rehabilitation activities.

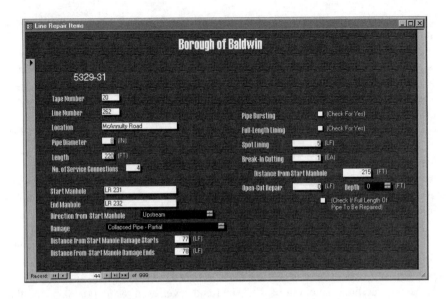

Figure 15.13   SRDMS line defect and repair entry form.

- Expands the content of the original GIS to include the location of service lateral connections identified during the televising process.
- Corrects inaccuracies in the original GIS based upon information obtained during the televising process.

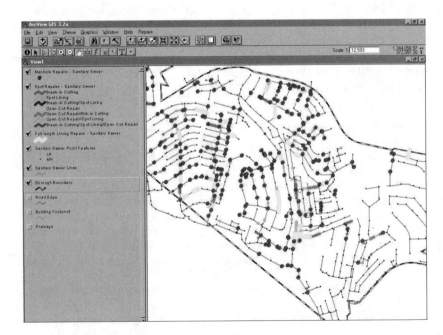

Figure 15.14   SRDMS line and manhole repair map.

Figure 15.13 shows the SRDMS line defect and repair entry form. The left side of this form receives input from the CCTV inspections database and automatically updates the GIS layer for pipes. The right side of the form, which recommends appropriate repairs, is completed by the project engineer or manager based on individual judgment and experience. Manhole and line repairs recommended by the user are added to the GIS as new layers. Figure 15.14 shows an ArcView screenshot of the line and manhole repair map. All the manholes targeted for repairs are clearly shown as large dots. All the sewers targeted for spot repairs or full-length lining repairs are shown in different colors.

USEFUL WEB SITES

Azteca Systems	www.azteca.com
Government Accounting Standards Board (GASB)	www.gasb.org
Pearpoint, Inc.	www.pearpoint.com
Peninsular Technologies, Inc.	www.pent.com
Red Hen Systems, Inc.	www.redhensystems.com
WinCan America, Inc.	www.wincanamerica.com

CHAPTER SUMMARY

This chapter provided an overview of GIS applications for the fourth M of the "4M approach" in the inspection and maintenance of water, wastewater, and stormwater systems. These applications include asset management and field inspections required by regulations such as GASB 34 and wet-weather overflow policies in the U.S. GIS also allows integrating work-order management and video inspections (e.g., CCTV inspections of sewers) with maps, which results in workflow efficiency improvement and cost savings for water and wastewater utilities. GIS can be used as a project management tool for system rehabilitation and repair work. Last but not least, GIS allows thematic mapping of inspection data to enable effective communication with decision makers and stakeholders. Real-world examples and case studies presented in this chapter show that GIS can be used for more efficient inspection and maintenance of water, wastewater, and stormwater systems.

CHAPTER QUESTIONS

1. What is the "infrastructure problem," and how can GIS help to solve this problem?
2. What is asset management, and how can GIS help in asset management of water, wastewater, and stormwater systems?
3. Describe two examples of GIS applications in regulatory compliance for water, wastewater, or stormwater systems.
4. What is water main isolation trace? How will you implement it on a mobile device?

CHAPTER 16

Security Planning and Vulnerability Assessment

Although water and wastewater systems are generally well prepared to deal with natural disasters and accidents, many utilities prior to September 11, 2001 had given little consideration to defending themselves against acts of terrorism. Read this chapter to see how GIS can help in safeguarding your water and sewer systems against the acts of terrorism.

Water distribution systems are especially vulnerable to acts of terror.

LEARNING OBJECTIVE

The learning objective of this chapter is to understand GIS applications in security planning and vulnerability assessment of water and sewer systems.

MAJOR TOPICS

- GIS applications in planning
- Security planning
- Vulnerability assessment
- Security modeling software

LIST OF CHAPTER ACRONYMS

AMSA Association of Metropolitan Sewerage Agencies
EPA (U.S.) Environmental Protection Agency
ERP Emergency Response Plan
OGC Open GIS Consortium
VSAT Vulnerability Self Assessment Tool

GIS APPLICATIONS IN PLANNING

GIS technologies have enabled planning professionals to make better and faster planning decisions. New software advances have revolutionized the preparation and use of master plans for water and sewer systems. GIS technologies provide various planning tools such as:

1. Remote-sensing data (described in Chapter 3)
2. Land-use/land-cover data (described in Chapter 3, Chapter 11, and Chapter 12)
3. Intelligent maps (described in Chapter 8 and Chapter 9)
4. Hydrologic and hydraulic models (described in Chapter 11 to Chapter 13)

With the help of case studies and examples, this chapter will focus on how to apply these tools for security planning purposes.

CITY OF ALBUQUERQUE'S WASTEWATER MASTER PLAN

In 1990, the City of Albuquerque, New Mexico, began a multiphase planning effort to evaluate its existing wastewater conveyance and treatment facilities, project future growth and flows, create programs for monitoring and rehabilitating existing infrastructure, and develop an organized approach to ensuring that future facilities keep pace with development. The City used GIS technology to develop a 40-year master plan for its wastewater collection and treatment system. The master plan was created by integrating the City's GIS data with a number of other computer applications including demographic modeling, land-use mapping, hydraulic analysis, and infrastructure rehabilitation planning. The GIS approach enabled the City to create a master plan that could be easily updated in future as new development takes place (American City & County, 1991).

SECURITY PLANNING

Water, wastewater, and stormwater systems face both natural (floods, droughts, earthquakes, fires, etc.) and man-made (terrorism, vandalism, sabotage, arson, cyber attacks, etc.) threats. Although water industry utilities have learned from natural disasters, their experience with human-caused threats is small (Grigg, 2002). The basic form of protection against attacks or threats of attacks is a security system. Security planning methods include hazard assessment, vulnerability assessment, mitigation, response planning, and crisis communications (Burns et al., 2001). Vulnerability assessment is an important step in disaster preparedness. In the U.S., many utilities are performing mandatory vulnerability assessments.

When terrorists attacked the World Trade Center complex in New York and the Pentagon building in Washington, D.C., on September 11, 2001, the U.S. witnessed one of the worst days of its 225-year history. The unfortunate event also witnessed the best of the American spirit as people everywhere, including the GIS community, rushed to help. GIS supported response efforts at the World Trade Center and the Pentagon by coordinating the search for survivors and identifying hazardous areas. Planning for collection of airborne imagery over Ground Zero began within hours of the terrorist attack. Only 3 hours after the attack, France's SPOT 4 satellite captured 20-m resolution infrared images of the fires blazing in Manhattan. On September 12, Space Imaging's IKONOS satellite acquired 1-m imagery of both the World Trade Center area and the Pentagon (Barnes, 2001b). LIDAR* images collected on September 19, 2001, at 5000 ft were used to create 3D depictions of the terrain and buildings of the World Trade Center complex (Logan, 2002). The media, to cover the stories related to the terrorist attack, frequently used these 3D renderings.

Since September 11, 2001, concerns over intentional contamination have resulted in new federal legislation in the U.S. requiring community water systems to develop Emergency Response Plans (ERPs) (Haestad Methods, 2003a). In the Public Health Security and Bioterrorism Preparedness and Response Act of 2002, U.S. Congress recognized the need for drinking water system utilities to undertake a more comprehensive view of water safety and security. It amended the Safe Drinking Water Act and specified actions that community water system utilities and the U.S. Environmental Protection Agency (EPA) must take to improve the security of the nation's drinking water infrastructure. In 2002, EPA issued $90-million grant monies to help U.S. water and wastewater utilities assess their security needs. In January 2003, the U.S. Government launched the Department of Homeland Security for coordinating the work of America's security and intelligence agencies.

Although water and wastewater systems are generally well prepared to deal with natural disasters and accidents, many utilities — prior to September 11 — had given

---

* For additional information on LIDAR data, please refer to Chapter 3 (Remote-Sensing Applications).

little consideration to defending themselves against acts of terrorism. Immediately after the September 11 terrorist attacks, many water and wastewater utilities in the U.S. quickly adopted or expanded security measures (Landers, 2002).

Vulnerability of Water Systems

Drinking water utilities today find themselves facing new responsibilities. Although their mission has always been to deliver a dependable and safe supply of water to their customers, the challenges inherent in achieving that mission have expanded to include security and counterterrorism. Water distribution systems are especially vulnerable to acts of terror. Water systems can be contaminated by injection of poisonous substances such as cyanide. Water system operators, therefore, should be able to monitor contaminant levels and to predict contaminant movement throughout the system. These are complex tasks that require real-time monitoring and distribution-system water quality modeling. Real-time monitoring of every contaminant that could be deliberately introduced in a water system is not practical. However, a change in indicator parameters can signal the possibility of intentional contamination. The operators should be equipped with the most effective means of interpreting the monitoring data, communicating the results to decision makers and the general public, and implementing the control measures quickly (Schreppel, 2003). At the present time, most applications are focusing on the security analysis of drinking water distribution systems.

Vulnerability of Sewer Systems

Sewer systems can be used to contaminate receiving waters that are used for drinking-water supply intakes. Terrorists can pour volatile matter in the sewer system to create fire hazards. Large combined and storm sewers can also be used to gain access to restricted areas that have been blocked for security reasons. Although much of the attention regarding the security of water infrastructure has focused on the drinking water side, wastewater professionals have been quick to acknowledge and attempt to address their vulnerabilities. For example, the Metropolitan Sewer District (MSD) of Greater Cincinnati, Ohio, turned its attention to its sewage collection system, analyzing the likelihood that its sewers — particularly some of its large combined sewers — might be used as conduits in a terrorist attack. MSD used a GIS to compare maps of its collection system with the maps of Cincinnati to conclude that there were no major targets of opportunity in close proximity to any of its large sewers (Landers, 2002). Analysis of using a sewer system to attack a wastewater treatment plant or pumping station still requires some research.

GIS APPLICATIONS IN VULNERABILITY ASSESSMENT

Water and wastewater utilities must remain vigilant and develop comprehensive engineering and management measures to protect against future threats (Grigg, 2003). Implementing homeland security necessitates understanding the systems, infrastructure, organizations, and vital interactions necessary for the well-being of communities. GIS is critical to this effort because it integrates all types of information and relates that information spatially. GIS can help in the areas of risk assessment, security planning, mitigation of the effects of attacks, preparedness activities, response measures, and recovery efforts.

GIS applications for community protection include modeling chemical releases in water and air, tracking and accessing information about hazardous storage sites, and managing hospital information. GIS applications can also reveal additional problems caused by proximity to hazards such as chemical storage. The modeling capabilities of GIS can simulate the effects of many types of attacks on critical resources, infrastructure, and populations. After identifying the hazards, assessing the risks, and prioritizing the values (i.e., assets of greatest value), both strategic and tactical plans can be formulated. GIS enhances response to acts of terror by helping managers to quickly assess the extent of damage and by speeding the decision-making process (ESRI, 2002d).

There is no one-size-fits-all security approach that will work for all water or sewer systems. Because each utility's situation is unique, security measures must be tailored to each system. In general, however, there are two basic requirements for the vulnerability assessment of water and sewer systems:

1. Network connectivity: This requirement pertains to what system components are connected to what and what is upstream (or downstream) of what. This capability would determine, for example, the area that would be impacted by a contaminated water storage tank. A GIS must have topology information to determine network connectivity. Network topology is inherent in many GIS packages and can easily provide this information. Network-tracing (described in Chapter 9 [Mapping Applications]) and isolation-tracing (described in Chapter 12 [Water Models]) functions are good examples of the network topological capabilities of GIS.
2. Contamination analysis: This requirement pertains to the strength and extent of the contaminants. GIS alone cannot provide this information. Hydraulic models capable of simulating both the quantity and the quality of water must be used in conjunction with a GIS.

GIS and hydraulic modeling can be used for infrastructure protection, vulnerability assessment, consequence management, and security planning to safeguard water, wastewater, and stormwater systems against potential sabotage activities and terrorist attacks. Lindley and Buchberger (2002) outlined a holistic approach that combined GIS and hydraulic modeling to integrate multiple risk factors so as to identify locations that may be vulnerable to contaminant intrusions. Intrusions were defined as the introduction (either accidental or deliberate) of an undesirable agent into the potable water distribution system. An intrusion pathway was defined as the connectivity route between the potable supply and the contaminant source. EPANET* software was used for hydraulic modeling. GIS was used for layering of the locations having adverse pressure, intrusion pathway, and source concerns.

---

* Additional information about EPANET is provided in Chapter 12 (Water Models).

SECURITY MODELING SOFTWARE

Sample security modeling software products are described in the following subsections.

H₂OMAP™ Protector

H₂OMAP Protector from MWH Soft, Inc. (Pasadena, California), is an add-on module for the H₂OMAP suite* that can be used for water security planning, infrastructure protection, and vulnerability assessment. Designed with the latest geospatial modeling technology, Protector uses a geodatabase for modeling various security scenarios. It can be used for estimating the consequences of a terrorist attack or a crisis event on a drinking-water supply infrastructure as well as formulating and evaluating sound emergency response, recovery, remediation and operations plans, and security upgrades. The program can be used to identify viable solutions before an incident or disaster occurs, or to assist in responding should it occur.

Protector allows users to model the propagation and concentration of naturally disseminated, accidentally released, or intentionally introduced contaminants and chemical constituents throughout water distribution systems; assess the effects of water treatment on the contaminant; and evaluate the potential impact of unforeseen facility breakdown (e.g., significant structural damage or operational disruption or both). It enables users to locate areas affected by contamination, calculate population at risk and report customer notification information, and identify the appropriate valves to close to isolate a contamination event. Finally, it helps users track contaminants to the originating supply sources, compute required purging water volume, develop efficient flushing strategies, determine the resulting impact on fire-fighting capabilities, and prepare data for eventual prosecution.

Figure 16.1 shows a Protector screenshot showing isolation analysis. It depicts affected and closed facilities color-coded according to their vulnerability (risk factors).

WaterSAFE™

WaterSAFE from Haestad Methods, Inc. (Waterbury, Connecticut), is an add-on component for WaterCAD and WaterGEMS** software products. WaterSAFE is designed to manage and safeguard water distribution systems. It is a water system security and emergency planning tool specifically created to study infrastructure vulnerability to terrorist attacks and natural events.

Harnessing the powers of ArcGIS, WaterSAFE enables water utilities to analyze the movement of multiple constituents and track multiple sources for a given period of time. With this information, water utilities can assess the impact in real time and swiftly relay the results to customers in contaminated areas to take the necessary precautions. These water quality analysis features can enable utilities to better respond, strategize, and implement a fast and effective recovery plan in the event of

---

* Additional information about H₂OMAP suit is provided in Chapter 12 (Water Models)
** Additional information about WaterCAD and WaterGEMS is provided in Chapter 12 (Water Models).

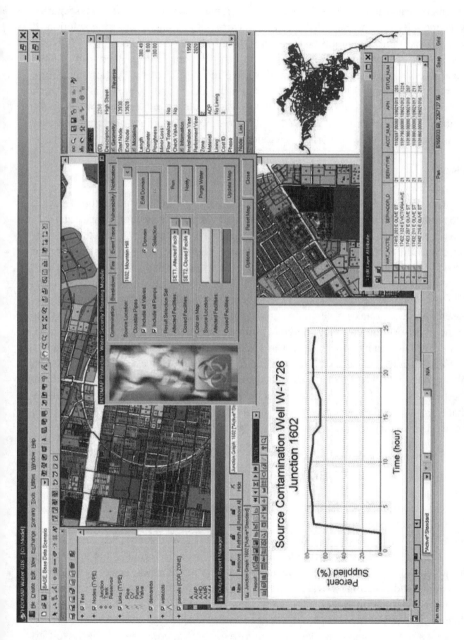

Figure 16.1 Protector screenshot showing isolation analysis.

a contamination emergency. WaterSAFE's extended water quality analysis features can also assist in conducting statistical analysis of water quality and evaluating contamination effects on a water treatment plant.

In 2003, a WaterCAD model was used to prepare an example ERP. Two scenarios were modeled: a rehearsed drill before a terrorist attack and a real-time solution. It was assumed that a contaminant was introduced into the tank of a typical midsized system and that complete mixing occurred in this tank. The vulnerability of the water system was assessed by monitoring the effects of injection location, the nature of the contaminant, the amount of contaminant added, and the period over which the contaminant was added. The modelers were able to reduce the spread of the contaminant, monitor the impact of flushing on the environment, and more quickly bring the system back to normal operation (Haestad Methods, 2003a).

VSAT™

Association of Metropolitan Sewerage Agency (AMSA) Washington, D.C., in cooperation with EPA, released a new water and wastewater system security analysis software in 2002. Known as Vulnerability Self Assessment Tool (VSAT), the software is available free of charge. Three versions, VSATWater, VSATWastewater, and VSATWater/Wastewater, are available for water, wastewater, and both water and wastewater systems, respectively. The VSAT products provide a structured, cost-effective approach for utilities to assess their vulnerabilities and to establish a risk-based approach to taking desired actions. The software allows utilities to assess the vulnerability of the complete range of utility assets including people (utility staff), physical plant, knowledge base, information-technology platform, and customers.

VSAT does not use GIS or hydraulic modeling. However, the information obtained from GIS and modeling can be used to run the VSAT program. For example, the first VSAT task is asset categorization and identification, which requires utility managers to conduct an inventory of utility assets. A GIS can be used to conduct an asset inventory for input to the VSAT program.

SECURITY PLANNING DATA ISSUES

The number of GIS data formats has increased exponentially with the growth in the GIS industry (Goodchild, 2002). According to some estimates, there might be more than 80 proprietary geographic data formats (Lowe, 2002b). Why are there so many geographic data formats? One reason is that a single format is not appropriate for all applications. For example, a single format cannot support both fast rendering in a command and control system and sophisticated topological analysis in a natural resource information system. Different data formats have evolved in response to diverse user requirements.

Users must possess considerable expertise to overlay, combine, or analyze different map layers or images. Converting from one format or type of data to another is cumbersome, time consuming, and error prone. Therefore, different GIS data formats or models used by different government organizations can lead to compatibility problems that can hamper data sharing needs during an emergency situation such as a terrorist attack. This can be avoided by adopting compatible data models and interoperable data formats throughout a city or county. Efforts are currently underway in the GIS industry to standardize data formats and database-management systems by promoting open platforms, data, and database management systems. For example, the Open GIS Consortium (OGC) has been created with the vision of "the complete integration of geospatial data and geoprocessing resources into mainstream computing." OGC was formed in 1994 to facilitate access to and geoprocessing of data held in systems or networks. It defines an open GIS as "open and interoperable geoprocessing" or "the ability to share heterogeneous geodata and geoprocessing resources transparently in a networked environment."

USEFUL WEB SITES

Association of Metropolitan Sewerage Agencies (AMSA)	www.amsa-cleanwater.org
Haestad Methods	www.haestad.com
MWH Soft	www.mwhsoft.com
Open GIS Consortium	www.opengis.org
Vulnerability Self Assessment Tool (VSAT)	www.vsatusers.net

CHAPTER SUMMARY

Water, wastewater, and stormwater systems can be vulnerable to acts of terrorism and sabotage. This chapter shows that GIS offers many useful applications in security planning and vulnerability assessment of these systems. GIS maps and hydraulic models can be used as effective security planning tools to evaluate the system vulnerability to natural and man-made disasters, and recommend solutions. For example, GIS and hydraulic models can be used to determine the areas of a water distribution system that can be affected by a contaminated water supply source. At the present time, most applications are focusing on the security analysis of drinking water distribution systems.

CHAPTER QUESTIONS

1. How are water and sewer systems vulnerable to acts of terrorism?
2. What can be done to protect water and sewer systems from acts of terrorism?
3. How can GIS applications help in security planning and vulnerability assessment of water and sewer systems?

CHAPTER 17

Applications Sampler

The use of GIS applications is growing throughout the world. This chapter shows how people around the world are applying GIS in their water, wastewater, and stormwater projects.

This chapter presents the latest examples of GIS applications in the water industry.

LEARNING OBJECTIVE

The learning objective of this chapter is to document GIS application projects around the world for water, wastewater, and stormwater systems.

MAJOR TOPICS

- Water system modeling
- Sewer system modeling
- Collection system rehabilitation and asset management
- Resource planning and capital improvement project (CIP) allocation
- Water quality management
- Water master planning

LIST OF CHAPTER ACRONYMS

BMP Best Management Practices
CCTV Closed-Circuit Television
CIP Capital Improvement Project
CIS Customer Information System
CSO Combined Sewer Overflow
DEM Digital Elevation Model
SCADA Supervisory Control and Data Acquisition
SSO Sanitary Sewer Overflow
SWAT Soil and Water Assessment Tool
TAZ Traffic Analysis Zone

This chapter presents a collection of recent case studies on GIS applications for water, wastewater, and stormwater systems. These case studies were written specially for publication in this book by 18 GIS and water professionals from 6 countries (Belgium, Bulgaria, Czech Republic, Denmark, Spain, and the U.S.). For the names and organizational affiliations of the case studies' authors, please see the Acknowledgments section. The case studies were submitted in response to the author's "Call for Case Studies" distributed to various Internet discussion forums.

DRAINAGE AREA PLANNING IN SOFIA

Application	Sewer system modeling
Author(s)	Milan Suchanek, Tomas Metelka
Project status	Completed in March 2003
Hardware	1000 MHz Pentium III personal computers
GIS software	Arc View 3.3
Other software	MOUSE 2002; MOUSE GM; MOUSE Gandalf; MOUSE LTS; AquaBase
GIS data	Subcatchment polygons, conduit lines, node points, CSO points, land use, population, property boundaries, buildings
Study area	City of Sofia, the capitol capital of Bulgaria, watershed area of about 400 km^2
Project duration	November 2001 to March 2003
Project budget	$460,000
Organization	Sofiyska Voda A.D.

The city of Sofia's sewerage system is a very complex combined drainage system dating back to early 1900. The drainage area is made up of about 400 km² of watershed area, draining wastewater from 1.2 million people. At the same time, the hydrological conditions of the city, which is located on the foothills of the Vitosha mountains, promote heavy rainfall, mainly during late spring.

By the end of March 2003, the first simulation model for the sewer drainage system was developed in Bulgaria. The skeletal planning model was built for the city of Sofia, having about 5000 manholes, 5000 pipes, 3000 catchments, and 140 CSO or diversion chambers. The study area was subdivided into seven main sub-catchments and their main trunk sewers. The respective submodels were built up and calibrated based on 6 weeks of flow survey at some 75 flow and 25 raingauge sites. Finally all seven submodels were merged into one combined model covering the whole drainage area.

The local GIS system, supported by comprehensive manhole and ancillary surveys, supplied the model with the system structural data. The overall data migration process was an important task in the project execution.

The Sofia drainage area planning project brings a new experience to the general view of project management. The management project was based on British standards, and the WaPuG *Code of Practice* handbook was followed. At the same time, Danish technology was applied, along with Czech and Bulgarian know-how. Figure 17.1 shows a screenshot of the Sofia model in MOUSE (Suchanek and Metelka, 2004).

PIPE RATING PROGRAM IN BUNCOMBE COUNTY

Application	Collection system rehabilitation and asset management
Authors	Ed Bradford, Roger Watson, Eric Mann, Jenny Konwinski
Project status	Implemented in 2004 and is being used to generate rehabilitation projects
Hardware	Standard desktop PCs (connected to network)
GIS software	ArcGIS 8.x
Other software	Microsoft Access XP, Windows Media Player, Microsoft Excel, Microsoft Word, etc., as necessary to generate reports
GIS data	Sewer line polylines, sewer structure points, 6-in., 1-ft, and 2-ft resolution DOQs, county parcel polygons, drainage basin polygons, digital video hyperlinks, CCTV inspection and defect-coded Microsoft Access tables, river and stream polylines, and road centerline polylines
Study area	Entire MSD service area (180 mi², 920 mi of sewer)
Project duration	Began in 2001. No end date scheduled
Project budget	No defined budget *per se* for the program itself. Construction projects have generated $700,000 in actual cost so far, which could increase in future
Organization	Metropolitan Sewerage District (MSD) of Buncombe County, North Carolina
Web site	www.msdbc.org

MSD's Pipe Rating Program is a new method of generating and prioritizing sewer rehabilitation projects. The typical approach of *reactive* planning is to define, develop, and complete a rehabilitation project after problems such as sanitary sewer overflows (SSO) or structural failures occur (Bradford et al., 2004a). Pipe Rating is a *proactive* tool that utilizes CCTV information, a GIS database, and real-world maintenance history to view, score, and rate pipe segments based on a number of

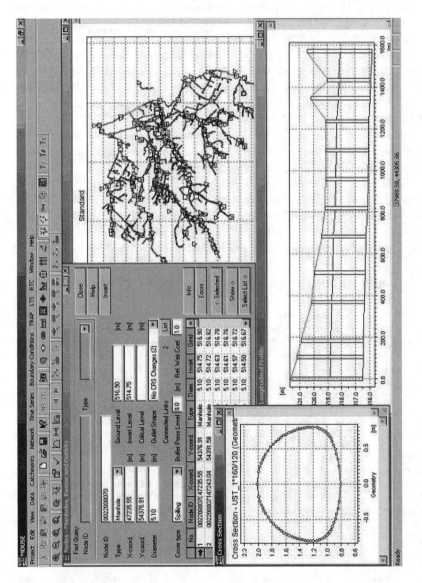

Figure 17.1   Screenshot of Sofia's MOUSE sewer system model.

factors. These factors, for any given manhole-to-manhole segment, include the number and severity of structural defects and the history of overflows on that segment. These are all combined to yield a rating, which may then flag a particular line segment for further investigation.

The data are gathered from a number of sources and incorporated into this program, which runs on the standard ArcGIS platform. CCTV video is captured by cameras traversing the pipes, and recorded on VCR tapes or in digital format (CD or DVD). First, each video is linked to its respective pipe segment within the District's GIS system so that it is immediately available to engineers and field responders. Second, a database is also created from the field data to record various features about the pipe, such as defect and structural information. This is accomplished by assigning to each defect a value pursuant to the standardized defect-rating manual developed by the District for this purpose. CCTV data are collected and embedded in a standard Microsoft Access database. When the technician identifies a defect along the segment, it is keyed in with the corresponding defect code. Each defect, and its corresponding severity score, is assigned to its pipe segment and used for future analysis performed in the GIS. Table 17.1 shows the defect structural scores used by MSD. These scores are based on MSD's standardized Sewer Condition Classification Manual. Scores are weighted according to MSD priorities.

Finally, all the information is used to calculate three pipe scores for each pipe: mean pipeline, mean defect, and peak defect scores. These scores allow users to visualize the severity in three categories for prioritization within the rehabilitation program. Once this has been accomplished, engineers can retrieve the videos from the GIS for confirmation and to evaluate them for rehabilitation procedures. Figure 15.1 shows an ArcGIS screenshot of mean pipeline ratings. Figure 17.2 and Figure 17.3

Table 17.1 Sewer Pipe Structural Defect Scores

Code	Description	Score
CC	Circumferential crack	20
CL	Longitudinal crack	20
CM	Multiple cracks	30
COH	Corrosion heavy	50
COL	Corrosion light	20
COM	Corrosion medium	30
D	Deformed sewer	65
FC	Circumferential fracture	40
FL	Longitudinal fracture	40
FM	Multiple fractures	60
HL	Hole large	50
HP	Hole patched	5
HS	Hole small	25
JDL	Joint displaced large	30
JDM	Joint displaced medium	5
OJL	Open joint large	30
OJM	Open joint medium	10
X	Collapsed pipe	100

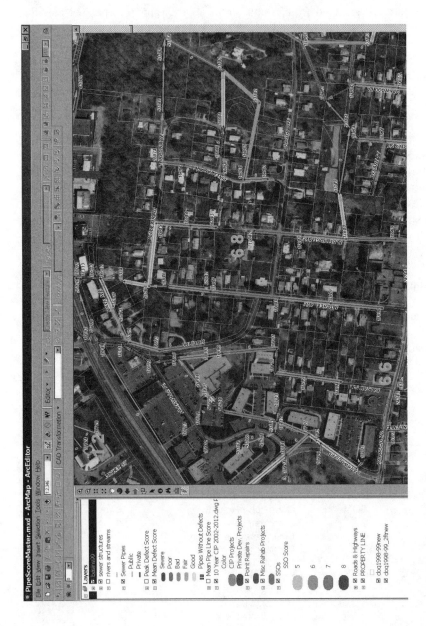

Figure 17.2 Map of mean defect ratings for sewer pipes.

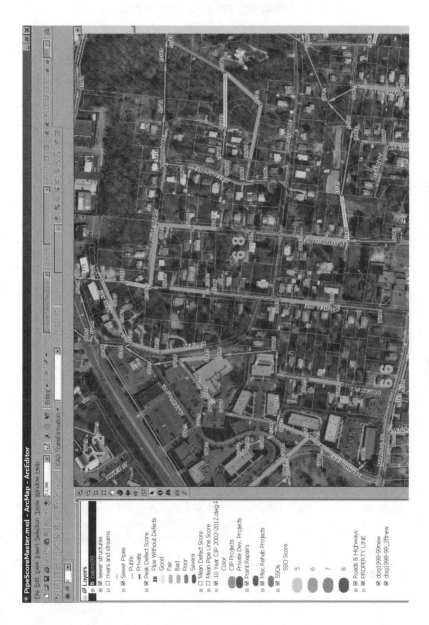

Figure 17.3 Map of peak defect ratings for sewer pipes.

show maps of mean defect and peak defect pipe ratings, respectively (Bradford et al., 2004a).

WATER SYSTEM MODELING IN TUCSON

Application	Resource planning and capital improvement project (CIP) allocation
Author	Dean Trammel
Project status	GIS and hydraulic modeling effort completed in 2003
Hardware	Dell personal computers (1495 MHz, 384 MB RAM) with Windows 2000 Professional
GIS software	ArcMap and ArcInfo
Other software	Haestad Methods' WaterGEMS, WaterCAD for AutoCAD, Microsoft Access, Microsoft Excel
GIS data	U.S. census–disaggregated population projection data in the form of polygon Shapefiles, with associated population projections for each year from 2000 through 2050
Study area	Tucson, Arizona
Project duration	Resource planning effort: March 2002 to early 2004, GIS and hydraulic modeling: about 9 months
Project budget	CIP-budgeted planning effort, part of regular business
Organization	Tucson Water, Tucson, Arizona

Tucson Water's potable system supplies over 180,000 services, maintains over 4,000 mi of pipe, and installs over 500 new meters per month. Tucson Water also serves reclaimed water to over 550 reclaimed services, maintains over 100 mi of reclaimed pipe, and installs between 75 and 100 new reclaimed meters per year. In addition to the GIS data listed above, the following data were also used:

- The Pima Association of Government (PAG) took U.S. census tracts for the year 2030 and broke them up further into smaller disaggregated population-projection polygons, or traffic analysis zones (TAZ).
- Engineers, planners, and administrators made up a Resource Planning Committee that determined a boundary for the water service provided by Tucson Water, which was a modified version of a boundary used for previous planning efforts.
- Polygon Shapefiles that describe pressure zones and water services areas (WSA) for the entire service area, based on a parcel base of Tucson Water's customers.
- Ten hydraulic models of the central water system called local area models and five smaller models that represent isolated systems outside the central system.
- Shapefiles of the proposed master plans. These files were used as a backdrop to compare population projections with prospective new developments planned for the near future, and to make sure that these areas were correctly represented in the 2030 and 2050 model.
- SCADA, billing, production records.
- Cost estimates.
- Hydrology supply scenarios.

Traffic and census tract population projections were analyzed using ArcGIS to estimate future potable water demands. The estimated demands were applied to Tucson Water's current hydraulic model using Haestad Methods' WaterGEMS software. The model was then used to determine what facility and piping modifications would be required to meet growth within current service areas and future development in out-lying areas. Using capital and energy cost estimation tools within the hydraulic mod-eling software, different planning scenarios can be compared as to their overall costs.

First, the WaterGEMS skeletonization tool, Skelebrator, was used to reduce the existing local area models down to a point where they could be combined into a complete system model. Second, within WaterCAD for AutoCAD, the import/export submodel tool was used to combine the ten local area models and the isolated system models into a complete water system model. Third, the new model was balanced and checked for calibration to peak-day conditions. Fourth, demands were allocated to the model using processed data to predict demands for years 2030 and 2050. The WaterGEMS demand allocation tool, LoadBuilder, was used to distribute this demand correctly to the model. Lastly, scenarios for new additions to the system were made using WaterGEMS and WaterCAD for AutoCAD, and the functionality, capital cost, energy costs, and operations and maintenance costs associated with these scenarios were evaluated using built-in tools in the hydraulic modeling soft-ware. Figure 17.4 shows a screenshot of WaterGEMS model layers in ArcMap (Trammel, 2004).

WATER SYSTEM MODELING IN THE CITY OF TRUTH OR CONSEQUENCES

Application	Development of a water distribution model in MIKE NET from GIS data
Authors	Eric Fontenot
Hardware	Standard Desktop PC, 128 MB RAM, 850 MHz Pentium III processor
GIS software	ArcView 3.2
Other software	AutoCAD, MIKE NET, Microsoft Access
GIS data	Water distribution pipes, network demands
Study area	Truth or Consequences, New Mexico
Project duration	4 months
Project budget	$42,000
Organization	DHI, Inc., Newton, Pennsylvania, and Hørsholm, Denmark
Web site	www.dhigroup.com

The City of Truth or Consequences' water supply system was strained due to population increase and real estate development. The City had concerns about the system's ability to deliver fire suppression flow to certain parts of the city at peak demand. The City, therefore, engaged in strategic planning to evaluate possible improvements.

A model of the existing water distribution system was created by DHI, Inc., from several data sources. The detailed, topologically oriented model of the water distribution network was created primarily from GIS data and updated from hard-copy maps of the water system. Water demands were determined and distributed, based on production and billing information contained in a customer information system (CIS).

The comprehensive modeling evaluated the current system, as well as those alternatives identified in a preliminary engineering report. The placement of two

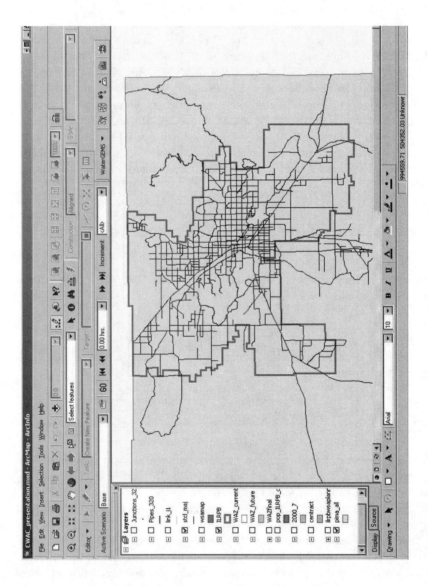

Figure 17.4   Screenshot of WaterGEMS model layers in ArcMap.

new storage tanks was evaluated, and recommendations were determined based on model results.

Background

Truth or Consequences is located east of Interstate 25 between Albuquerque and Las Cruces, adjacent to Elephant Butte Lake. The water system serves a combined population of approximately 7300 in Truth or Consequences and Williamsburg. There were 2820 residential and commercial connections, 268 of which were located in Williamsburg. The water system had 5 groundwater wells that fed a 200,000-gal clearwell, a 1,000,000-gal tank serving a low-pressure zone, a 3,000,000-gal tank serving a high-pressure zone, and 2 booster pump stations. Disinfection via chlorination was provided prior to the clearwell.

The five groundwater wells, ranging in age from 2 to 50 years, pumped into a 200,000-gal tank. Water was pumped from the 200,000-gal tank to the 1,000,000-gal storage tank, which supplied the low-pressure zone. Some of the piping in the low-pressure zone was quite old, dating back to the 1940s when the water system was originally constructed. Around 1970, the water system was divided into high- and low-pressure zones with the construction of the 3,000,000-gal tank. Water from the low-pressure zone was pumped into a storage tank to supply the high-pressure zone.

Due to increased demands and an aging system, the City was planning on constructing two additional tanks, one in each pressure zone. The current system did not maintain adequate pressure at peak demand, and the City had some concerns about the system's ability to deliver fire suppression flow to certain parts of the city. The 1,000,000-gal tank needed to be drained and rehabilitated, which could be done only after the construction of additional storage. The 3,000,000-gal tank also needed maintenance. Because it was the only tank in the high-pressure zone, an additional tank or booster station capable of providing adequate fire suppression flow was required to permit the draining and painting of the tank.

Building the MIKE NET Model from Various Data Sources

The City provided GIS data for the water distribution system, historical SCADA information for water tanks and booster pumps, billing information from the City's CIS, fire suppression flow data, and a hard copy of the water distribution system map. The GIS information included the network geometry, pipe diameter, and pipe material.

The GIS information supplied by the City was a mixture of AutoCAD files (DWG extension) and ArcView Shapefiles (SHP extension). The pipe locations and diameters were all contained in one AutoCAD drawing file. It was determined that modeling pipes 4 in. and larger would be adequate to represent the water distribution system and determine the locations of the problem areas in the system, and evaluate the proposed alternatives.

The drawing file was separated by layer and exported to the DXF format, so that it could be imported to MIKE NET. Each exported layer contained network geometry as well as pipe material and pipe diameter information. The following nine DXF files were created and imported into separate MIKE NET project files:

- 4-, 8-, and 12-in. PVC pipes
- 4-, 8-, and 12-in. AC pipes
- 4-, 8-, and 12 in. CI pipes

These files were checked for connectivity errors and merged together to form the City's water distribution network in MIKE NET. Figure 17.5 shows the network model built from GIS data in MIKE NET.

The GIS data, water system drawings, fire suppression flow data, billing data, and SCADA data were used for preliminary model building. The GIS data were used to build the network topology and connectivity. The GIS data was found to be about 85% accurate. Errors in the GIS data included incorrect network connectivity, missing pipes, omissions in separation between the high- and low-pressure zones, incorrect pipe material labels, and incorrect pipe diameters. The network connectivity was checked against the water system drawings and updated to match the drawings in which differences occurred.

Street names, from an ArcView Shapefile, were geocoded to each pipe in the model. The billing data supplied by the city were summed by street name, and the demand for that street was distributed to those pipes in MIKE NET bearing that street name.

Elevation data were supplied in 2-ft contour intervals in the City's GIS data. These data were extracted to an X, Y, Z triplet text file using a DHI-developed ArcView extension, and the node elevations were assigned by using the Generate Node Elevation tool in MIKE NET.

Tank and pump data were added manually from the water system drawings and information obtained during interviews. Upon completion of the model, these data were exported from MIKE NET to ArcView Shapefiles and supplied to the City as an updated GIS coverage of their water distribution assets (Fontenot, 2004).

ARCGIS AND ARCFM INTEGRATION IN BELGIUM

Application	ArcGIS/ArcFM 8 for water distribution network management
Authors	Reynaert et al. (2003)
Project status	In use since January 2002
Hardware	Citrix application servers & standard Dell desktop PCs
GIS software	ArcInfo/ArcView 8 , ArcSDE 8, ArcIMS 4, ArcFM 8
Other software	SQL Server 2000, Citrix Metaframe XP
GIS data	Water distribution network data and topographic data
Study area	Antwerp province, Belgium
Project duration	1998 to 2002
Organization	Pidpa
Web site	www.pidpa.be

Pidpa is a European company that produces and distributes drinking water to more than 1.1 million people throughout Antwerp province in Belgium (Flanders).

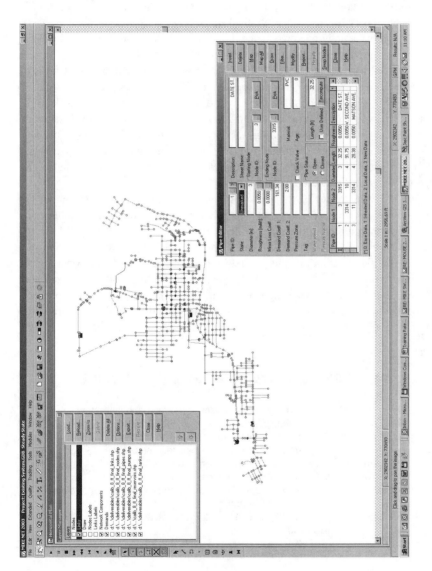

Figure 17.5 The city of Truth or Consequences' MIKE NET model developed from GIS.

Pidpa used ESRI's ArcGIS 8.2 software and Miner & Miner's ArcFM 8.1.3 software to develop a data-centric GIS solution for the management of its drinking water network, including functions for water main isolation trace and pressure zone trace (Horemans and Reynaert, 2003).

Pidpa selected ESRI as the strategic partner to implement an open and generic GIS solution for the management of its drinking water network data. The ArcGIS 8/ArcFM 8 implementation is based on a centrally located SQL Server 2000 database (with ArcSDE 8 on top) containing newly entered and converted vector data as well as large amounts of raster data at scales 1:1000 and 1:5000. All data are combined into a seamless hybrid map that is used in ArcInfo, ArcView, and ArcIMS on a daily basis by about 15 editors and a large group of viewers. Figure 17.6 shows a screenshot of ArcInfo 8 and ArcFM 8 being used on a seamless hybrid map with raster and vector data.

One of the major project goals was to create a strong generic water distribution data model that could be shared and further enriched with input from other European drinking water companies. Together with the choice for commercial off-the-shelf (COTS) software, this approach ensures a feasible future evolution of the imple-mented GIS and can create a large user base.

Pidpa was well aware of the fact that much of the data used throughout the company has a geographical element. With this in mind, Pidpa started integrating the GIS with many other information systems for use in the office as well as in the field. At the time of this writing, the main operational links were to CIS and to a library of 90,000 synoptic drawings. A first set of interfaces to SAP was operational, and a link to the SCADA system and hydraulic modeling software was under development. It was anticipated that by sharing data between these environments and optimizing workflow over the boundaries, users would be able to get required data in a user-friendly fashion (Reynaert et al., 2004).

WATER SYSTEM MASTER PLANNING IN PRAGUE

Application	Conceptual hydraulic modeling of water supply system
Authors	Peter Ingeduld, Zdenek Svitak, Josef Drbohlav
Project status	Conceptual model complete
Hardware	Standard Pentium III and IV desktop computers
GIS software	ArcView, LIDS (MicroStation and ORACLE based)
Other software	MIKE NET, Microsoft Access
GIS data	Water supply objects layer of Prague's water supply system
Study area	Prague, Czech Republic
Project duration	January 2001 to June 2002 (conceptual model)
Project budget	$165,000
Organization	Pražská vodohospodářská spole čnost a.s. (Prague stockholding company)

Hydraulic modeling of water supply and water distribution networks was used for planning and linking consumers to the network, evaluating the remaining capacity of the network, and planning for network breakdown. Modeling was also used to simulate various loading scenarios, such as fire suppression flow analysis and its impact on the water supply system, reconstruction of existing pipes and planning for new pipes,

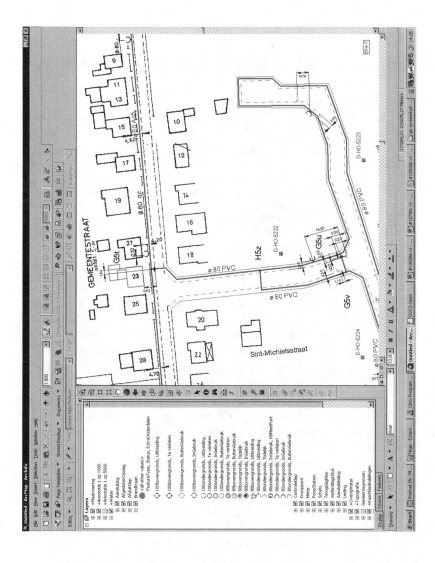

Figure 17.6 Screenshot of ArcInfo 8 and ArcFM 8.

pressure zone evaluation, and for comparing measured and simulated data. A conceptual model with all important water tanks, pumping stations, and water sources was developed for the whole Prague water supply system. This model will greatly assist in formulating and answering questions regarding supply strategies and in providing high system reliability. Hydraulic modeling was done using DHI's MIKE NET software. Figure 17.7 shows a screenshot of MIKE NET for the Prague model (Ingeduld et al., 2004).

WATER QUALITY MANAGEMENT IN MECKLENBURG COUNTY

Application	Water quality management planning
Authors	Carl W. Chen and Curtis Loeb
Project status	Completed in August 2003
Hardware	Personal computers
GIS software	Included in Watershed Analysis Risk Management Framework (WARMF)
Other software	SWAT
GIS data	10-m DEM, land-use Shapefiles for existing and future conditions
Study area	The Palisades golf course community, Mecklenburg County, North Carolina, at the eastern shore of Lake Wylie; 5.7 mi² for The Palisades and 1137 mi² of watershed surrounding Lake Wylie
Project duration	2002 to 2003 (12 months)
Project budget	$145,000
Organization	Systech Engineering, Inc., San Ramon, California

Rhein Interests, LLC, and Crescent Resources, LLC (North Carolina), proposed to develop The Palisades as a new community comprising homes, a golf course, open spaces, an equestrian center, offices, and stores. The Mecklenburg County required the project to be designed with best management practices (BMP) to meet the water quality criteria of total suspended sediment, total phosphorus, total nitrogen and algae (chlorophyll a) in the eight coves of Lake Wylie. The County approved the use of two GIS-based watershed models, SWAT and WARMF, for the project. SWAT simulated the nonpoint loads of pollutants from The Palisades, which were converted to point source loads for input to WARMF. WARMF simulated nonpoint source loads of pollutants from other areas and the water quality responses in the coves that not only receive the nonpoint source loads from The Palisades but also exchange water with the main body of Lake Wylie. Two levels of resolution were used to delineate the watershed; the finer resolution of The Palisades was nested with the coarser resolution of other subwatersheds. The lake was divided into 29 stratified sections, including coves. As a part of the project planning and design, a monitoring program was established to collect meteorology data at four stations, stream flow and water quality data at four stream stations, and water quality data at eight coves during the spring and summer months of 2002. The data were used to calibrate the models. The calibrated models were used to determine the BMP train necessary to meet the water quality criteria for coves. The water quality management plan for The Palisades was approved by the County. Figure 17.8 shows a screenshot of the WARMF application for The Palisades golf course community (Chen and Loeb, 2004).

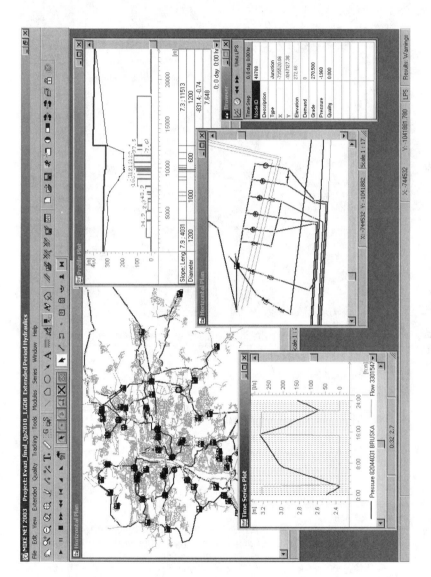

Figure 17.7 Screenshot of Prague's MIKE NET water system model.

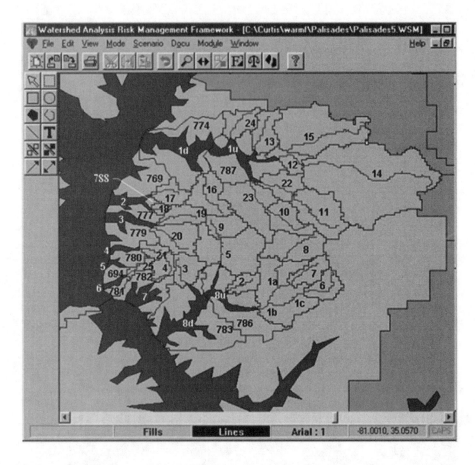

Figure 17.8   Screenshot of the WARMF application for The Palisades golf course community.

WATER MASTER PLANNING IN SUECA, SPAIN

Application	Quantity and quality modeling of water distribution systems
Authors	Hugo Bartolin and Fernando Martinez
Project status	Completed
Hardware	Standard desktop PC (Intel Pentium IV)
GIS software	ArcView GIS 3.2a, Spatial Analyst (Optional)
Other software	EPANET 2
GIS data	Three AutoCAD files with cartographic information (pipe polylines, valves, water sources, street segments, building blocks, future zones of development)
Study area	Town of Sueca (about 26,000 inhabitants) and its tourist beach area (about 30,000 in summer), both supplied by independent networks. The town is located in Valencia on the east coast of Spain.
Project duration	3 months (December 2002 to February 2003)
Project budget	$6000 (consultancy)
Organization	Polytechnic University of Valencia. Grupo REDHISP — IIAMA (Institute of Water and Environmental Engineering)
Web site	www.redhisp.upv.es www.redhisp.upv.es/software/gisred/gisred_eng.htm

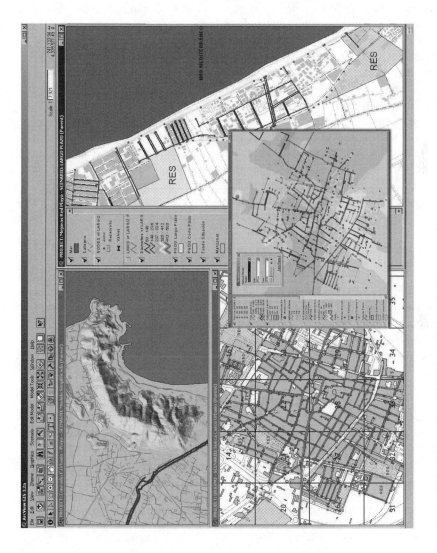

Figure 17.9 Screenshot of GISRed extension for Sueca master plan.

GISRed is a customized extension of ArcView GIS 3.2 software for modeling and calibration of water distribution networks. It basically integrates hydraulic modeling tools, the widely used hydraulic simulation software EPANET (Version 2), and a hydraulic calibration module based on genetic algorithms.

The extension is essentially a tool to assist technicians in the task of modeling water distribution networks and supporting decision-making issues based on the model, all within a GIS environment. It is useful for performing complex tasks such as importing a whole or partial network from an external data source, creating a hydraulic network model, and automatically calibrating it.

One of the tasks GISRed can assist in is master planning issues. In the Sueca master plan project, a general procedure using the GISRed extension was carried out. The main idea was to define a scenario for both independent network models and give a diagnosis of the performance of each real system. Based upon this, new short- and long-term scenarios were proposed, taking into account the variation of the population, the new growing areas, and, therefore, the future demand projection. Finally, network system improvements were planned and simulated to check the feasibility of each alternative and support all the action strategies. In addition to that, the models were connected by two mains in order to enhance the quality of service and prevent emergencies. Figure 17.9 shows a screenshot of GISRed extension for Sueca master plan (Bartolin and Martinez, 2004).

CHAPTER SUMMARY

This chapter presented eight case studies on GIS applications for water, wastewater, and stormwater systems in Belgium, Bulgaria, Czech Republic, Denmark, Spain, and the U.S. The case studies indicate that GIS is being used in all aspects of water, wastewater, and stormwater management, from planning and H&H modeling to mapping and asset management.

CHAPTER QUESTIONS

1. On the basis of the case studies presented in this chapter, which GIS application appears to be the most common?
2. On the basis of the case studies presented in this chapter, which GIS software appears to be the most commonly used?
3. Using the case study presentation format used in this chapter, write a case study summary of a GIS project completed in your organization. Optionally, submit this case study to the author (info@GISapplications.com) for potential publication at GISapplications.com Web site.

Acronyms

2-D Two-Dimensional
3-D or 3D Three-Dimensional
ACE Army Corps of Engineers
ACP Asbestos Cement Pipe
ADRG Arc Digitized Raster Graphics
AIRS Aerometric Information Retrieval System
AM/FM Automated Mapping/Facilities Management
AM/FM/GIS Automated Mapping/Facilities Management/Geographic Information
    System
AML Arc Macro Language
AMSA Association of Metropolitan Sewerage Agencies
API Application Programming Interface
ASCE American Society of Civil Engineers
ASCII American Standard Code for International Interchange
ASP Application Service Provider
AVI Audio Video Interleaved (digital movie video format)
AWRA American Water Resources Association
AWWA American Water Works Association
BASINS Better Assessment Science Integrating Point and Nonpoint Sources
BIL Band Interleaved by Line
BIP Band Interleaved by Pixel
BMP Best Management Practice
BRS Biennial Reporting System
BSQ Band SeQuential
BYU Brigham Young University
CAD Computer-Aided Drafting/Computer-Aided Design
CADD Computer-Aided Drafting and Design
CAFM Computer-Aided Facility Management
CAM Computer-Aided Mapping
CCTV Closed-Circuit Television
CD Compact Disc
CD-ROM Compact Disc-Read Only Memory
CGM Computer Graphic Metafile

CIFM Computer Integrated Facility Management
CIP Cast Iron Pipe
CIP Capital Improvement Project
CIS Customer Information System
CMMS Computerized Maintenance Management System
CMOM Capacity, Management, Operations, and Maintenance
CMP Corrugated Metal Pipe
COE Corps of Engineers
COGO Coordinate Geometry
COM Component Object Model
CORBA Common Object Request Broker Architecture
COTS Commercial Off-the-Shelf
CSO Combined Sewer Overflow
CTG Composite Theme Grid
DAK Data Automation Kit
DBMS Database Management System
DDE Dynamic Data Exchange
DEM Digital Elevation Model
DEP Department of Environmental Protection
DFIRM Digital Flood Insurance Rate Map
DGPS Differential Global Positioning System
DHI Danish Hydraulic Institute
DIB Device Independent Bitmap
DIME Dual Independent Map Encoding
DIP Ductile Iron Pipe
DLG Digital Line Graph
DLL Dynamic Link Library
DOP Digital Orthophoto
DOQ Digital Orthophoto Quadrangle
DOQQ Digital Orthophoto Quarter Quadrangle
DR3M Distributed Routing Rainfall Runoff Model
DRG Digital Raster Graphics (USGS topographic maps)
DSS Data Support System
DSS Decision Support System
DTED Digital Terrain Elevation Model
DTM Digital Terrain Model
DVD Digital Versatile Disc or Digital Video Disc
DXF Drawing Exchange Format
EDC EROS Data Center
EMF Enhanced Metafile Format
EOS Earth Observation System
EPA (U.S.) Environmental Protection Agency
ERDAS Earth Resource Data Analysis System
EROS Earth Resources Observation Systems
ERP Emergency Rresponse Plan
ESDLS EPA Spatial Data Library System
ESRI Environmental Systems Research Institute

FAP File of Arcs by Polygon
FAT Feature Attribute Table
FEMA Federal Emergency Management Agency
FGDC Federal Geographic Data Committee
FIRM Flood Insurance Rate Map
FIS Flood Insurance Study
FMSIS Flood Map Status Information Service
FTP File Transfer Protocol
GAAP Generally Accepted Accounting Principales
GASB Government Accounting Standards Board
GEMS Geographic Engineering Modeling Systems
GIF Graphics Interchange Format
GIRAS Geographic Information Retrieval and Analysis System
GIS Geographic Information Systems
GLONASS Global Navigation Satellite System (Russian Systemsystem)
GML Geography Markup Language
GPD Gallons Per Day
GPS Global Positioning System
GRASS Geographic Resources Analysis Support System
GUI Graphical User Interface
H&H Hydrologic and Hydraulic
HEC Hydrologic Engineering Center (U.S. Army Corps of Engineers)
HGL Hydraulic Gradient Line
HoLIS Honolulu Land Information System
HRAP Hydrologic Rainfall Analysis Project
HSG Hydrologic Soil Group
HSPF Hydrologic Simulation Program — Fortran
HTML Hypert Text Markup Language (a file extension)
HTTPS Secure Hypert Text Transmission Protocol
HUC Hydrologic Unit Code
HUD Housing and Urban Development
IAC Interapplication Communication
ICR Information Collection Rule
IFSAR Interferometric Synthetic Aperture Radar
IMS Internet Map Server
ISE Integral Square Error
IT Information Technology
JFIF JPEG File Interchange Format
JPEG Joint Photographic Experts Group
JSP Java Server Pages
LAN Local Area Network
LANDSAT Land Satellite
LBS Location-Based Services
LIDAR Laser Imaging Detection and Ranging
LOMC Letter of Map Change
LTCP Long-Term Control Plan
LULC Land Use/Land Cover

MA Metropolitan Area
MGD Million Gallons per Day
MIXEL Mixed Cells
MMS Maintenance Management System
MMS Multimedia Mapping System
MPEG Moving Picture Experts Group (digital movie video format)
MrSID Multir Resolution Seamless Image Database
MS4 Municipal Separate Storm Sewer System
MSC Map Service Center
NAD-27 North American Datum of 1927
NAD-83 North American Datum of 1983
NASA National Aeronautics & and Space Administration (U.S.)
NAVSTAR Navigation System by Timing and Ranging
NCDC National Climatic Data Center (U.S.)
NED National Elevation Dataset
NEXRAD Next Generation Weather Radar
NGDC National Geospatial Data Clearinghouse
NGVD National Geodetic Vertical Datum
NHD National Hydrography Dataset
NLCD National Land Cover Database
NMAS National Map Accuracy Standards
NMC Nine Minimum Controls (for CSOs)
NOAA National Oceanic and Atmospheric Administration (U.S.)
NPDES National Pollution Discharge Elimination System
NPS Nonp Point Source
NQL Network Query Language
NRCS Natural Resources Conservation Service
NSDI National Spatial Data Infrastructure
NSGIC National States Geographic Information Council (U.S.)
NWI National Wetlands Inventory
NWS National Weather Service
O&M Operation and Maintenance
ODBC Open Database Connectivity
OGC Open GIS Consortium
OIRM Office of Information Resources Management
OLE Object Linking and Embedding
PASDA Pennsylvania Spatial Data Access
PC Personal Computer
PCS Permit Compliance System
PDA Personal Digital Assistant (an electronic handheld information device)
PDF Portable Document Format (Adobe Acrobat)
PDOP Positional Dilution of Precision
PFPT Peak Flow Presentation Table
PIXEL Picture Element
POTW Publicly Owned Treatment Works
PPS Precise Positioning Service
PRR Practical Release Rate

PRV Pressure Regulating Valve
PVC Polyvinyl Chloride
QA/QC Quality Assurance/Quality Control
RBE Rule Base Engine
RCP Reinforced Concrete Pipe
RCRIS Resource Conservation and Recovery System
RDBMS Relational Database Management System
RF Reach File
RGB Red Green Blue (scientific hues or primary colors)
RMSE Root Mean Square Error
ROM Read Only Memory
RPC Remote Procedure Call
RR Release Rate
RTB Retention Treatment Basin
RTK Real Time Kinematic
SA Selective Availability
SAR Synthetic Aperture Radar
SCADA Supervisory Control and Data Acquisition
SCS Soil Conservation Service
SDE Spatial Database Engine
SDTS Spatial Data Transfer Standard
SHC System Hydraulic Characterization (for CSOs)
SHE System Hydrologique Europeen
SML Simple Macro Language
SMS Short Message Service
SPC State Plane Coordinate (map projection system)
SPS Standard Positioning Service
SQL Structured Query Language
SSL Secure Socket Layer
SSO Sanitary Sewer Overflow
SSURGO Soil Survey Geographic
STATSGO State Soil Geographic
STORM Storage Treatment Overflow Runoff Model
STP Sewage Treatment Plant
SVG Scalable Vector Graphics
SWAT Soil and Water Assessment Tool
SWMM Storm Water Management Model
TAZ Traffic Analysis Zone
TCP Terra-Cotta Pipe
TCP/IP Transmission Control Protocol/Internet Protocol
TIF Tag Image File
TIFF Tag Image File Format
TIGER Topologically Integrated Geographic Encoding and Referencing System (U.S. Census Bureau mapping system)
TIN Triangular Irregular Network/Triangulated Irregular Network
TM Thematic Mapper
TMDL Total Maximum Daily Load

TMS TIGER Map Service
TR Technical Release
TR-20 Technical Release 20
TRI Toxic Release Inventory
TRR Theoretical Release Rate
TVP Topological Vector Profile
UAV Unmanned Aerial Vehicles
UCGIS University Consortium of Geographic Information Science
UPS Uninterruptible Power Supply
URL Uniform (or Universal) Resource Locator
USACE United States Army Corps of Engineers
USACERL United States Army Construction Engineering Research Laboratory
USBC United States Bureau of Census
USGS United States Geological Survey
UTM Universal Transverse Mercator
VB Visual Basic
VBA Visual Basic for Applications
VCP Vitrified Clay Pipe
VCR Video Cassette Recorder
VHS Video Home System (video cassette format)
VRML Virtual Reality Modeling Language
VSAT Vulnerability Self Assessment Tool
WAAS Wide Area Augmentation System
WAP Wireless Application Protocol
WCS Web Coverage Service
WEF Water Environment Federation
WFS Web Feature Service
WGS84 World Geodetic System of 1984
WIMS Wastewater Information Management System
WISE Watershed Information System
WML Wireless Markup Language
WMS Watershed Modeling System
WMS Web Map Server
WSR Weather Surveillance Radar
WSDL Web Service Description Language
WWW World Wide Web
XLST Extensible Stylesheet Language Transformation
XML Extensible Markup Language

Conversion Factors

Conversion Factors from Customary U.S. Units to Système International (SI) Units

To Covert	To	Multiply by
Acres	Square meters (m^2)	4047
Acres	Hectares (ha)	0.4047
Cubic feet	Liters (l)	28.32
Cubic feet per second (cfs)	Cubic meters per second	0.02832
Cubic yards	Cubic meters (m^3)	0.7646
Feet (ft)	Meters (m)	0.3048
U.S. gallons (water)	Liters (l)	3.7843
U.S. gallons (water)	Cubic centimeters (cm^3)	3785
U.S. gallons (water)	Milliliters (ml)	3785
U.S. gallons (water)	Grams (g)	3785
U.S. gallons (water)	Cubic meters (m^3)	0.0038
Gallons per minute (GPM)	Liters per second	0.0631
Gallons per minute (GPM)	Liters per minute	63.08
Inches (in.)	Centimeters (mm)	2.54
Inches (in.)	Millimeters (mm)	25.4
Miles (mi)	Kilometer (km)	1.6093
Millions gallons per day (MGD)	Cubic meters per day	3785
Millions gallons per day (MGD)	Cubic meters per second	0.044
Pounds (lb)	Kilograms (kg)	0.4536
Square feet (ft^2)	Square meters (m^2)	0.0929
Square miles (mi^2)	Square kilometers (km^2)	2.5921
Yards	Meters (m)	0.9144
Square miles (mi^2)	Square kilometers (km^2)	2.5899

References

Alston, R. and Donelan, D. (1993). Weighing the benefits of GIS. *American City & County*, October 1993, Vol. 108, No. 11, 14.

American City & County (1994). GIS assists Texas town in meeting NPDES regulations. *American City & County*, March 1994, 62.

American City & County (1993). CAGIS restores order to Cincinnati infrastructure. *American City & County*, August 1993, 24.

American City & County (1991). Albuquerque integrates GIS to meet needs. *American City & County*, May 1991, 30.

Anderson, F. (2000). Water resources planning and management in the Internet era. *Water resources impact*, American Water Resources Association, Vol. 2, No. 5, September 2000, 5–8.

Anderson, G. (1998). Program yields wastewater piping hydraulic model, *Water World*, October, 1998.

ArcNews (2001a). Colorado utility maps in the Rockies with imagery. *ArcNews*, Vol. 23, No. 3, Fall 2001, ESRI, 20–21.

ArcNews (2001). GIS goes with the flow. *ArcNews*, Vol. 23, No. 1, Spring 2001, ESRI, 34–35.

ArcNews (2000). ArcGIS 8.1 nears release. *ArcNews*, Vol. 22, No. 4, Winter 2000/2001, ESRI, Spring 2000, 1–2.

ASCE (2003). Little progress made on infrastructure, ASCE report finds. *ASCE News*, American Society of Civil Engineers, Vol. 28, No. 10, October 2003, 1.

ASCE (2001). New systems make surveying easier more accurate. *Civil Engineering*, May 2001, 74–76.

ASCE (1999). GIS Modules and Distributed Models of the Watershed, Task Committee Report, Paul A. DeBarry and Rafael G. Quimpo Committee Co-chairs, ASCE.

ASCE (1998). New poll shows voters worried about aging infrastructure, *ASCE News* American Society of Civil Engineers, November 1998.

Azteca Systems (1999). White Paper: Pipeworks Data Model Concepts, Azteca Systems Web site at http://www.azteca.com/data_model.htm.

Babbitt, L.M. (2002). Lehigh County Authority, Allentown, PA. *Water Writes*, ESRI, Fall 2002, 6–7.

Barnes, S. (2001a). ASPRS 2001: Remote sensing opportunities soar. *Geospatial Solutions*, Vol. 11, No. 6, June 2001, 16.

Barnes, S. (2001b). United in purpose — spatial help in the aftermath. *Geospatial Solutions*, Vol. 11, No. 11, November 2001, 34–39.

Barnes, S. (2001). ESRI and Sun Launch LBS portal. *Geospatial Solutions*, February 2001, 16.

Barnes, S. (2000). One last thing — GIS market revenues top $1.5 billion. *Geospatial Solutions*, Vol. 10, No. 6, June 2000, 58.

Bartolin, H. and Martinez, F. (2004). GISRed, an assistant tool to develop master planning projects of water distribution system. Case study submission in response to a call for case studies distributed by the author (U. M. Shamsi) for this book.

Bathurst, J.C., Wicks, J.M., and O'Connell, P.E. (1995). The SHE/SHESED Basin scale water flow and sediment transport modelling system, Chapter 16 in *Computer Models of Watershed Hydrology*, V.P. Singh (Ed.), Water Resources, 563–594.

Battin, A., Kinerson, R., and Lahlou, M. (1998). EPA's Better Assessment Science Integrating Point and Nonpoint Sources (BASINS) — A Powerful Tool for Managing Watersheds. Internet file retrieved 11/6/98 from The Center for Research in Water Resources, The University of Texas at Austin. http://www.crwr.utexas.edu/gis/gishyd98/epa/batin/p447.htm.

Bell, J. (1999a). ArcView GIS. *Professional Surveyor*, March 1999, 48–50.

Bell, J. (1999). AutoCAD Map 3. *Professional Surveyor*, January/February 1999, 52–53.

Bernhardsen, T. (1999). *Geographic Information Systems — An Introduction*, 2nd ed., John Wiley & Sons, New York, 372.

Binford, M.W. (2000). Clark lab's Idrisi32. *Geo Info Systems*, April 2000, 40–42.

Booth, R. and Rogers, J. (2001). Using GIS technology to manage infrastructure capital assets. *Journal AWWA*, American Water Works Association, November, 2001, Vol. 93, No. 11, 62–68.

Bradford, E., Watson, R., Mann, E., and Konwinski, J. (2004a). MSD Pipe Rating Program. Case study submission in response to a call for case studies distributed by the author (U.M. Shamsi) for this book.

Bradford, E., Watson, R., Mann, E., and Konwinski, J. (2004). Pipe rating. *Water Environment & Technology*, Water Environment Federation, February, 2004, Vol. 16, No. 2, 39–43.

Bufe, M. (2003). Going digital. *Water Environment & Technology*, Water Environment Federation, October 2003, Vol. 15, No. 10, 24–27.

Burns, N.L. et al. (2001). Security analysis and response for water utilities. AWWA, Denver, CO.

Butler, D. and Maksimovic, C. (Eds.) (1998). *UDM'98 4th International Conference on Developments in Urban Drainage Modeling*. Imperial College of Science, Technology & Medicine, London.

California Water Districts Link GIS to Document Management System (1998). *ArcNews*, Spring 1998, 19.

Cannistra, J.R. (1999). Converting utility data for a GIS. *Journal AWWA*, American Water Works Association, February 1999, 55–64.

Cannistra, J.R., Leadbeater, R., and Humphries, R. (1992). Washington suburban sanitary commission implements GIS. *Journal AWWA*, AWWA, July 1992, 62–67.

CE News (2001). Group seeks Congressional aid for water infrastructure. Vol. 13, No. 5, June 2001, 13–16.

Cerasini, V. and Herskowitz, J. (2001). City streamlines customer service. *Imaging Notes*, Space Imaging, Thornton, CO, Vol. 16, No. 5, September/October 2001, 26–27.

Chang, T., Hsu, M., Teng, W., and Huang, C. (2000). A GIS-assisted dstributed watershed model for simulating flooding and inundation. *Water Resources Bulletin*, American Water Resources Association, Vol. 36, No. 5, October 2000, 975–988.

Charnock, T.W., Hedges, P.D., and Elgy, J. (1996). Linking multiple process models with GIS. In K. Kovar and H.P. Nachtnebel (Eds.), *HydroGIS'96: Application of Geographic Information Systems in Hydrology and Water Resources Management*. International Association of Hydrologic Sciences Publication No. 235. IAHS, Wallingford, U.K. 29–36.

Chen, C.W. and Loeb, C. (2004). Watershed Analysis Risk Management Framework (WARMF) Case Study. Case study submission in response to a call for case studies distributed by the author (U.M. Shamsi) for this book.

Cheves, M. (2000). GIS and surveying. *Professional Surveyor*, Vol. 20, No. 9, October 2000, 24.

Chien, P. (2000). Endeavour maps the world in three dimensions. *GEOWorld*, April 2000, Vol. 13, No. 4.

Christian, S.M. and Yoshida, G.K. (2003). Mapping out success. *Water Environment & Technology*, Vol. 15, No. 8, August 2003, 50–55.

City of Glendale Database Design (1996). ESRI, Redlands, CA, July 1996.

Coates, J. (2003). Water main shutdown application prevents costly mistakes. *ArcUser*, ESRI, January–March, 2003, Vol. 6, No. 1, 20–21.

Computational Hydraulics Int. (1995). *PCSWMM*. User's manual, 36 Stuart Street, Guelph, Ontario, Canada N1E 4S5.

Corbley, K.P. (2001). Adding value to data. *Imaging Notes*, Vol. 16, No. 2, March/April 2001, 27–28.

Corbley, K.P. (2000). Plugging in. *Imaging Notes*, Space Imaging, Thornton, CO, March/April 2000, 14–15.

Criss, Cori. (2000). WinCan simplifying data gathering, *Trenchless Technology*, November 2000, 64–65.

Curtis, T.G. (1994). SWMMDUET: Enabling EPA SWMM with the ARC/INFO paradigm. *ArcNews*, Environmental Systems Research Institute, Redlands, CA, Spring 1994, 20.

Dalton, L.M. (2001). Prospecting spatial database offerings. *Geospatial Solutions*, Vol. 11, No. 10, October 2001, 40–44.

Daratech (2000). Geographic Information Systems — Markets and Opportunities, Cambridge, MA.

Davis, J. (2000). Keys to success. *Geo Info Systems*, February 2000, 32–35.

Davis, K. and Prelewicz, G. (2001). Concerned about CMOM? *Water Environment & Technology*, Vol. 13, No. 3, March 2001, 33–36.

Day, J.M. (1998). Rainfall Quantity Analysis from a Watershed Perspective: Using the Virtual Rain Gauge, International User Conference, ESRI, San Diego, CA, July 27–31, 1998.

DeGironimo, E. and Schoenberg, J. (2002). The key to the vault. *Water Writes*, ESRI, Fall 2002, 1–3.

Denver Water Department (Colorado, USA) installed one of the first AM/FM/GIS system "Synercom" in USA in 1986 (Lang, 1992).

DeVantier, B.A. and Feldman, A.D. (1993). Review of GIS applications in hydrologic modeling, *Journal of Water Resources Planning and Management*, 119 (2): 246–261.

Djokic, D., Meavers, M.A., and Deshakulakarni, C.K. (1993). *ARC/HEC2: An ArcInfo — HEC-2 Interface*. User's manual, Center for Research in Water Resources, The University of Texas at Austin, Austin, TX.

Doan, J.H. (1999). Hydrologic Model of the Buffalo Bayou Using GIS, International User Conference, ESRI, San Diego, CA, July 26–30, 1999.

Dorris, V.K. (1989). Systems link geography and data, *ENR*, June 1, 1989, 30–37.

Dougherty, L. (2000). Census 2000, census sense, part 2, geographic products for one and all. *Geospatial Solutions*, September 2000, 36–39.

Doyle, G. and Grabinsky, M. (2003). Applying GIS to a water main corrosion study. *Journal AWWA*, AWWA, May 2003, Vol. 95, No. 5, 90–104.

Doyle, M.J. and Rose, D. (2001). Protecting your assets. *Water Environment & Technology*, Water Environment Federation, July, 2001, Vol. 13, No. 7, 43–47.

Dueker, K.J. (1987). Geographic information systems and computer-aided mapping, *APA Journal*, Summer issue, 383–390.

Edelman, S., Dudley, T., and Crouch, C. (2001). Much ado about data, *CE News*, January 2001, 52–55.

Elroi Consulting, Inc. (2002). Stormwater Utility GIS. Company literature at www.elroi.com.

Engelhardt, J. (2001a). Mapping the future. *Imaging Notes*, Space Imaging, Thornton, CO, May/June 2001, 30.

Engelhardt, J. (2001). GITA's technology report. *Geospatial Solutions*, February 2001, 14.

Engman, E.T. (1993). Remote sensing. Chapter 24 in *Handbook of Hydrology*, D.R. Maidment (Ed.), McGraw Hill, NY, 24.1–24.23.

EPA (2002). *SWMM Redevelopment Project Plan.* Version 5, Water Supply and Water Resources Division, National Risk Management Research Laboratory, Office of Research and Development, U.S. Environmental Protection Agency, Cincinnati, OH.

EPA (2000). Environmental Planning for Communities — A Guide to the Environmental Visioning Process Utilizing a Geographic Information System (GIS). EPA/625/R-98/003, Technology Transfer and Support Division, Office of Research and Development, U.S. Environmental Protection Agency, Cincinnati, OH, September 2000.

ERDAS (2001a). Ventura capitalizes with imagery. *ERDAS News*, ERDAS, April 2001, Vol. 2, No. 2, 18.

ERDAS (2001b). IMAGINE OrthoBASE Pro-Automatic DEM extraction made easy. *ERDAS News*, ERDAS, April 2001, Vol. 2, No. 2, 5.

ERDAS (2001). There's more to a Pixel than meets the eye. *ERDAS News*, ERDAS, January 2001, Vol. 2, No. 1, 18–19.

ESRI (2004a). Inventorying Infrastructure with ArcPad. *ArcUser*, ESRI, January–March, 2004, Vol. 7, No. 1, 38–39.

ESRI (2004). Taking GIS on the road. *ArcUser*, ESRI, January–March, 2004, Vol. 7, No. 1, 30–32.

ESRI (2003). GIS standards and interoperability. *ArcNews*, ESRI, Vol. 25, No. 1, Spring 2003, 1.

ESRI (2002). AAA's TripTik/Traveler rated the best cartographer, *ArcNews*, ESRI, Vol. 24, No. 3, Fall 2002, 1.

ESRI (2002a). Elsinore Valley municipal water district goes live with ArcGIS. *ArcNews*, ESRI, Vol. 23, No. 4, Winter 2001–2002, 20.

ESRI (2002b). City of Germantown, Tennessee, implements water inventory, modeling. *ArcNews*, Vol. 24, No. 1, ESRI, Spring 2002, 16.

ESRI (2002c). Developer samples for water/wastewater, *Water Writes*, ESRI, Winter 2002, 9.

ESRI (2002d). A common vision for homeland security, *ArcUser*, ESRI, Vol. 5, No. 1, January–March 2002, 8–11.

ESRI (2000). Sharing geographic knowledge: providing access to geographic information, *ArcNews*, ESRI, Spring 2000, 6.

ESRI. (1999). Utility market embraces ArcFM GIS solution, *ArcNews*, ESRI, Spring 1999, 4 p.

ESRI (1994). Massachusetts water resources authority's sewer management project. *The GIS People*, June 1994.

ESRI (1992). ArcView User's Guide, 2nd ed., Environmental System Research Institute, Redlands, CA.

ESRI Educational Services (1995). *Introduction to Avenue*, Environmental System Research Institute, Redlands, CA.

Estes-Smargiassi, S. (1998). Massachusetts water resources authority uses GIS to meet objectives cost-effectively. *Water Writes*, ESRI.

EWRI (2001). ASCE's 2001 report card for America's infrastructure. *EWRI Currents*, Environmental and Water Resources Institute, ASCE, Vol. 3, No. 2, June 2001, 9–10.

Farkas, A.L. and Berkowitz, J.B. (2001). State-of-the-industry report. *Environmental Engineer*, American Academy of Environmental Engineers, Vol. 37, No. 4, October 2001, 22–26.

Fillmore, L., Doyle, M., Khan, I., Russell, T., and Osei-Kwadwo, G. (2001). CMOM to see efficiency. *Water Environment & Technology*, Vol. 13, No. 3, March 2001, 39–43.

Fontenot, E. (2004). Development of a Water Distribution Model in MIKE NET from GIS Data. Case study submission in response to a call for case studies distributed by the author (U.M. Shamsi) for this book.

Frank, A. (2001). Building better maps. *Imaging Notes*, Space Imaging, Thornton, CO, May/June 2001, 18–19.

Gaines, J.B. (2001). The power of preventive maintenance. *Water Environment & Technology*, Vol. 13, No. 8, August 2001, 60–65.

Garbrecht, J. and Martz, L. 2000. TOPAZ, Topographic Parameterization Software, Grazing Labs Research Laboratory Web site at http://grl.ars.usda.gov/topaz/TOPAZ1.HTM.

Gay, S. (2003). Remote sensing, GIS, and a taxing dilemma. *Geospatial Solutions*, April 2003, Vol. 13, No. 4, 34–38.

Geographic Information Systems, (2000). USGS Web site at www.usgs.gov/research/gis/title.html.

Geospatial Solutions (2004). Rand report: Federal spatial Web sites don't help terrorists. *Geospatial Solutions*, Vol. 14, No. 5, May 2004, 16.

Geospatial Solutions (2001). Gaze into the future — hot trends 2001. January 2001, 20–29.

Geospatial Solutions (2001a). Innovative, imaginative, intricate - top Apps 2001. August 2001, 16.

GEOTELEC (2004). Computerizing Facility Management Activities. Company Web site, www.geotelec.com/cfma.html.

GEOWorld (2001). Wireless fleet management improves. September 2001, Vol. 14, No. 9, 36.

Gilbrook, M.J. (1999). GIS paves the way. *Civil Engineering*, ASCE, Vol. 69, No. 11, November 1999, 34–39.

Gish, B.W. (2001). Resolution demonstration. *Local Government Handbook for GIS Implementation within the Commonwealth of Pennsylvania CD-ROM*, May 23, 2001, Pennsylvania Mapping and Geographic Information Consortium, Lewisburg, Pennsylvania, www.pamagic.org.

GITA (2001). *Geospatial Technology Report 2000*. Geospatial Information & Technology Association, www.gita.org, January 2001.

GITA (2003). *Geospatial Technology Report 2002: A Survey of Organizations Implementing Information Technologies*. Geospatial Information & Technology Association, www.gita.org.

Goldstein, H. (1997). Mapping convergence: GIS joins the enterprise, *Civil Engineering*, ASCE, June, 1997, 36–39.

Goodchild, M.F. (1996). The Application of Advanced Technology in Assessing Environmental Impacts. In D.L. Corwin and K. Loague (Eds.), *Application of GIS to the Modeling of Non-point Source Pollutants in the Vadose Zone*. Madison, WI, Soil Science Society of America, 1–17.

Goodchild, M. (1998). Uncertainty: the Achilles heels of GIS, *Geo Info Systems*, November 1998, 50.

Goodchild, M. (2002). MapFusion for GIS interoperability. *Geospatial Solutions*, Vol. 12, No. 4, April 2002, 48–51.

Gorokhovich, Y., Khanbilvardi, R., Janus, L., Goldsmith, V., and Stern, D. (2000). Spatially distributed modeling of stream flow during storm events. *Journal of the American Water Resources Association*, Vol. 36, No. 3, June 2000, 523–539.

Goubert, D. and Newton, R. (1992). Small utility GIS, *Civil Engineering*, ASCE, November 1992, 69–71.

Graybill, G.R. (1998). Database Design Issues for the Integration of ARC/INFO with Hydraulic Analysis Applications, International User Conference, ESRI, San Diego, CA, July 27–31, 1998.

Greene, R.G. and Cruise, J.F. (1995). Urban watershed modeling using Geographic Information System, *Journal of Water Resources Planning and Management*, Vol. 121, No. 4, 318–325.

Griffin, C.B. (1995). Data quality issues affecting GIS use for environmental problem-solving. *Proceedings of the National Conference on Environmental Problem-Solving with Geographic Information Systems*, EPA/625/R-95/004, Office of Research and Development, U.S. Environmental Protection Agency, Cincinnati, OH, September 1995, 15–30.

Grigg, N.S. (2003). Surviving disasters: learning from experience. *Journal AWWA*, American Water Works Association, Vol. 95, No. 9, September 2003, 64–75.

Grigg, N.S. (2002). Surviving disasters in water utilities: learning from experience. AWWA Research Foundation, Denver, CO.

Guetti, B. (1998). Developing a Water Distribution System Maintenance Application, International User Conference, ESRI, San Diego, CA, July 27–31, 1998.

Haestad Methods, (2004). Bethlehem SCADA connects waterGEMS. *Client Care Newsletter*, 11, 7, Haestad Press, Waterbury, CT.

Haestad Methods (2003). One-on-one with Jack Dangermond. *ClientCare Newsletter*, March/April 2003, Haestad, Waterbury, CT.

Haestad Methods (2003a). Securing water systems. *ClientCare Newsletter*, June/July 2003, Haestad, Waterbury, CT.

Haestad Methods (2002). Water, a demanding solution. *ClientCare Newsletter*, Oct/Nov 2002, Haestad, Waterbury, CT.

Haestad Methods, Walski, T., Chase, D., Savic, D., Grayman, W., Beckwith, S., and Koelle, E. (2003). *Advanced Water Distribution Modeling and Management*, Haestad, Waterbury, CT.

Hallam, C.A., Salisbury, J.M., Lanfear, K.J., and Battaglin, W.A. (Eds.) (1996). *Proceedings of the AWRA Annual Symposium: GIS and Water Resources*. American Water Resources Association, Herndon, VA.

Hamid, R. and Nelson, R.E. Models, GIS, and databases — the essential tools in wastewater planning, Collection System Odyssey, *2001 Collection Systems Specialty Conference*, WEF, Bellevue, WA, July 8–11, 2001.

Hammond, M. (1999). GIS Lands on the Map, *PCWeek*, January 4, 1999, 42–47.

Hansen, R. 1998. Importing U.S. Census TIGER Files into ArcView GIS, *ArcUser*, ESRI, July–September 1998, 46.

Harlin, J.M. and Lanfear, K.J. (Eds.) (1993). *Proceedings of the symposium on Geographic Information Systems and water resources*. American Water Resources Association, Bethesda, MD.

Hay, L.E. and Knapp, L.K. (1996). Integrating a Geographic Information System, A scientific visualization system, and a precipitation model. *Water Resources Bulletin*, American Water Resources Association, Vol. 32, No. 2, April 1996, 357–369.

He, C., Shi, C., Yang, C., and Agosti, B.P. (2001). A windows–based GIS-AGNPS interface. *Journal of American Water Resources Association*, Vol. 37, No. 2, April 2001, 395–406.

Heaney, J.P., Sample, D., and Wright, L. (1999). *Geographical Information Systems, Decisions Support Systems, and Urban Stormwater Management.* Cooperative Agreement Report No. CZ826256-01-0, U.S. Environmental Protection Agency, Edison, NJ.

Heineman, M. (2000). SWMMTools, ArcScripts Web site (arcscripts.esri.com), ESRI.

Henstridge, F. (1998). Developing the strategic plan for a GIS. *Professional Surveyor*, May/June 1998, 47–48.

Henstridge, F. (1998a). Developing the strategic plan for a GIS — Part 2. *Professional Surveyor*, November/December 1998, 48–49.

Herrlin, J. (2000). Go with the flow — delineating sewerage tributaries. *Geo Info Systems*, March 2000, 34–39.

Hoffman, J. and Crawford, D. (2000). Using comprehensive mapping and database management to improve urban flooding modeling. *Proceedings of the Conference on Stormwater and Urban Water Systems Modeling*, Computation Hydraulics International, Toronto, Canada, February 24–25, 2000.

Horemans, R. and Reynaert, B. (2003). A technical overview: How Pidpa links GIS with ERP, CIS, and SCADA. *ArcNews*, ESRI, Winter, 2003/2004.

Horemans, R. and Reynaert, B. (2002). ArcFM 8 Water: The Solution for the European Drinking Water Company. 17th ESRI European User Conference, Bruges, Belgium, October 14–16, 2002.

Hsu, Y.M. (1999). 3D.Add_XYZ Script, ArcScripts Web site, www.esri.com/arcscripts/scripts.cfm, ESRI.

Huber, W.C. and Dickinson, R.E. (1988). *Storm Water Management Model.* User's manual, Version 4, Environmental Research Laboratory, U.S. Environmental Protection Agency, Athens, GA.

Huey, D.M. (1998). Managing NPDES Permit Compliance: Development of an Integrated GIS Management System, International User Conference, ESRI, San Diego, CA, July 27–31, 1998.

Hutchinson, S. and Daniel, L. (1995). *Inside ArcView.* Onward Press, 329.

Hydrosphere Data Products, 1996. *GeoSelect Pennsylvania*, CD-ROM and User's guide, NCDC Hourly Precipitation GIS, Data Vol. 6, Boulder, CO.

IDRISI 32, (2000). Image Gallery, Clark Labs Web site at http://www.clarklabs.org/03prod/idrisi.htm.

Imwalle, S. (1998). The Four Stages of Water/Sewer GIS Development, International User Conference, ESRI, San Diego, CA, July 27–31, 1998.

Ingeduld, P., Svitak, Z., and Drbohlav, J. (2004). Prague Master Plan of Water Supply System. Case study submission in response to a call for case studies distributed by the author (U.M. Shamsi) for this book.

Irrinki, S. 2000. The digital utility, *Water Environment & Technology* Vol. 12, No. 12, December 2000, Water Environment Federation, 29–33.

Jackson, T.J. (2000). Global soil moisture monitoring with satellite microwave remote sensing, *Water Resources IMPACT*, American Water Resources Association, Vol. 2, No. 5, September 2000, 15–16.

James, W. (1993). Introduction to SWMM. Chapter 1 in *New Techniques for Modelling the Management of Stormwater Quality Impacts.* W. James (Ed.), Lewis, Boca Raton, FL, 1–28.

Jenkins, D. (2002). Making the leap to ArcView 8.1. *Geospatial Solutions*, January 2002, 46–48.

Jenson, S.K. (1991). Application of hydrologic information automatically extracted from digital elevation models. *Hydrologic Processes*, Vol. 5, 31–44.

Jenson, S.K. and Dominique, J.O. (1988). Extracting topographic structure from digital elevation data for Geographic Information System analysis. *Photogrammetric Engineering and Remote Sensing*, Vol. 54, No.11, 1593–1600.

Joffe, B.A., Burnes, T., Jackson, M., Perry, L., Slocum, W., and Wuthrich, D. (2001). All for one and one for all — Consensus builds a successful GIS, *GEOWorld*, June, 2001, 46–49.

Karimi, H.A. and Hammad, A. (2004). *Telegeoinformatics — Location-Based Computing and Services*, CRC, Boca Raton, FL.

Kavaliunas, J. and Dougherty, L. (2000). Census 2000, census sense, part 1, what geospatial pros need to know. *Geospatial Solutions*, July 2000, 40–43.

Kindleberger, C. (1992). Tomorrow's GIS. *American City & County*, April 1992, 38–48.

Kite, G.W. (1995). The SLURP model, Chapter 15 in *Computer Models of Watershed Hydrology*, V.P. Singh (Ed.), Water Resources, 521–562.

Kopp, S.M. (1996). Linking GIS and hydrologic models: Where have we been, Where are we going? In K. Kovar and H.P. Nachtnebel (Eds.), *HydroGIS'96: Application of Geographic Information Systems in Hydrology and Water Resources Management*. International Association of Hydrologic Sciences Publication No. 235. IAHS Press, Wallingford, U.K., 133–139.

Kovar, K., and Nachtnebel, H.P. (Eds.) (1996). *HydroGIS'96: Application of Geographic Information Systems in Hydrology and Water Resources Management*. International Association of Hydrologic Sciences Publication No. 235. IAHS, Wallingford, U.K.

Kuppe, J.B. (1999). Imagery technology helps county make GIS data accessible, *ArcUser*, ESRI, October–December, 1999, 10–11.

Lahlou, M., Shoemaker, L., Paquette, M., Bo, J., Choudhury, S., Elmer, R., and Xia, F. (1996). *Better Assessment Science Integrating Point and Nonpoint Sources BASINS*, User's manual, Version 1, EPA-823-R-96-001, Exposure Assessment Branch, Office of Water, U.S. Environmental Protection Agency, Washington, D.C.

Lamb, K.B., Quinn, P., Romanowicz, R., and Freer, J. (1995). TOPMODEL, Chapter 18 in *Computer Models of Watershed Hydrology*, V.P. Singh (Ed.), Water Resources Publications, 627–668.

Landers, J. (2002). Safeguarding Water Utilities, *Civil Engineering*, Vol. 72, No. 6, June 2002, 48–53.

Lanfear, K.J. (2000). The Future of GIS and Water Resources, Water Resources Impact, American Water Resources Association, Vol. 2, No. 5, September 2000, 9–11.

Lang, L. (1992). Water's New World, *Civil Engineering*, ASCE, June 1992, 48–50.

Laurent, F., Anker, W., and Graillot, D. (1998). Spatial modelling with geographic information systems for determination of water resources vulnerability application to an area in Massif Central (France). *Journal of the American Water Resources Association*, Vol. 34, No. 1, February 1998, 123–134.

Lee, S.G. (1998). Routing Meter Readers ... Leveraging an Investment in Data Conversion! International User Conference, ESRI, San Diego, CA, July 27–31, 1998.

Lewis, P. (1985). Beyond description. *Annals of the Association of American Geographers*, Vol. 75, No. 4, 465–477.

Lillesand, T.M. and Kiefer, R.W. (1999). Remote Sensing and Image Interpretation, 4th ed., John Wiley & Sons, New York, 724.

Limp, W.F. (2001). Millennium moves — raster GIS and image processing products expand functionality in 2001. *GEOWorld*, April 2001, Vol. 14, No. 4, 39–42.

Lindley, T.R. and Buchberger, S.G. (2002). Assessing intrusion susceptibility in distribution systems, *Journal AWWA*, AWWA, Vol. 94, No. 6, June 2002, 66–79.

Lior, S., Miller, K., and Thomas, C. (1998). Automating Philadelphia Water Department's Shutoff Program, International User Conference, ESRI, San Diego, CA, July 27–31, 1998.

Logan, B. (2002). The lessons of 9/11. *Geospatial Solutions*, September 2002, 26–30.

Longley, P.A., Goodchild, M.F., Maguire, D.J., and Rhind, D.W. (1999). *Geographic Information Systems*, 2nd ed., John Wiley & Sons.

Lopes, J., Bell, E., and Corriveau, A. (2002). Asset Management Rehaul. *Water Environment & Technology*, Vol. 14, No. 12, December 2002, 40–43.

Lowe, J.W. (2002). Mobile enterprise — act two: the props. *Geospatial Solutions*, April 2002, Vol. 12, No. 4, 44–47.

Lowe, J.W. (2002a). Mobile enterprise — act one: the players. *Geospatial Solutions*, March 2002, Vol. 12, No. 3, 44–47.

Lowe, J.W. (2000). Building a spatial Web site? Know your options. *Geo Info Systems*, April 2000, 44–49.

Lowe, J.W. (2000a). Utilities databases Internet. *Geo Info Systems*, March 2000, 46–48.

Lowe, J.W. (2002b). GIS meets the mapster. *Geospatial Solutions*, Vol. 12, No. 2, February 2002, 46–48.

Luce C.H. (2001). Remote sensing change detection. Book review, *Journal of the American Water Resources Association*, AWRA, Vol. 37, No. 1, 238–239.

Luijten, J.C. (2000). Dynamic hydrological modeling. *ArcUser*, ESRI, July–September 2000, 21–22.

Lundeen, M. (2003). Get Polygon Info Script, ArcScripts Web site, arcscripts.esri.com, ESRI.

Lunetta, R.S. and Elvidge, C.D. (1998). *Remote Sensing Change Detection*, Ann Arbor Press, 121 South Main Street, Chelsea, MI 48118. 318 pages, 1998, ISBN 1-57504-037-9.

Lyman, T.G. (2001). A guide to GPS — learn to obtain better accuracy. *CE News*, May 2001, Vol. 13, No. 4, 52–55.

Maguire, D.J. (1999). ARC/INFO version 8: object-component GIS, Vol. 20, No. 4, Winter 1998/1999, ESRI, 1–2.

Maidment, D.R. (2000). ArcGIS Hydro Data Model, GIS Hydro 2000 preconference seminar, 20th Annual User Conference, ESRI, San Diego, CA, June 2000. http://www. crwr. utexas.edu/giswr/.

Maidment, D.R. (1998). GIS Hydro'98 — Introduction to GIS Hydrology, GIS Hydro 1998 preconference seminar, 18th Annual User Conference, ESRI, San Diego, CA, July 1998.

Maidment, D.R. (1997). Integration of GIS and Hydrologic Modeling, preconference seminar, GIS Hydro97 CD ROM, 1997 User Conference, ESRI.

Maidment, D.R. (1993). GIS and Hydrologic Modeling. Chapter 13 in *GIS and Environmental Modeling*, M.F. Goodchild, B.O. Parks, and L.F. Steyaert (Eds.), Oxford University Press, New York, 147–167.

Martin, L., Castllo, J.G., and Ackerman, D. (1998). Innovative technology combines the components of a sewer system to create virtual computer models. *Proceedings of the 1998 National Conference on Environmental Engineering*, T.E. Wilson (Ed.), ASCE, June 7–10, 1998, Chicago, IL, 669–674.

Maslanik, J.M. and Smith, P.A. (1998). *The City of Harrisburg Water Distribution System Hydraulic Model*, USFilter Engineering and Construction, Pittsburgh, PA.

McKinnon, C. and Souby, J. (1999). Preserving the Wild West, *Imaging Notes*, Space Imaging, Thornton, CO, November/December 1999, 26–27.

Mehan, G.T., III (2003). Monitoring is the key. *Water Environment & Technology*, Vol. 15, No. 11, November 2002, Water Environment Federation, 23–27.

Merry, C.J. (2000). Role of technology in the future of water resources remote sensing developments, *Water Resources Impact*, American Water Resources Association, Vol. 2, No. 5, September 2000, 13–14.

Meyer, S.P., Salem, T.H., and Labadie, J.W. (1993). Geographic information systems in urban storm-water management, *Journal of Water Resources Planning and Management*, Vol. 119, No. 2, 206–228.

Meyers, J. (2002). The worst mistakes in GIS project history (and how to avoid them). *Geospatial Solutions*, Vol. 12, No. 7, July 2002, 36–39.

Michael, J. (1994). Creating digital orthophotos requires careful consideration of project design elements. *Earth Observation Magazine*, Vol. 3 No. 2, 34–37.

Miller, D. (2000). Fine-tuning digital orthophoto quarter quads, *ArcUser*, ESRI, July–September, 2000, 45–47.

Miller, R.C., Guertin, D.P., and Heilman, P. (2004). Information technology in watershed management decision making. *Journal of the American Water Resources Association*, AWRA, April, 2004, Vol. 40, No. 2, 347–357.

Miotto, A. (2000). A better image, *Civil Engineering*, January 2000, 42–45.

Mische, M.R. (2003). Surveying with GPS. *CE News*, November 2003, Vol. 15, No. 10, 40–41.

Mitchell, A. (1997). Zeroing in, ESRI, Redlands, CA, 57–64.

Moglen, G.E. (2000). Effect of orientation of spatially distributed curve numbers in runoff calculations. *Journal of the American Water Resources Association*, AWRA, Vol. 36, No. 6, 1391–1400.

Monmonier, M. (1996). How to lie with maps, University of Chicago Press, 2nd ed.

Morgan, T.R. and Polcari, D.G. (1991). Get set! Go for mapping, modeling, and facility management. *Proceeding of Computers in the Water Industry Conference*, AWWA, Houston, TX.

Morrison, D.A. and Ruiz, M.O. (2001). Instant replay–Red hen's multimedia GIS. *Geospatial Solutions*, October 2001, Vol. 11, No. 10, 45–47.

Murphy, C. (2001). How the discontinuation of SA has affected survey-quality GPS. *CE News*, May 2001, Vol.13, No. 4, 54–55.

Murphy, C. (2000). GIS trends. *CE News*, December 2000, 62–65.

Murphy, C. (2000a). Digital cartographic data — the bottom line, *CE News*, July 2000, 52–56.

Nelson, J., Jones, N., and Miller, W. (1993). Integrated hydrologic simulation with TINs, advances in hydro-science and -engineering, *Proceedings of the 1st International Conference on in Hydro-Science and -Engineering*, Sam S.Y. Wang (Ed.), Center for Computational Hydroscience and Engineering, Washington, D.C., June 7–11, 1993, Vol. I, 571–578.

Neslon, J.E. (1997). From a grid or coverage to a hydrography: unlocking your GIS data for hydrologic applications. *Proceedings of the 1997 ESRI User Conference*, ESRI, San Diego, CA, July 8–11, 1997.

Obermeyer, N.J. (1998). The evolution of public participation GIS, *Cartrography and Geographic Information Systems*, 25(2), 65–66.

Olivera, F. and Bao, J. 1997. ArcView Extension for estimating Watershed Parameters and Peak Discharges According to the TxDOT Statewide Regional Rural Regression Equations, preconference seminar, GIS Hydro97 CD ROM, 1997 User Conference, ESRI.

Ono, T., Cote, L., Dodge, S., and Pier, E. The Wastewater Management System (WIMS), ArcUser calendar insert, ESRI, July 1998.

PaMAGIC (2001). *Local Government Handbook for GIS Implementation within the Commonwealth of Pennsylvania, Part II: GIS Data*. Interim Draft Document, March 2001, Pennsylvania Mapping and Geographic Information Consortium, Lewisburg, PA, www.pamagic.org.

Payraudeau, S., Tournoud, M.G., Cernesson, F., and Picot, B. (2000). Annual nutrients export modelling by analysis of land use and geographic information: case of a small Mediterranean catchment. *Proceedings of the First World Water Congress of the International Water Association*, Book 5, Water Resources and Waste Management, Paris, France, July 3–7, 2000, 320–327.

Peng, Z. (1998). Internet GIS, *CE News*, December 1998, 42–47.

Pfister, B., Burgess, K., and Berry, J. (2001). What's a map? *GEOWorld*, Vol. 14, No. 5, May 2001, 40–43.

Phipps, S.P. (1995). Using raster and vector GIS data for comprehensive storm water management, *Proceedings of the 1995 User Conference*, ESRI. May 22–26.

Pimpler, E. and Zhan, B. (2003). Drop by drop, modeling water demand on the lower Colorado. *Geospatial Solutions*, Vol. 13, No. 5, May 2003, 38–41.

PipeTech Brochure (2000). Peninsular Technologies, 555 Ada Dr., Ada, MI 49301-0728.

PipeTech Website (2001). Peninsular Technologies, www.pent.com.

Prins, J.G., Reil, J., Madison, M., and Swimley, C. (1998) ArcView GIS for City of Stockton's Sewer Model, International User Conference, ESRI, San Diego, CA, July 27–31, 1998.

Quimpo, R.G. and Al-Medeij, J. (1998). Automated watershed calibration with Geographic Information Systems. *Proceedings of the 1st Interagency Hydrologic Modeling Conference*, Las Vegas, NV, 765–772.

Quimpo, R.G. and Shamsi, U.M. (1991). Reliability-based distribution system maintenance, *Journal of Water Resources Planning and Management*, ASCE, Vol. 117, No. 3, May/June 1991, 321–339.

Reichardt, M. (2004). Building the sensor web. *Geospatial Solutions*, March 2004, Vol. 14, No. 3, 36–47.

Reynaert, B., Horemans, R., and Vercruyssen, P. (2004). ArcGIS/ArcFM 8 For Water Distribution Network Management. Case study submission in response to a call for case studies distributed by the author (U. M. Shamsi) for this book.

Rindahl, B. (1996). Analysis of Real-Time Raingage and Streamgage Flood Data Using ArcView 3.0, International User Conference, ESRI, Palm Springs, CA, May 20–24, 1996.

Ritter, K. and Gallie, E.A. (1998). Runoff97: a GIS-based, distributed, surface-runoff model, *Proceedings of the Conference on Stormwater and Related Modeling*: *Management and Impacts*, Computation Hydraulics International, Toronto, Canada, February 19–20, 1998.

Robertson, J.R. (2001). Feeding the flames — airborne imagery fuels GIS. *GeoWorld*, Vol. 14, No. 3, March 2001, 36–38.

Roesner, L.A., Aldrich, J.A., and Dickinson, R.E. (1988). *Storm Water Management Mode EXTRAN Addendum*. User's manual, Version 4, Environmental Research Laboratory, U.S. Environmental Protection Agency, Athens, GA.

Rosenberg, M., Schwind, B., and Lewis, A. (2001). Mapping and monitoring California's vegetation. *Imaging Notes*, Space Imaging, May/June 2001, 24–25.

Rossman, L.A. (2000). *EPANET 2* User's manual, EPA/600/R-00/057, National Risk Management Research Laboratory, U.S. Environmental Protection Agency, Cincinnati, OH.

Rossman, L.A. (2000a). Computer models/EPANET, Chapter 12 in *Water Distribution Systems Handbook*, L.W. Mays (Ed.), McGraw-Hill, 12.1–12.23.

Rugaard, L.C. (2001). CMOM++ — improving collection system reliability to achieve a goal of no overflows. *Conference Proceedings, 2001 Collection System Odyssey: Combining Wet Weather and O&M Solutions*, Water Environment Federation.

Samuels, W.B., Pickus, J.M., Field, M., Ingman, G., and Welch V. (2000). Watershed data information management in Montana — finding the trouble spots, *ArcNews*, Vol. 22, No. 3, Fall 2000, ESRI, 10.

Schell, T.T., Acevedo, M.F., Bogs, F.C., Newell, J., Dickson, K.L., and Mayer, F.L. (1996). Assessing pollutant loading to Bayou Chico, Florida by integrating an urban storm-water runoff and fate model with GIS. *CD-ROM Proceedings of The 3rd International Conference/Workshop on Integrating GIS and Environmental Modeling*, U.S. National Center for Geographic Information and Analysis, Santa Fe, NM, January 21–25, 1996.

Schilling, K.E. and Wolter, C.F. (2000). Application of GPS and GIS to map channel features in Walnut Creek, Iowa. *Journal of American Water Resources Association*, AWRA, December 2000, 1423–1434.

Schock, M.R. and Clement, J.A. (1995). You can't do that with these data! or: uses and abuses of tap water monitoring analyses. *Proceedings of the National Conference on Environmental Problem-Solving with Geographic Information Systems*, EPA/ 625/R-95/004, Office of Research and Development, U.S. Environmental Protection Agency, Cincinnati, Ohio, September 1995, 31–41.

Schreppel, C.K. (2003). Distribution system security — setting the alarm for an early warning, *Opflow*, AWWA, Vol. 29, No. 6, June 2003, 1–6.

Schultz, G.A. (1988). Remote sensing in hydrology, *Journal of Hydrology*, Vol. 100, 239–265.

Seiker, F. and Verworn, H.R. (Eds.) (1996). IAHR/IAWQ *Proceedings of the 7th Annual Conference On Urban Storm Drainage*, Hannover, Germany. Vols. I, II, and III.

Sevier, C. and Basford, C. (1996). Hydraulic modeling helps district plan for future, react to challenges. *WaterWorld*, May 1996, 46–47.

Seybert, T.A. (1996). Effective Partitioning of Spatial Data for Use in a Distributed Runoff Model. Ph.D. dissertation, Department of Civil and Environmental Engineering, Pennsylvania State University, State College, Pennsylvania.

Shamsi, U.M. (2002). *GIS Tools for Water, Wastewater, And Stormwater Systems*, ASCE Press, Reston, VA. www.GISApplications.com.

Shamsi, U.M. (2002a). A new pipe replacement approach. *Journal of American Water Works Association*, AWWA, Vol. 94, No. 1, January 2002, 52–58.

Shamsi, U.M. (2001). GIS and Modeling Integration. CE News, Vol. 13, No. 6, July 2001, 46–49.

Shamsi, U.M. (2000). Internet GIS for water industry, Chapter 8 in *Practical Modeling for Urban Water Systems*, Monograph 11, W. James (Ed.), Computational Hydraulics International, Toronto, Ontario, 139–153.

Shamsi, U.M. (1999). GIS and Water Resources Modeling: State-of-the-Art. Chapter 5 in *New Applications in Modeling Urban Water Systems*, W. James (Ed.), Computational Hydraulics International, Guelph, Ontario, Canada, 93–108.

Shamsi, U.M. (1998). ArcView applications in SWMM modeling. Chapter 11 in *Advances in Modeling the Management of Stormwater Impacts*, W. James(Ed.), Vol. 6. Computational Hydraulics International, Guelph, Ontario, Canada. 219–233.

Shamsi, U.M. (1997). SWMM graphics. Chapter 7 in *Advances in Modeling the Management of Stormwater Impacts*. W. James (Ed.), CHI, Guelph, Ontario, 129–53.

Shamsi, U.M. (1996). Stormwater management implementation through modeling and GIS. *Journal of Water Resources Planning and Management*, ASCE, Vol. 122, No. 2, 114–127.

Shamsi, U.M. (1991). Three dimensional graphics in distribution system modeling. *Proceedings of the Specialty Conference on Computers in the Water Industry*, AWWA, Houston, TX, April 1991, 637–647.

Shamsi, U.M., Benner, S.P., and Fletcher, B.A. (1996). A computer mapping program for sewer systems. Chapter 7 in *Advances in Modeling the Management of Stormwater Impacts*, W. James (Ed.), Computation Hydraulics International, Guelph, Ontario, Canada, 97–114.

Shamsi, U.M., Benner S.P., and Fletcher, B.A. (1995). A computer mapping program for water distribution systems, *Distribution System Symposium*, American Water Works Association, Nashville, TN, September 10–13, 1995.

Shamsi, U.M. and Fletcher, B.A. (2000). AM/FM/GIS applications for stormwater systems, Chapter 7 in *Applied Modeling of Urban Water Systems*, Monograph 8, W. James (Ed.), Computational Hydraulics International, Toronto, Ontario, 123–139, 2000.

Shamsi, U.M. and Scally, C. (1998). *An Application of Continuous Simulation to Develop Rainfall versus CSO Correlations*, WEFTEC Asia, Conference and Exhibition on International Wastewater and Water Quality Technology, WEF, Singapore, March 7–11, 1998.

Shamsi, U.M., and Smith, P. (2004). ArcGIS and SWMM5: perfect for each other. *Proceedings of Stormwater and Urban Water Systems Modeling Conference*, Computational Hydraulics International, Toronto, Ontario, February 19–20, 2004.

Singh, V.P. (Ed.) (1995). *Computer Models of Watershed Hydrology*. Water Resources Publications, 1130.

Slawecki, T., Theismann, C., and Moskus, P. (2001). GIS-A-GI-GO. *Water Environment & Technology*, Vol. 13, No. 6, June 2001, Water Environment Federation, 33–38.

Somers, R. (1999). Project milestones: Measuring success, managing expectations. *Geo Info Systems*, October 1999, 20–26.

Somers, R. (1999a). Hire, rent, or train? Making effective investments in GIS expertise. *Geo Info Systems*, May 1999, 20–25.

Space Imaging (2001). Diverse data sources shed light on global change. *Imaging Notes*, Vol. 16, No. 2, March/April 2001, 24–25.

Sponemann, P., Beeneken, L., Fuchs, L., Prodanovic, D., and Schneider, S. (1996). Criteria for geographic information systems used in urban drainage. In F. Seiker and H.R. Verworn (Eds.) (1996) *IAHR/IAWQ Proceedings of the 7th Annual Conference on Urban Storm Drainage*, Hannover, Germany. Vol. III, 1689–1694.

Stalford, R.N. and Townsend, J.A. (1995). Water dstribution system modeling, GIS, and facility management systems, what do they have in common? A case study. *Proceedings of the 1995 International User Conference*, ESRI, San Diego, CA.

Stern, C.T. and Kendall, D.R. (2001). Mobile solutions deliver improved utility infrastructure and operations management. *Journal AWWA*, American Water Works Association, November, 2001, Vol. 93, No. 11, 69–73.

Stone, D. (1999). County reaps rewards from airborne laser survey, *CE News*, November 1999, 58–60.

Stupar, D., Heise, S. and Schielbold, J. (2002). Ocean county utilities authority takes a business-driven approach to GIS design and implementation. *Water Writes*, ESRI, Winter 2002, 6–7.

Suchanek, M. and Metelka, T. (2004). Sofia Drainage Area Plan. Case study submission in response to a call for case studies distributed by the author (U. M. Shamsi) for this book.

Swalm, C., Gunter, J.T., Miller, V., Hodges, D.G., Regens, J.L., Bollinger, J., and George, W. (2000). GIS blossoms on the mighty Mississippi. *Geo Info Systems*, April 2000, 1998, 30–34.

Taher, S.A. and Labadie, J.W. (1996). Optimal design of water-distribution networks with GIS, *Journal of Water Resources Planning and Management*, ASCE, Vol. 122, No. 4, July/August 1996, 301–311.

TenBroek, M.J. and Roesner, L.A. (1993). MTV — Analysis tool for review of computer models. *Proceedings of the Conference on Computers in the Water Industry*, Water Environment Federation, Santa Clara, CA, August 8–11, 1993, 173–181.

Teo, Frederick K.S. (2001). For Tacoma water GIS is a wireless wireless wireless world. *Geospatial Solutions*, 40–45, February 2001.

The San Diego Union Tribune (1998). GIS — New Technology's Scope is Almost Infinite. Sunday, July 26, 1998, A-12.

Thoen, B. (2001). Maximizing your search for geospatial data. *GEOWorld*, Vol. 14, No. 4, April 2001, 34–37.

Thomas, G. (1995). A Complete GIS-based Stormwater Modeling Solution, *Proceedings of the 1995 User Conference*, ESRI, May 22–26.

Thompson, S. (1991). Mapping the latest trends in GIS. *American City & County*, May 1991, 26–38.

Thrall, G.I. (1999). Low cost GIS, *Geo Info Systems*, April 1999, 38–40.

Thrall, G.I. (1999b). Want to think spatially? Raster GIS with MFworks, *Geo Info Systems*, September 1999, 46–50.

Thrall, G.I. (2003). Laptops for GIS warriors, *Geospatial Solutions*, Vol. 13, No. 10, October 2003, 49–52.

Thuman, A. and Mooney, K. (2000). TMDLs: How did we get here and where are we going? Environmental Perspectives, *The HydroQual Newsletter*, Fall 2000, Vol. 5, 1–4.

Tiewater, P. (2001). GASB statement 34: Not just for accountants and engineers. *Presentation in the 3rd Annual Sewer Rehabilitation Conference*, Three Rivers Wet Weather Demonstration Program, Warrendale, PA, September 25–26, 2001.

TOPAZ (2000). Topographic parameterization software, Grazing Labs Research Laboratory Website at http://grl.ars.usda.gov/topaz/TOPAZ1.HTM.

Trammel, D. (2004). Resource Planning and CIP Allocation. Case study submission in response to a call for case studies distributed by the author (U. M. Shamsi) for this book.

Turner, A.K. (2001). The future looks bright (in any spectrum) for GIS data. *GeoWorld*, Vol. 14, No. 5, May 2001, 30–31.

Turner, B.G., Arnett, C.J., and Boner, M.C. (2000). Wet weather monitoring and calibration of USEPA BASINS model provides basis for watershed management framework compliance with TMDL allocations. *Proceedings of the 1st World Water Congress of the International Water Association*, Book 5, Water Resources and Waste Management, Paris, France, July 3–7, 2000, 439–446.

Understanding GIS — The ARC/INFO Method, 1990. ESRI, Redlands, CA.

U.S. Department of Agriculture (1986). *Urban Hydrology for Small Watersheds*. Technical Release 55, Soil Conservation Service, U.S. Department of Agriculture, Washington, D.C., 2.5–2.8.

USGS (2000). USGS Digital Elevation Model Data, USGS Web site at http://edcwww.cr.usgs.gov/glis/hyper/guide/usgs_dem.

Venden, E. and Horbinski, T. (2003). Gurnee streamlines notification process. *Geospatial Solutions*, January 2003, Vol. 13, No. 1, 26–27.

Walski, T. (2001). Importance and accuracy of node elevation data. *Current Methods*, Vol. 1, No. 1, Haestad, Waterbury, CT.

Walski, T.M. and Male, J.W. (2000). Maintenance and rehabilitation replacement. In *Water Distribution Systems Handbook*, L.W. Mays Ed., McGraw-Hill, 17.1–17.28.

Walski, T., Toothill, B., Skoronski, D., Thomas, J., and Lowry. S. (2001). Using digital elevation models. *Current Methods*, Haestad, Vol. 1, No. 1, 91–99.

Wang, M., Hjelmfelt, A.T., and Garbrecht., J. (2000). DEM aggregation for watershed modeling. *Water Resources Bulletin*, American Water Resources Association, Vol. 36, No. 3, June 2000, 579–584.

Water Resources Bulletin, American Water Resources Association, Vol. 36, No. 3, June 2000, 579–584.

Waters, N. (2001). Network query language is the new, new thing in GIS. *GEOWorld*, April 2001, Vol. 14, No. 4, 30.

Wells, E. (1991). Needs analysis: Pittsburgh's first step to a successful GIS. *Geo Info Systems*, October 1991, 30–38.

Whittemore, R.C. and Beebe, J. (2000). EPA's BASINS model: Good science or serendipitous modeling? *Journal of the American Water Resources Association*, June 2000, Vol. 36, No. 3, 493–499.

Wild, M. and Holm, D. (1998). Consensus building at superfund sites using GIS. *Proceedings of the 1998 National Conference on Environmental Engineering*, T.E. Wilson (Ed.), ASCE, June 7–10, 1998, Chicago, IL, 326–331.

Wilson, J.D.P. (2001). The next frontier—GIS empowers a new generation of mobile solutions. *GEOWorld*, June 2001. Vol. 14, No. 6, 36–40.

Wilson, J.P., Mitasova, H., and Wrightang, D. (2000). Water resources applications of GIS. Articles currently under peer review by the URISA Journal, University Consortium for Geographic Information Science (UCGIS) Web site, www.ucgis.org/apps_white/water.html.

WinCan Web site (2001). WinCan America, www.wincanamerica.com.

WMS (2000). Watershed Modeling System Web site, http://emrl.byu.edu/wms.htm, Environmental Modeling Research Laboratory (EMRL), Department of Civil and Environmental Engineering, Brigham Young University, Provo, UT.

Zeiler, M. (1999). *Modeling Our World*, ESRI Press, ESRI, Redlands, CA, 199 p.

Zimmer, R.J. (2001). GIS vision. *Professional Surveyor*, November 2001, Vol. 21, No. 10, 38–39.

Zimmer, R.J. (2001a). DEMs — A discussion of digital elevation data and how that data can be used in GIS. *Professional Surveyor*, Vol. 21, No. 3, March 2001, 22–26.

Zimmer, R.J. (2001b). GPS resource mapping for GIS. *Professional Surveyor*, Vol. 21, No. 4, April 2001, 44–47.

Zimmer, R.J. (2000). Parcel mapping part I. *Professional Surveyor*, Vol. 20, No. 9, October 2000, 52–55.

Index

Page numbers marked with the suffix n indicate that the index entry is located in the note on that page

G

I